STUDENT PERSPECTIVES IN MILITARY GEOGRAPHY AND GEOSCIENCES

Navigating Environmental, Urban, and Security Challenges

EDITORS

Ivan Henrico
Susan Henrico
Rikus le Roux

SUN PRESS

Student Perspectives in Military Geography and Geosciences: Navigating Environmental, Urban, and Security Challenges

Published by AFRICAN SUN MeDIA under the SUN PReSS imprint

This publication was subjected to an independent double-blind peer evaluation by the publisher.

First edition 2026

ISBN: 978-0-6398894-2-9
ISBN: 978-0-6398894-3-6 (eBook)
https://doi.org/10.52779/9780639889436

Set in Times New Roman 9/12pt

Cover design, typesetting and production by AFRICAN SUN MeDIA
Cover image: © AFRICAN SUN MeDIA

SUN PReSS is an imprint of AFRICAN SUN MeDIA. Scholarly, professional and reference works are published under this imprint in print and electronic formats.

This publication can be ordered from:
orders@africansunmedia.co.za
Takealot: bit.ly/2monsfl
Google Books: bit.ly/2k1Uilm
africansunmedia.store.it.si (eBooks)
Amazon Kindle: amzn.to/2ktL.pkL

Visit africansunmedia.co.za for more information.

TABLE OF CONTENTS

Preface I

Contributors and Affiliations III

SECTION 1

Environmental Management and Monitoring (Natural Environment) 1

1. Accuracy assessment of Ordinary Kriging and Inverse Distance Weighting based on estimation error for predicting naturally occurring radioactive materials 3

   B. Mtshawu, J. Bezuidenhout, and R.K. Botlholo

2. Mapping and determining land degradation using GIS and remote sensing: A case of General De La Rey Military Training Area 23

   N.V. Mathebula and R.R. le Roux

3. Temporal analysis of the effects of warfare on vegetation in Gaza 47

   T. Henrico, D. Naidoo, Z. Lourens, and Z. Münch

4. Investigating the distribution of sediments using natural radionuclides through in-situ measurements on the Kilindini Harbour, Mombasa, Kenya 67

   R.K. Botlholo, J. Bezuidenhout, and B. Mtshawu

SECTION 2

Urban Dynamics, Crime, and Security (Human-Environmental Interactions) 87

5. Spatio-temporal analysis for monitoring residential expansion: A case of Middelpos settlement in Saldanha from 2004–2024 89

   N.L. Monyamane and L.P. Welman

6. Using hotspot analysis to geo-visualise criminal activity in the Saldanha Military Area: 2017–2023 101

   T.M. Tlhekwe and H.A.P. Smit

7. Evaluating crime patterns in South Africa: The imperative for enhanced police presence in Mamelodi East 119

   L.D. Pila and P. Schmitz

8. Evolving approaches to human security: Geographic insights into contemporary challenges 143

   A.R. Govender and I. Henrico

SECTION 3
Military, Maritime, and Security Infrastructure (Planning and Decision-making) 175

9. Geospatial analysis of risk and consequences of an explosion incident at Koeberg nuclear power plant 177
T.L. Molefe and D. Putter

10. Enhancing utility management through GIS and GPS integration: A case of the Saldanha Military Area 197
S. Nyembe and I. Henrico

11. Maritime surveillance in Saint Helena Bay using a multisource approach 215
J.M. Mametja and D. Putter

SECTION 4
Sustainable and Renewable Energy Solutions 231

12. Photovoltaic solar farm site selection using the Analytical Hierarchy Process and Geographic Information Systems: A case in the Saldanha Military Area 233
K.T. Kortman and S. Henrico

13. Modelling the maximum potential wave energy zones in North Bay, Saldanha, for small-scale green energy 257
L.L. Matiwane and R.L. Uys

14. Geospatial tools and techniques in military exercise planning: Insights from the South African Military Academy's Exercise Trans Enduro 269
M. Mathoho and H.A.P. Smit

About the Editors 287

PREFACE

This edited volume, *Student Perspectives in Military Geography and Geosciences: Navigating Environmental, Urban, and Security Challenges*, emerged from an innovative student-centred initiative that reflects a strong culture of academic mentorship and collaborative research within Stellenbosch University's Faculty of Military Science. Most of the chapters are based on postgraduate research conducted in the Department of Military Geography at the Saldanha Campus. Additional contributions originate from the Departments of Physics (Mil) and Nautical Science. One chapter (Chapter 3) reflects collaborative undergraduate research undertaken by students from the Department of Geography and Environmental Studies at the main campus of Stellenbosch University, while Chapter 8 represents an undergraduate research project developed within the Faculty of Military Science. Many of the student contributions were developed through mentorship and co-authorship with senior academics from these departments, as well as from the Department of Geography at the University of South Africa (UNISA), highlighting the cross-institutional collaboration that underpins this volume.

Each chapter originated from undergraduate coursework or honours-level research conducted under structured academic supervision. Through this process of academic guidance and refinement, these student-led projects evolved into publishable scholarly contributions. Many chapters were further developed following their presentation at the 15th International Conference on Military Geosciences (ICMG 2024), held in Somerset West, South Africa, from 9 to 14 June 2024. Participation in this international forum provided student researchers with an opportunity to engage with global scholars, present their findings, and further strengthen their academic work. Collectively, the chapters presented in this volume demonstrate rigorous research, critical analysis, and the practical application of geospatial methods within contemporary military and environmental contexts. At a time when geospatial technologies are increasingly central to understanding environmental change, urban dynamics, and security challenges, this volume highlights the important role that emerging scholars can play in advancing applied research within military geography and geosciences.

All chapters in this volume underwent a formal academic review process in accordance with the policies of African Sun Media. Academic reviewers were independently selected, including two national reviewers from outside the authors' institutions and one international reviewer. The peer review followed a double-blind process: manuscripts were sent anonymously to reviewers, and their feedback was returned anonymously to the authors. Based on the reviewers' recommendations, the authors revised their work to ensure academic rigour and overall quality.

The book is thematically organised into four key sections, reflecting the diverse areas of focus within this interdisciplinary field.

The first section, *Environmental Management and Monitoring (Natural Environment)*, explores advanced geospatial methods for analysing environmental processes and risks. Chapters in this section include the assessment of radionuclide distribution in sediments (Chapters 1 and 4), the evaluation of land degradation in military training areas (Chapter 2), and the analysis of vegetation changes in conflict-affected regions (Chapter 3).

The second section, *Urban Dynamics, Crime, and Security (Human–Environmental Interactions)*, examines the spatial dimensions of urban development and security challenges. The chapters in this section explore residential expansion in informal settlements (Chapter 5), spatially targeted crime analysis in military and civilian contexts (Chapters 6 and 7), and the application of geospatial technologies to enhance human security strategies in the face of climate change, migration, and conflict (Chapter 8).

The third section, *Military, Maritime, and Security Infrastructure (Planning and Decision-Making)*, focuses on the strategic application of geospatial intelligence in infrastructure management and surveillance. Chapters 9, 10, and 11 provide insights into nuclear facility risk assessment (Chapter 9), utility infrastructure management in military contexts (Chapter 10), and maritime surveillance to detect illegal activities (Chapter 11).

The final section, *Sustainable and Renewable Energy Solutions*, addresses the growing importance of energy security and operational sustainability in defence contexts. Chapters 12, 13, and 14 examine photovoltaic solar farm site selection (Chapter 12), wave energy potential in coastal environments (Chapter 13), and the use of geospatial tools to support efficient military exercise planning (Chapter 14).

This volume showcases the academic capabilities of student researchers while highlighting the value of structured academic mentorship. By transforming coursework and honours research into peer-reviewed scholarly outputs, the contributors demonstrate how student-led research can meaningfully contribute to advancing knowledge in military geography and geosciences.

We commend the student authors and their academic mentors for their dedication, innovation, and scholarly commitment. We hope that this volume will encourage continued research, collaboration, and intellectual curiosity within the evolving fields of military geography and geosciences.

EDITORS:

Ivan Henrico, Susan Henrico, and Rikus le Roux

CONTRIBUTORS AND AFFILIATIONS

Chapter	Student author(s)	Mentor / co-author(s)	Institutional affiliation(s) of mentors / co-authors
1	R.K. Botlholo	Prof. J. Bezuidenhout and Dr. B. Mtshawu	Department of Physics (Mil) & Military Geography, Faculty of Military Science, Stellenbosch University
2	N.V. Mathebula	Assoc. Prof. R.R. le Roux	Department of Military Technology, Faculty of Military Science, Stellenbosch University
3	T. Henrico, D. Naidoo, Z. Lourens	Dr. Z. Münch	Department of Geography and Environmental Studies, Stellenbosch University
4	R.K. Botlholo	Prof. J. Bezuidenhout and Dr. B. Mtshawu	Department of Physics (Mil) & Military Geography, Faculty of Military Science, Stellenbosch University
5	N.L. Monyamane	Dr. L.P. Welman	Department of Military Geography, Faculty of Military Science, Stellenbosch University
6	T.M. Tlhekwe	Prof. H.A.P. Smit	Department of Geography, University of South Africa
7	L.D. Pila	Prof. P. Schmitz	Department of Geography, University of South Africa
8	A.R. Govender	Assoc. Prof. I. Henrico	Department of Military Geography, Faculty of Military Science, Stellenbosch University

9	T.L. Molefe	Dr. D. Putter	Department of Defence Intelligence Studies, Faculty of Military Science, Stellenbosch University
10	S. Nyembe	Assoc. Prof. I. Henrico	Department of Military Geography, Faculty of Military Science, Stellenbosch University
11	J.M. Mametja	Dr. D. Putter	Department of Defence Intelligence Studies, Faculty of Military Science, Stellenbosch University
12	K.T. Kortman	Dr. S. Henrico	Department of Defence Intelligence Studies, Faculty of Military Science, Stellenbosch University
13	L.L. Matiwane	Mr. R.L. Uys	Department of Nautical Science, Faculty of Military Science, Stellenbosch University
14	M. Mathoho	Prof. H.A.P. Smit	Department of Geography, University of South Africa

SECTION 1

Environmental Management and Monitoring Natural Environment

This section examines the intersection of environmental risk, geospatial science, and military geoscience through spatial interpolation, remote sensing, and field-based assessments. The four chapters present rigorous approaches to monitoring environmental hazards – natural and conflict-induced – by leveraging advanced geospatial techniques to support defence readiness, ecosystem monitoring, and resource protection.

Chapter 1 assesses the accuracy of two spatial interpolation methods – Ordinary Kriging and Inverse Distance Weighting (IDW) – for predicting the spatial distribution of naturally occurring radioactive materials in sediments from the Port Reitz–Kilindini hydraulic system in Mombasa, Kenya. Using cross-validation and error statistics, the study found that kriging provides the most accurate results, offering critical insights for environmental risk mapping in coastal and port-adjacent zones.

Chapter 2 shifts the focus to land degradation in the General De La Rey Military Training Area in South Africa. Using multi-temporal Landsat imagery and vegetation indices such as NDVI, SAVI, and TCB, the authors mapped changes in vegetation cover and soil conditions between 2013 and 2023. Their analysis highlighted the spatial effects of military activities, such as live-fire exercises and vehicle movement, on vegetation health, supporting long-term monitoring in military ecosystems.

Chapter 3 explores the environmental impacts of the conflict in Gaza, a highly sensitive and geopolitically significant region. Using Sentinel-2A satellite imagery, the chapter quantifies vegetation change over a one-year period (2023–2024) using NDVI, SAVI, and MSAVI. Although no statistically significant differences in vegetation were detected, the study revealed strong index correlations and reinforced the value of vegetation monitoring in conflict-affected regions.

Chapter 4 builds on the work presented in Chapter 1 by analysing the sediment transport in Mombasa harbour through in situ measurements of uranium, thorium, and potassium concentrations using the Delta Underwater Gamma-ray System (DUGS). The spatial distribution of radionuclides was visualised using kriging techniques, identifying natural sediment tracers and linking radionuclide concentrations to geomorphological and hydrodynamic processes in the harbour.

Together, these chapters demonstrate the value of geospatial tools in capturing environmental change, whether caused by natural sedimentation, anthropogenic degradation, or warfare. They underscore the importance of spatial accuracy and multisource data integration for decision-making in civilian and military contexts.

CHAPTER 1

Accuracy assessment of Ordinary Kriging and Inverse Distance Weighting based on estimation error for predicting naturally occurring radioactive materials

B. Mtshawu, J. Bezuidenhout, and R.K. Botlholo

Abstract

Spatial interpolation requires two basic inputs: known points and an interpolation method. However, the choice of spatial interpolation method can greatly impact the quality of the prediction and spatial variability of site-specific natural radionuclides in sediments. Ordinary Kriging (OK) is often proposed as a statistical technique with superior mathematical properties, such as a minimum error variance. However, due to its robustness and simplicity, Inverse Distance Weighting (IDW) continues to be widely used. This study aimed to compare the prediction accuracy of the two most frequently used interpolation techniques (OK and IDW) in determining the degree of spatial variability of natural radionuclides in the sediments of the Port Reitz–Kilindini hydraulic system in Mombasa, Kenya. The comparison of these two interpolation methods was achieved using the total error (TE) of cross-validation and validation statistics. The mean error (ME) and root mean square error (RMSE) were calculated and combined to determine which interpolator produced the lowest TE. The interpolator that produced the lowest TE portrayed the most accurate predictions of the study area. The natural radionuclide (^{40}K, ^{232}Th, ^{238}U) dataset was collected in Port Reitz, Mombasa using the Delta Underwater Gamma System (DUGS), which consists of a 3"×3" NaI(Tl) scintillation detector, a TB-5 multichannel analyser (MCA) from Amptek®, and an external computer that interfaces with the hardware for data logging and processing. A total of 8,674 sample points were collected across the study area. The preliminary results obtained from the comparison of the two applied interpolation methods indicated that kriging was the most suitable method for predicting and mapping the spatial distribution of natural radionuclides in the sediments of Port Reitz, Mombasa, Kenya.

Keywords: Inverse Distance Weighting (IDW), naturally occurring radioactive materials (NORMs), Ordinary Kriging (OK), prediction accuracy, spatial interpolation

Introduction

The selection of an appropriate interpolation method for mapping radiometric measurements of naturally occurring radionuclides (Potassium [^{40}K], Thorium [^{232}Th], and Uranium [^{238}U]) is complex because of spatial heterogeneity, the complexities of suboptimal sampling densities, and the presence of significant noise or discontinuities in the data (Alasadi & Abojassim, 2022; Sajid et al., 2013; Le Roux et al., 2023). According to MacEachren and Davidson (1987), the accuracy of interpolation methods is often determined by the accuracy of data measurement, data density, data distribution, and spatial variability in a specific study area. There are no specific standards, tests, or established guidelines for selecting the most suitable interpolation method (Sajid et al., 2013). Typically, the performance of an interpolation method is often evaluated and compared using cross-validation techniques (Tomczak, 1998; Gong et al., 2014; Singh & Verma, 2019). The primary objective of cross-validation is to increase reliability, gain sufficient confidence in an interpolation method, and ensure that predictions accurately represent spatial variability. The performance of the interpolation method is deemed acceptable when it exhibits a minimum error and low variance (Ikechukwu et al., 2017). Lam (1983) reviewed various interpolation techniques and found that each interpolation technique has advantages and disadvantages, and the choice of an interpolation model depends largely on the characteristics and quality of the original data, the level of accuracy desired, and the available computational resources. The overall conclusion of Lam's study was that the performance of interpolation techniques relies greatly on the density, spatial arrangement, and complexity of the data.

Many interpolation techniques have been developed to predict the values of spatial phenomena at unsampled locations based on the regionalised variable theory (Journel & Huijbregts, 1978; Goovaerts, 1999a; Goovaerts, 1999b; Meng et al., 2013). Henly (1981) provides a concise review of the theory, stating that the regionalised variable can be defined as a standard random function of a probability distribution law, which it follows; however, it must be normally distributed and characterised by a mean and a variance. Therefore, if this random function conforms to a standard statistical distribution, it will be fully described by a limited set of parameters. Burrough and McDonnell (1998: 133) further explain that the "regionalised variable theory assumes that spatial variation of any variable can be expressed as the sum of three components". The first is the standard component with a constant mean, and the second is a random but spatially correlated component, known as the regionalised variable. The third component is the residual component or error term.

The foundation of spatial interpolation is based on the First Law of Geography proposed by Waldo Tobler, which asserts that everything is related to everything else, but near things are more related than distant things (Waters, 1988). Spatial autocorrelation is a formal property that measures the degree to which near and distant things are related (Smith et al., 2017). Most interpolation methods utilise spatial autocorrelation by weighting nearby sample points more heavily than those farther away. Therefore, interpolators can be defined as mathematical functions used in the construction of continuous surfaces from a set of collected points (Childs, 2004). It is used for the densification of a sample that does not cover the entire area of interest.

The interpolation techniques commonly used for naturally occurring radionuclides (^{40}K, ^{232}Th, and ^{238}U) include Inverse Distance Weighting (IDW) and Ordinary Kriging (OK) (Bezuidenhout, 2019; Le Roux et al., 2023; Henrico et al., 2023). Both methods estimate the value of a parameter at unsampled locations based on the measurements from the surrounding locations with specific weights assigned to each measurement. IDW is simpler to implement, whereas kriging is more complex and time-consuming. However, kriging generally provides a more precise depiction of the spatial structure of the data and provides more useful information about the distribution of estimation errors (Kravchenko & Bullock, 1999). For both interpolation methods, the measured and predicted means, standard deviation, mean error (ME) and/or root mean square error (RMSE) are reported. The use of ME and RMSE provides an evaluation of the error in the units of the variable, which often offers more insight into the model's performance than what can be obtained from relative error alone or goodness-of-fit measures (Sajid et al., 2013). The difference between the two interpolators is that OK is a stochastic interpolator, whereas IDW is a deterministic interpolator. According to Chang (2019), a stochastic interpolation method considers the presence of randomness in its variables and offers an assessment of prediction errors with estimated variances. A deterministic interpolation method estimates values at unsampled locations based on known sampled locations using a weight function in a search neighbourhood and provides no assessment of errors with predicted values. However, the underlying theory for both methods is based on the similarity of near points to create a spatially continuous surface.

Although the difference in performance between stochastic and deterministic interpolation methods remains unclear, the interpolation method should be used in practice. While some researchers argue that stochastic methods are more powerful, others advocate using deterministic methods over stochastic methods (Ferreira, 2015). Several studies (conducted by Tabios & Salas, 1985; Laslett et al., 1987; Gotway & Hartford, 1996; Mueller et al., 2001; Junior et al., 2006; Bezuidenhout, 2019; Le Roux et al., 2023; Henrico et al., 2023) have compared the performance of kriging and IDW methods. However, the results are rather contradictory. Some researchers have demonstrated that kriging provides a more precise description of the spatial structure of the studied phenomenon. Some studies have shown that IDW produces better results, while others have not found significant differences between these methods.

However, little research has been conducted on standardising interpolation methods for mapping the spatial distribution and ratios of naturally occurring radionuclides (^{40}K, ^{232}Th, and ^{238}U) in sediments (Mtshawu et al., 2023). This study aimed to use ArcGIS Pro 3.3 to compare the prediction accuracy of OK and IDW based on the cross-validation assessment criterion (i.e. scatter plot, ME, RMSE, and TE) to measure the degree of accuracy in mapping the spatial variability of naturally occurring radionuclides in the sediments of the Port Reitz–Kilindini hydraulic system in Mombasa, Kenya.

Study area

Mombasa is Kenya's second-largest city, positioned on a coral island southeast of Kenya, overlooking the Indian Ocean, and connected to the mainland by a series of causeways and bridges (Rees et al., 1996). The Makupa Causeway links the island to the mainland and separates the two inland lagoons. Port Tudor, located to the north of the island, is connected to the ocean via a channel through the Port of Mombasa. The smaller, narrower channel called Port Kilindini connects the larger inland lagoon and Port Reitz to the ocean on the island's southern side (Figure 1.1). The Port of Mombasa is one of East Africa's largest and busiest ports, ranked 5th in Africa and 120th in the world for having handled 1.01 Twenty-Foot Equivalent Units (TEUs) in 2014 (Otoi et al., 2015). The Port of Mombasa is important to Kenya for several reasons. First, due to its central geographical location in the East Africa region, this port is a gateway and exit point for cargo belonging to the vast hinterland that includes Kenya, Uganda, Burundi, the Democratic Republic of Congo, South Sudan, Tanzania, Ethiopia, and Somalia. Second, the port is a significant economic driver through maritime-related industries, creating employment and attracting foreign direct investments (Njeru, 2009). Finally, the Port of Mombasa creates an interface between maritime and inland transport systems, thus facilitating regional and international trade (Kimeli et al., 2018). The Kenya Maritime Authority (KPA) manages the

Port of Mombasa, which includes Kilindini Harbour, Port Reitz, Port Tudor, and the whole of the tidal waters (Njeru, 2009).

The study area (Figure 1.1) in which in-situ measurements were conducted is the Port Reitz–Kilindini hydraulic system, located at 39°38' E, 4°04' S. This area features a relatively straight morphology without significant prominences along its length, making it more navigable for large ocean vessels than other areas. According to Onundo and Mwema (2017), this area is a large natural deep-water channel characterised by a complex system of constrictions and sills. The bathymetry of this Port Reitz–Kilindini hydraulic system progressively narrows in width entering the port, from nearly 1,000 m in the south to approximately 100–600 m in the north of Port Reitz, reaching depths of up to 50 m (Onundo, 2017). The estimated freshwater that flows into the Reitz–Kilindini stream systems is on average 4.4 m^3/s, which is due to the rainfall patterns significantly affecting the variability in freshwater flow within the stream systems (Rees et al., 1996). Mombasa experiences a tropical climate system characterised by high temperatures ranging from 20 °C to 34 °C year-round and two main rainy seasons: the long rains (from April to June) and the short rains (from October to December) (Olivera Viciedo et al., 2021).

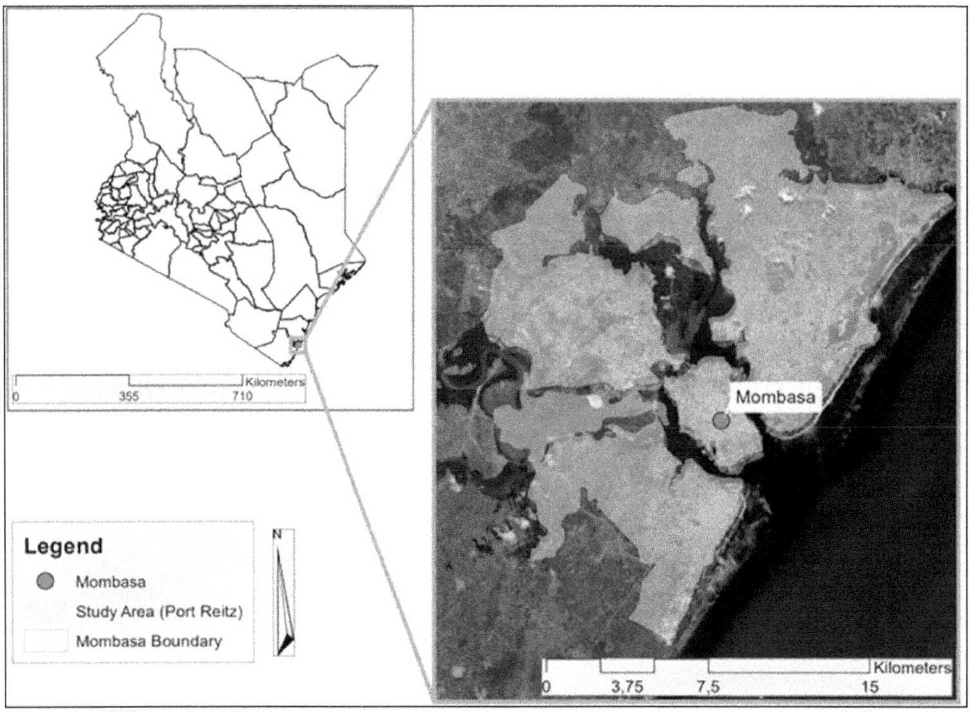

Figure 1.1: Map of the study area showing the Port Reitz–Kilindini hydraulic system. (*Source:* Authors' own work using ArcGIS Pro 3.3)

Methods

The Delta Underwater Gamma System (DUGS) consists of a 3"×3" NaI(Tl) scintillation detector, an Amptek® TB-5 multichannel analyser (MCA), and an external computer for data logging and processing. All components were housed in a watertight stainless-steel cylinder with delta-shaped wings for stability and obstacle avoidance on the seabed. Measurements were conducted using a Trimble® Yuma rugged tablet computer. Figure 1.2 shows the components and assembly of the system.

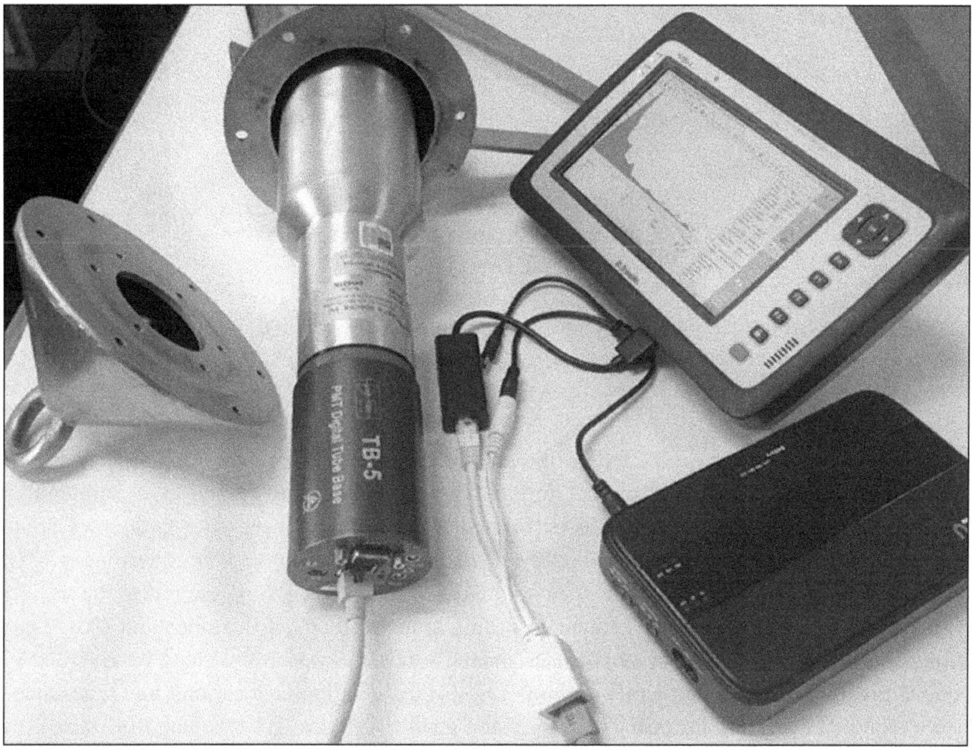

Figure 1.2: NaI(Tl) scintillation detector, TB-5 MCA, power bank, and computer with the housing of the DUGS detector.
(*Source:* Mtshawu et al., 2023)

The DUGS was engineered to continuously collect gamma-ray data, which were recorded alongside the geographical location and acquisition time. Figure 1.3 illustrates a standard gamma-ray spectrum captured by the DUGS, showing peaks corresponding to various natural radionuclides and their respective decay products. To handle fluctuations in the water temperature, the system incorporates an automated drift correction technique developed by Bezuidenhout et al. (2023).

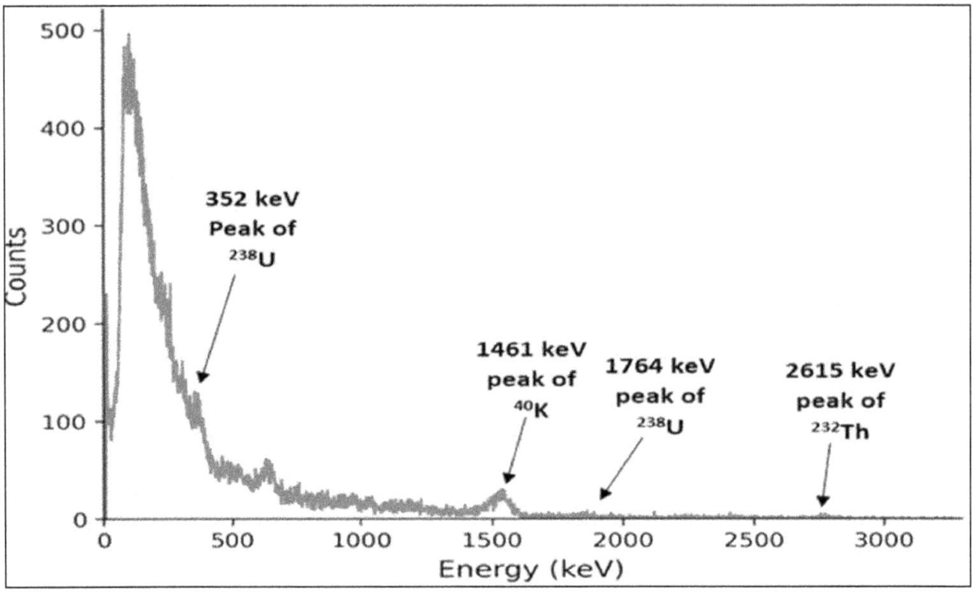

Figure 1.3: An in-situ gamma-ray spectrum acquired by the DUGS with the peaks of the natural radionuclides and their progeny.
(*Source:* Mtshawu et al., 2023)

Energy calibrations were conducted on the system before and after each measurement period using natural and anthropogenic nuclide gamma-ray peaks, as described by Bezuidenhout (2023). The efficiency calibration was performed at the Nuclear Energy Corporation of South Africa (NECSA) using three standard reference pads containing known concentrations of K_2O, U_3O_8, and ThO_2. The resulting efficiency parameters, which convert the count rate to the activity concentration (Bq/kg), were determined according to the methods of Bezuidenhout (2015) and Chiozzi et al. (2000). For the in-situ measurements, the DUGS was towed along the riverbed by a boat travelling at a speed of 1 m/s. Gamma-ray spectra, geographic locations, and timestamps were recorded every 5 s for post-processing and analysis. Figure 1.4 is a map illustrating the route taken while sampling the study area with a total of 8,674 sample points gathered. Each point provided the gamma-ray spectra, geographical position, and time log intervals of ^{40}K, ^{232}Th, and ^{238}U.

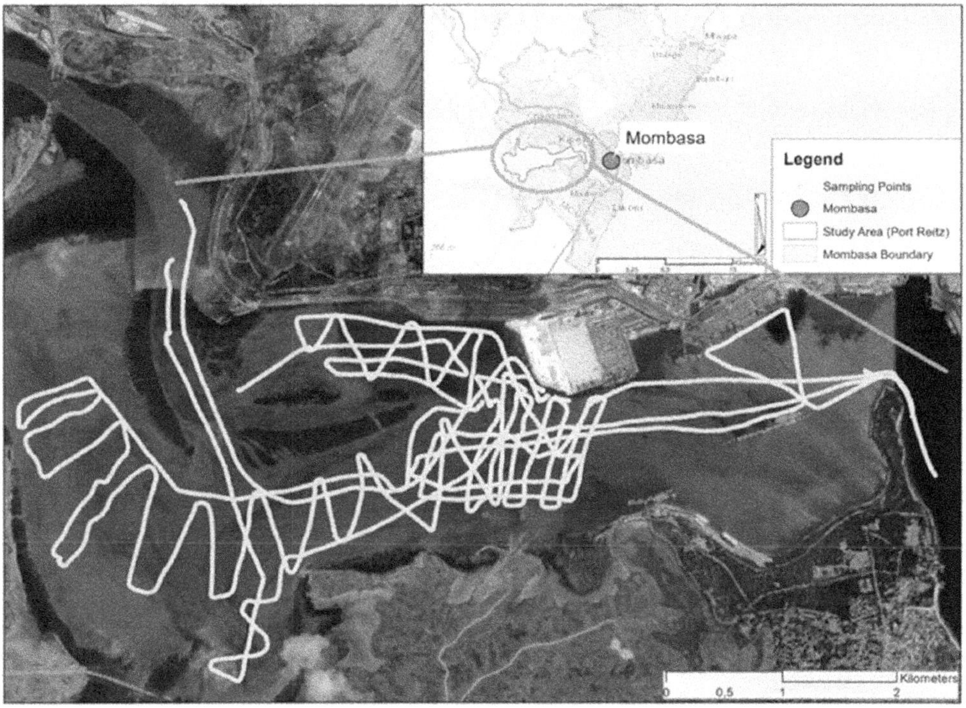

Figure 1.4: A map of the study area showing the route taken during data collection, with a total of 8,674 sampling points collected.
(*Source:* Authors' own work using ArcGIS Pro 3.3)

Spatial analysis

Inverse Distance Weighting (IDW)

IDW is a deterministic interpolation technique in which values at unsampled locations are estimated based on known sampled locations using a weight function in a search neighbourhood (Moeletsi et al., 2016). IDW assumes that the rate of correlations and similarities between neighbours is proportional to the distance between them, which is described as a function inversely proportional to the distance of every neighbouring point (Setianto & Triandini, 2013). It is important to note that the definition of the neighbouring radius and the related power to the function inversely proportional to the distance are considered important problems in this method (Ferreira et al., 2017).

This interpolation technique is recognised as not only deterministic, but also local and exact (Mtshawu, 2014). A local interpolation technique uses a sample of unknown points to estimate an unknown value, whereas an exact interpolator refers to an interpolation method that predicts a value at the point location that is the same as its known value (Chang, 2019). IDW is a straightforward interpolation method because it does not require pre-modelling, such as kriging. The formula and weighted function for the IDW were provided by Setianto and Triandini (2013).

This can be seen in the following equation:

$$Z_0 = \frac{\sum_{i=1}^{N} Z_i \left(d_i^{-n}\right)}{\sum_{i=1}^{N} Z_i \left(d_i^{-n}\right)}$$

where:

Z_0 is the estimation value of variable Z in point i.

Z_i is the sample value in point i.

d_i is the distance from the sample point to the estimated point.

N is the coefficient that determines the weight based on the distance.

n is the total number of predictions for each validation case.

Kriging

Kriging is a local interpolation technique that utilises a geostatistical method developed by G. Matheron in 1960 based on a master's thesis by D.G. Krige in 1951 (Burrough & McDonnell, 1998). A local interpolator uses the information in the vicinity of the point being estimated only (Singh & Verma, 2019). Kriging is exact and stochastic (Chang, 2019). It is exact because it predicts a value at the point location that is the same as the known value, thus minimising the prediction error at that point. It is stochastic because it considers the presence of randomness in its variables and offers an assessment of prediction errors with estimated variances.

OK is one of the most basic kriging methods (Setianto & Triandini, 2013). It offers an estimate for a location where the variable Z is not directly observed, based on the weighted average of adjacent observed sites within a given area. The theory is derived from regionalised variables. According to Burrough and McDonnell (1998: 133), "regionalised variable theory assumes that spatial variation of any variable can be expressed as the sum of three components". The first is the standard component with a constant mean; the second is a random but spatially correlated component, also known as the regionalised variable; and the last is the residual component, also referred to as the error. OK can be briefly described by considering an intrinsic random function denoted by:

$$\widehat{Z}\left(S_0\right) = \sum_{i=1}^{N} \lambda_i Z\left(S_i\right)$$

where:

$Z(S_i)$ The measured value at the ith.

λ_i is an unknown weight for the measured value at the ith location.

S_0 is the prediction location.

N is the number of measured values.

Semi-variogram

Kriging uses a semi-variogram to predict expected differences in values between pairs of data with a similar orientation (Collins & Woodcock, 1999). The semi-variogram can then be described as a representation of the average rate of change of a property with distance (Lam, 1983). According to Esri (2024), the semi-variogram can be defined as:

$$y(S_i, S_j) = \frac{1}{2}\, Variance\left(Z(S_i) - Z\left(S_j\right)\right)$$

If two locations S_i and S_j are close to one another in terms of distance measured $d\,(S_i, S_j)$ one should expect them to be similar. Therefore, the difference in their values $(Z(S_i) - Z(S_j)$ is small. As S_i and S_j become farther apart, they become increasingly different. Hence, the difference in their values $(Z(S_i) - Z(S_j)$ will become larger. Figure 1.5 shows a schematic representation of the semi-variogram.

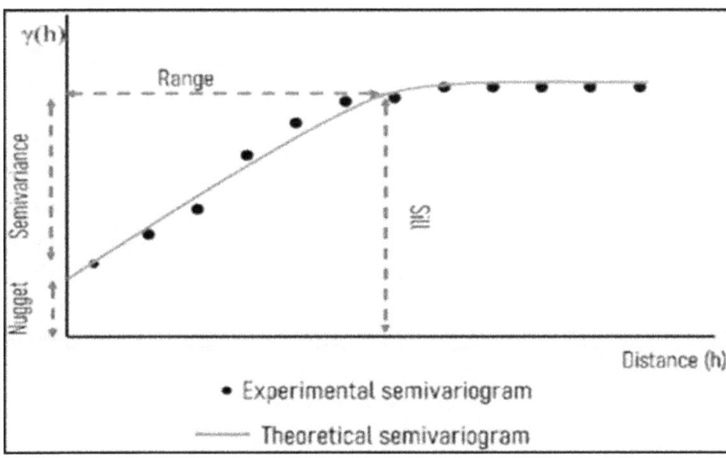

Figure 1.5: Semi-variogram.
(*Source:* Rocha et al., 2021)

The semi-variogram is important because it provides comprehensive information about a regionalised variable, including the size of the zone of influence around the sample, the isotropic or anisotropic nature of the variable, and the variable's spatial consistency (Cressie, 1993).

Validation and cross-validation

Validation and cross-validation refer to statistical techniques for assessing and comparing learning algorithms. It involves dividing the data into two segments: one is used for learning and training the model, and the other is used for validating the model's performance (Cressie, 1993). Each interpolation technique was then compared based on the ME and RMSE. The main aim of this comparative effort was to determine which interpolator produced the lowest TE, which is the combination of RMSE and ME Kravchenko and Bullock (1999). According to Kravchenko and Bullock (1999), the interpolator that produces the lowest TE portrays the most accurate predictions of the study area.

The ME is the average difference between the measured and predicted values. The value should be as close to zero as feasible. According to Al-Shboul et al. (2018), the ME measures the model's bias, where a positive ME indicates overprediction and a negative ME indicates underprediction. The formula for ME was described by Esri (2024).

$$ME = \frac{\sum_{i=1}^{n}\left(\widehat{Z}(S_i) - z(S_j)\right)}{n}$$

The RMSE indicates how closely the model predicts the measured values. The smaller the error, the better. The statistic (i.e. RMSE) measures the prediction accuracy and indicates the average deviation of the predicted values from the measured values. The values are expressed in the units of the data. For instance, for potassium interpolation measured in Bq/kg, a RMSE value of 1.5 means that the predictions differed from the measured values by approximately 1.5 Bq/kg on average (Esri, 2024).

$$RMSE = \sqrt{\frac{\sum_{i=1}^{n}\left(\widehat{Z}(S_i) - z(S_j)\right)^2}{n}}$$

Cross-validation scatter plot

ArcGIS Pro also offers a range of graphs and summaries comparing measured values with predicted values. Among these graphs, a scatter plot of the predicted versus measured natural radionuclides was produced. These scatter plots should be on a line of approximately 1:1 (grey-blue line, which is the scatter plot regression line). However, this is not always the case because the slope is usually less than 1. The fitted line through the scatter of points is shown in blue. If all the data are independent and there is no autocorrelation, the blue line should be horizontal. With autocorrelation and a good interpolation method, the blue line should be closer to the grey line in a 1:1 manner. It is important to note that all interpolation surfaces, cross-validation scatter plots, and cross-validation ME and RMSE were computed using the default parameters of the ArcGIS Pro 3.3 geostatistical analyst tool.

Results and discussions

The accuracy of the two interpolators in predicting naturally occurring radionuclides (^{40}K, ^{232}Th, and ^{238}U) was performed on the data collected at the Port Reitz–Kilindini hydraulic system, using all 8,674 sampled datasets. A total of six interpolation surfaces (Figures 1.6, 1.7, and 1.8) and six cross-validation scatter plots (Figures 1.9, 1.10, and 1.11) were produced. For the OK method, all the scatter plot graphs for ^{40}K, ^{232}Th, and ^{238}U were closer to the 1:1 line, indicating that OK provides better spatial autocorrelation. The IDW scatter plots showed a poorer autocorrelation than that of OK. This is because IDW uses a simpler logarithm than the complex logarithm for kriging, which fits a model. Validation was performed for each dataset, and the ME and RMSE values produced by each interpolation technique were tabulated. The TE was then calculated (Table 1.1). The interpolator that produced the lowest TE was considered the most accurate interpolator between the two (OK and IDW). All OK MEs between the measured and predicted values were closer to zero, and the RMSE was less than that produced by IDW. Subsequently, OK produced a smaller TE than IDW. This means that OK produced unbiased predictions.

Therefore, OK is the most suitable method for predicting and mapping the spatial distribution of ^{40}K, ^{232}Th, and ^{238}U.

Table 1.1: Evaluation performance of Ordinary Kriging (OK) and Inverse Distance Weighting (IDW) of ^{40}K, ^{232}Th, and ^{238}U through cross-validation and validation statistics

	OK			IDW		
	ME	RMSE	TE	ME	RMSE	TE
^{40}K	0.24	90.41	90.65	0.53	93.30	93.82
^{232}Th	-0.42	17.47	17.05	-0.05	18.30	18.25
^{238}U	-0.50	25.64	25.14	0.27	26.71	26.98

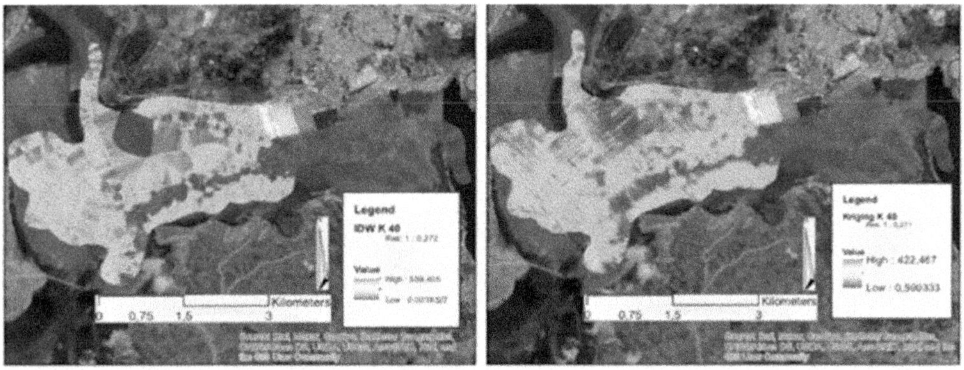

Figure 1.6: Inverse Distance Weighting and Ordinary Kriging prediction surface maps for ^{40}K. (*Source:* Authors' own work using ArcGIS Pro 3.3)

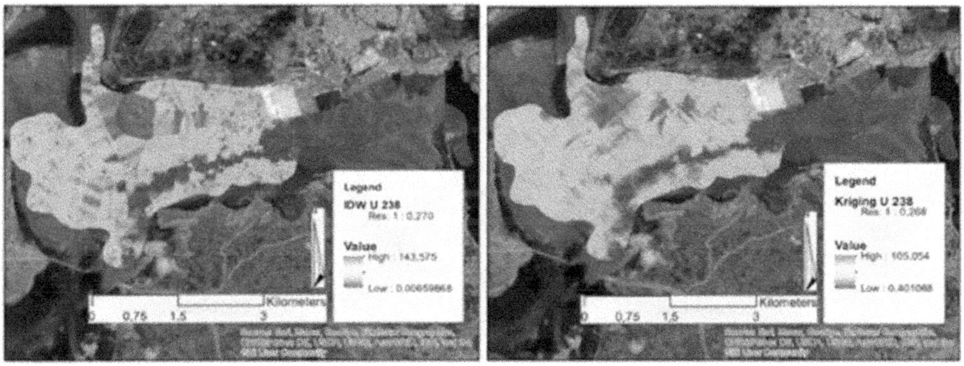

Figure 1.7: Inverse Distance Weighting and Ordinary Kriging prediction surface maps for ^{238}U. (*Source:* Authors' own work using ArcGIS Pro 3.3)

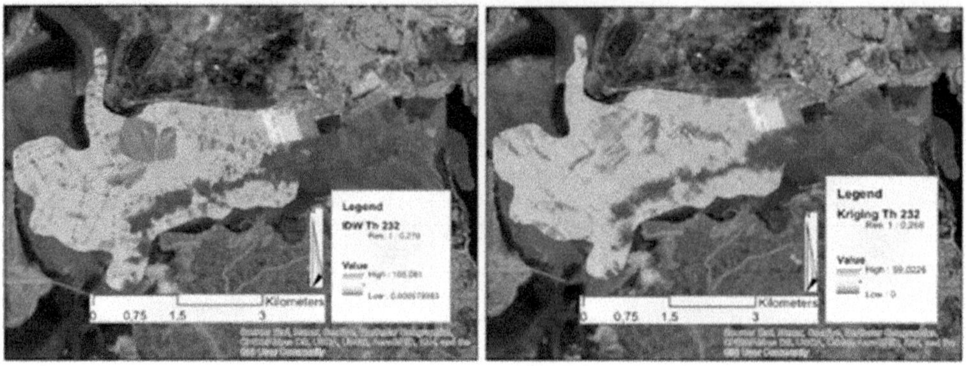

Figure 1.8: Inverse Distance Weighting and Ordinary Kriging prediction surface maps for ^{232}Th.
(*Source:* Authors' own work using ArcGIS Pro 3.3)

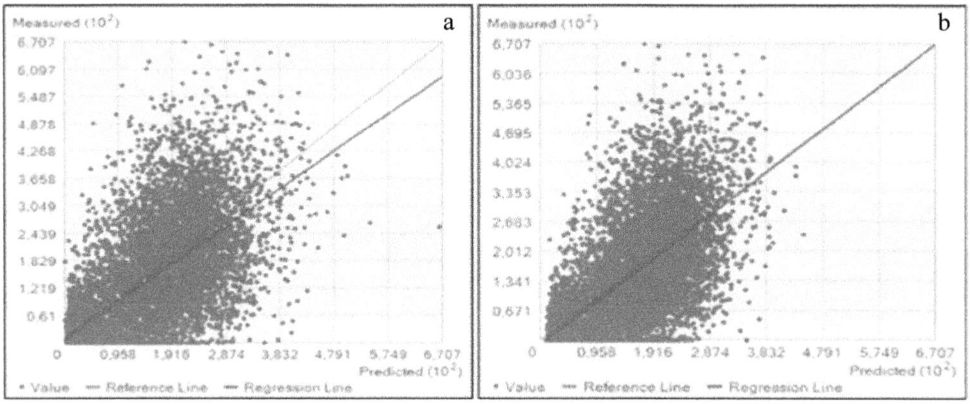

Figure 1.9: Cross-validation scatter plot of prediction errors for ^{40}K using Inverse Distance Weighting (a) and Ordinary Kriging (b).

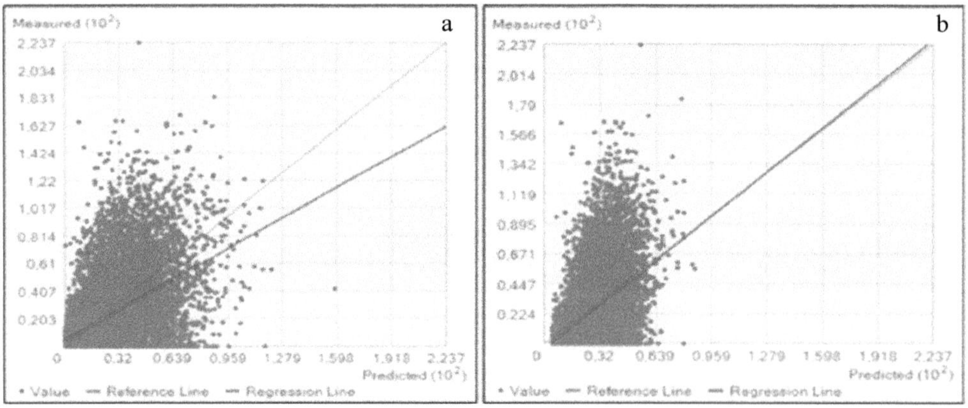

Figure 1.10: Cross-validation scatter plot of prediction errors for ^{238}U using Inverse Distance Weighting (a) and Ordinary Kriging (b).

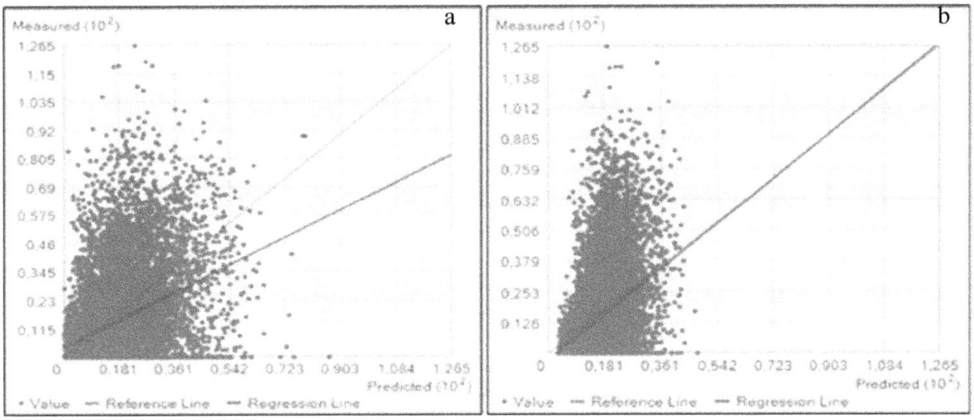

Figure 1.11: Cross-validation scatter plot of prediction errors for ^{232}Th using Inverse Distance Weighting (a) and Ordinary Kriging (b).

Conclusion

In this study, IDW was demonstrated to be a simpler/straightforward interpolation technique with fewer procedural steps compared to kriging. As outlined in the previous sections, the IDW estimates cell values by averaging the values of nearby sample data points, giving more weight to points closer to the centre of the cell being estimated. IDW tends to be more effective when sample data points representing the minimum and maximum values of the surface are known. Because IDW is an averaging process, all interpolated values are within the sample range. Conversely, for more detailed datasets, such as those presented in this study, kriging is the preferred method. Similar to IDW, kriging forms weights from surrounding measured values to predict values at unmeasured locations. As with IDW interpolation, the closest measured values usually have the greatest influence. However, kriging uses a more complex weighting system derived from a semi-variogram that reflects the spatial structure of the data, in contrast to IDW's simpler distance-based algorithm.

The primary aim of this study was to evaluate and compare the effectiveness of OK and IDW for interpolation using radiometric measurements of the natural radionuclides ^{40}K, ^{232}Th, and ^{238}U obtained from the Port Reitz–Kilindini hydraulic system using cross-validation and validation statistics. The interpolation methods that produced the ^{40}K, ^{232}Th, and ^{238}U surface maps in this study adhered precisely to the procedures outlined in the methods section. Both cross-validation and validation (ME and RMSE) were calculated and combined to determine which interpolator produced the lowest TE. A comparative analysis of the two interpolation methods demonstrated that kriging was the most suitable method for predicting and mapping the spatial distribution of radiometric measurements of the natural radionuclides ^{40}K, ^{232}Th, and ^{238}U. Although kriging was demonstrated to be a more reliable interpolator in this study, IDW also has notable advantages. The creation of maps is a crucial initial step in site-specific management because these maps illustrate spatial variability and provide the basis for controlling it. From this study, it can be concluded that, although OK performed slightly better than IDW in mapping the spatial distribution of radiometric measurements of the natural radionuclides ^{40}K, ^{232}Th, and ^{238}U, both interpolation methods can provide good-quality spatial distribution maps for radiometric measurements of the natural radionuclides.

Acknowledgements

This work was funded by the IAEA under the coordinated research project (CRP: F22074) and RAF 0056.

References

Alasadi, L.A. & Abojassim, A.A. 2022. Mapping of natural radioactivity in soils of Kufa districts, Iraq using GIS technique. *Environmental Earth Sciences*, 81(279). https://doi.org/10.1007/s12665-022-10407-8

Al-Shboul, K.F., Alali, A.E., Al-Shurafat, A.W., Arrasheed, A.A. & Al-Shboul, S.A. 2018. Spatial evaluation of radionuclide concentrations and the associated radiation hazards using the Kriging method. *Journal of Radioanalytical and Nuclear Chemistry*, 317: 1285–1297. https://doi.org/10.1007/s10967-018-6015-x

Bezuidenhout, J. 2015. In situ gamma ray measurements of radionuclides at a disused phosphate mine on the West Coast of South Africa. *Journal of Environmental Radioactivity*, 150: 1–8. https://doi.org/10.1016/j.jenvrad.2015.07.030

Bezuidenhout, J. 2019. Estimation of radon potential through measurement of uranium concentrations in granite geology. *South African Journal of Science*, 115(7–8): 1–4. https://doi.org/10.17159/sajs.2019/5768

Bezuidenhout, J. 2023. Investigating naturally occurring radionuclides in sediment by characterizing the catchment basin geology of rivers in South Africa. *Journal of Applied Geophysics*, 213. https://doi.org/10.1016/j.jappgeo.2023.105037

Bezuidenhout, J., le Roux, R.R., & Kilel, K. K.2023. The characterization and optimization of an Underwater Gamma-Ray Detection System (DUGS). *Journal of Marine Science and Engineering*, 11(1). https://doi.org/10.3390/jmse11010171

Burrough, P.A. & McDonnell, R.A. 1998. *Principles of Geographical Information Systems*. New York, NY: Oxford University Press.

Chang, K.T. 2019. *Introduction to Geographic Information Systems*. Ninth edition. New York, NY: McGraw-Hill.

Childs, C. 2004. Interpolating surfaces in ArcGIS spatial analyst. *ArcUser*, July–September: 32–35.

Chiozzi, P., Pasquale, V. & Verdoya, M. 2002. Naturally occurring radioactivity at the Alps-Apennines transition. *Radiation Measurements*, 35(2): 147–154. https://doi.org/10.1016/S1350-4487(01)00288-8

Collins, J.B. & Woodcock, C.E. 1999. Geostatistical estimation of resolution-dependent variance in remotely sensed images. *Photogrammetric Engineering and Remote Sensing*, 65(1): 41–50.

Cressie, N. 1993. *Statistics for Spatial Data*. New York, NY: John Wiley & Sons. https://doi.org/10.1002/9781119115151

Esri [Environmental Systems Research Institute]. 2024. How kriging works. https://pro.arcgis.com/en/pro-app/latest/tool-reference/3d-analyst/how-kriging-works.htm [17 July 2024].

Ferreira, C.J. 2015. A probabilistic approach to quantify the value of information of 4D seismic projects. Paper presented at the SPE Annual Technical Conference and Exhibition, Houston, Texas, USA, September 2015. https://doi.org/10.2118/178731-STU

Ferreira, I.O., Rodrigues, D.D., Santos, G.R.D. & Rosa, L.M.F. 2017. In bathymetric surfaces: IDW or Kriging? *Boletim de Ciências Geodésicas*, 23(3): 493–508. https://doi.org/10.1590/s1982-21702017000300033

Gong, G., Mattevada, S. & O'Bryant, S.E. 2014. Comparison of the accuracy of kriging and IDW interpolations in estimating groundwater arsenic concentrations in Texas. *Environmental Research*, 130: 59–69. https://doi.org/10.1016/j.envres.2014.01.007

Goovaerts, P. 1999a. Geostatistics in soil science: State-of-the-art and perspectives. *Geoderma*, 89(1–2): 1–45. https://doi.org/10.1016/S0016-7061(98)00078-0

Goovaerts, P. 1999b. Using elevation to aid the geostatistical mapping of rainfall erosivity. *Catena*, 34(3–4): 227–242. https://doi.org/10.1016/S0341-8162(98)00116-7

Gotway, C.A. & Hartford, A.H. 1996. Geostatistical methods for incorporating auxiliary information in the prediction of spatial variables. *Journal of Agricultural, Biological, and Environmental Statistics*, 1(1): 17–39. https://doi.org/10.2307/1400558

Henrico, I., Henrico, S., le Roux, R. & Bezuidenhout, J. 2023. Radiometric mapping of the Berg River estuary. *Transactions in GIS*, 27(1): 105–114. https://doi.org/10.1111/tgis.13011

Ikechukwu, M.N., Ebinné, E., Idorenyin, U. & Raphael, N.I. 2017. Accuracy assessment and comparative analysis of IDW, spline, and kriging in spatial interpolation of landform (topography): An experimental study. *Journal of Geographic Information System*, 9(3): 354–371. https://doi.org/10.4236/jgis.2017.93022

Journel, A.G. & Huijbregts, C.J. 1978. *Mining Geostatistics*. London: Academic Press.

Junior, H.K., Scolforo, J.R., de Mello, J.M. & de Oliveira, A.D. 2006. Avaliação de interpoladores estatísticos e determinísticos como instrumento de estratificação de povoamentos clonais de *Eucalyptus sp. Cerne*, 12(2): 123–136.

Kimeli, A., Thoya, P., Ngisiang'e, N., Ong'anda, H. & Magori, C. 2018. Satellite-derived bathymetry: A case study of Mombasa Port Channel and its approaches, Kenya. *Western Indian Ocean Journal of Marine Science*, 17(2): 93–102. https://doi.org/10.4314/wiojms.v17i2.8

Kravchenko, A. & Bullock, D.G. 1999. A comparative study of interpolation methods for mapping soil properties. *Agronomy Journal*, 91(3): 393–400. https://doi.org/10.2134/agronj1999.00021962009100030007x

Lam, N.S.N. 1983. Spatial interpolation methods: A review. *The American Cartographer*, 10(2): 129–150. https://doi.org/10.1559/152304083783914958

Laslett, G.M., McBratney, A.B., Pahl, P.J. & Hutchinson, M.F. 1987. Comparison of several spatial prediction methods for soil pH. *Journal of Soil Science*, 38(2): 325–341. https://doi.org/10.1111/j.1365-2389.1987.tb02148.x

Le Roux, R., Henrico, S., Bezuidenhout, J. & Henrico, I. 2023. Inverse Distance Weighting as an alternative interpolation method to create radiometric maps of natural radionuclide concentrations using QGIS. *Proceedings of the ICA (ICA-Proc)*, 5(10). https://doi.org/10.5194/ica-proc-5-10-2023

MacEachren, A.M. & Davidson, J.V. 1987. Sampling and isometric mapping of continuous geographic surfaces. *The American Cartographer*, 14(4): 299–320. https://doi.org/10.1559/152304087783875723

Meng, Q., Liu, Z. & Borders, B.E. 2013. Assessment of regression kriging for spatial interpolation – Comparisons of seven GIS interpolation methods. *Cartography and Geographic Information Science*, 40(1): 28–39. https://doi.org/10.1080/15230406.2013.762138

Moeletsi, M.E., Shabalala, Z.P., de Nysschen, G.A. & Walker, S. 2016. Evaluation of an inverse distance weighting method for patching daily and dekadal rainfall over the Free State province, South Africa. *Water SA*, 42(3): 466–474. http://dx.doi.org/10.4314/wsa.v42i3.12

Mtshawu, B. 2014. *Optimising interpolation as a tool for use in soil property mapping.* Master's dissertation. Bloemfontein: University of the Free State.

Mtshawu, B., Bezuidenhout, J. & Kilel, K.K. 2023. Spatial autocorrelation and hotspot analysis of natural radionuclides to study sediment transport. *Journal of Environmental Radioactivity*, 264. https://doi.org/10.1016/j.jenvrad.2023.107207

Mueller, T.G., Pierce, F.J., Schabenberger, O. & Warncke, D.D. 2001. Map quality for site-specific fertility management. *Soil Science Society of America Journal*, 65(5): 1547–1558. https://doi.org/10.2136/sssaj2001.6551547x

Njeru, G.N. 2009. Performance of the Mombasa Port: An empirical analysis. *Kenya Institute for Public Policy Research and Analysis. KIPPRA Discussion Paper* no. 101. https://repository.kippra.or.ke/bitstream/handle/123456789/2678/DP101.pdf?sequence=1

Olivera Viciedo, D., de Mello Prado, R., Martinez, C.A., Habermann, E., Branco, R.B.F., de Cássia Piccolo, M., Calero Hurtado, A., Peña Calzada, K. & Lata Tenesaca, L.F. 2021. Water stress and warming impact nutrient use efficiency of Mombasa grass (*Megathyrsus maximus*) in tropical conditions. *Journal of Agronomy and Crop Science*, 207(1): 128–138. https://doi.org/10.1111/jac.12452

Onundo, L.P. 2017. *Estimating tidal energy resource potential for power production along the Kenyan coastline.* Doctoral dissertation. Nairobi: University of Nairobi.

Onundo, L.P. & Mwema, W.N. 2017. Estimating marine tidal power potential in Kenya. *International Journal of Energy and Environmental Engineering*, 10(7): 1013–1017.

Otoi, O.S., Park, Y.S., Mutisya, K.M. & Park, J.S. 2015. The status of marine traffic on Port of Mombasa and maritime activities in Kenya. Asia Navigation Conference, Kitakyushu, Fukuoka, Japan.

Rees, J.G., Williams, T.M., Nguli, M.M., Kairu, K.K. & Yobe, A.C. 1996. Contaminant transport and storage in the estuarine creek systems of Mombasa, Kenya. Nottingham: British Geological Survey.

Rocha, S.S., Pitombo, C.S., Costa, L.H.M. & de França Marques, S. 2021. Applying optimisation algorithms for spatial estimation of travel demand variables. *Transportation Research Interdisciplinary Perspectives*, 10. https://doi.org/10.1016/j.trip.2021.100369.

Sajid, A.H., Rudra, R.P. & Parkin, G. 2013. Systematic evaluation of kriging and inverse distance weighting methods for spatial analysis of soil bulk density. *Canadian Biosystems Engineering*, 55: 1.1–1.10. https://doi.org/10.7451/CBE.2013.55.1

Setianto, A. & Triandini, T. 2013. Comparison of kriging and inverse distance weighted (IDW) interpolation methods in lineament extraction and analysis. *Journal of Applied Geology*, 5(1): 21–29. https://doi.org/10.22146/jag.7204

Singh, P. & Verma, P. 2019. A comparative study of spatial interpolation technique (IDW and Kriging) for determining groundwater quality. In: S. Venkatramanan, M.V. Prasanna & S.Y. Chung (eds.). *GIS and Geostatistical Techniques for Groundwater Science*. Amsterdam: Elsevier. 43–56. https://doi.org/10.1016/B978-0-12-815413-7.00005-5

Smith, T.B., Smith, N. & Weleber, R.G. 2017. Comparison of nonparametric methods for static visual field interpolation. *Medical & Biological Engineering & Computing*, 55: 117–126. https://doi.org/10.1007/s11517-016-1485-x

Tabios III, G.Q. & Salas, J.D. 1985. A comparative analysis of techniques for spatial interpolation of precipitation. *JAWRA Journal of the American Water Resources Association*, 21(3): 365–380. https://doi.org/10.1111/j.1752-1688.1985.tb00147.x

Tomczak, M. 1998. Spatial interpolation and its uncertainty using automated anisotropic inverse distance weighting (IDW) – Cross-validation/Jackknife approach. *Journal of Geographic Information and Decision Analysis*, 2(2): 18–30.

Waters, N.M. 1988. Expert systems and systems of experts. In: W.J. Coffey (ed.). *Geographical Systems and Systems of Geography: Essays in Honour of William Warntz*. Ontario: Department of Geography, University of Western Ontario. 173–187.

CHAPTER 2

Mapping and determining land degradation using GIS and remote sensing: A case of General De La Rey Military Training Area

N.V. Mathebula and R.R. le Roux

Abstract

Military training areas are crucial for maintaining operational readiness. However, intensive activities such as live ammunition exercises and heavy vehicle movements can contribute to land degradation through vegetation loss and soil disturbance. This study assessed land degradation in the General De La Rey Military Training Area, a key South African National Defence Force (SANDF) training site, using cost-effective geographic information systems (GIS) and remote sensing techniques. By analysing vegetation cover and changes in 2013 and 2023, Landsat 8 OLI/TIRS-derived vegetation indices – Normalised Difference Vegetation Index (NDVI), Soil-Adjusted Vegetation Index (SAVI), and Tasselled Cap Brightness (TCB) – were used to evaluate vegetation cover and soil conditions. Spectral profiles of the mean surface reflectance were used to compare the changes in dams, buildings, bare soil, and vegetation in the area. Maps were compiled using the calculated vegetation indices to visualise vegetation cover and detect changes in soil conditions between 2013 and 2023, providing a spatial representation of potential land degradation trends. Contrary to expectations, the results indicated an overall increase in vegetation cover with no significant signs of soil erosion or degradation. Spectral analysis also confirmed these results, revealing distinct differences between the features in band 5 near-infrared (NIR) to band 7 shortwave infrared 2 (SWIR2), which show variations in vegetation health and land cover. An accuracy assessment was conducted to evaluate how effectively the data points were classified into the correct categories. The results indicated an overall classification accuracy of 71%, demonstrating the reliability of the methodology for long-term environmental monitoring in military landscapes.

Keywords: geographic information system, land degradation, military training area, remote sensing, vegetation indices

Introduction

On a global scale, land degradation has emerged as one of the most significant environmental challenges (Wessels et al., 2007). Land degradation is defined as a process in which there is a reduction in the value of biological and economic productivity (Hoffman & Todd, 2000). Land degradation is considered a severe and persistent environmental problem in Southern Africa, driven by a combination of climatic and human factors. Consequently, land resources can no longer provide the ecosystem with the essential services required for its survival. Modifications are made to the soil and vegetation, resulting in increased environmental vulnerability and loss of biodiversity. In essence, land degradation can intensify soil erosion, and if prolonged, can decrease the soil's nutrient supply, causing vegetation vulnerability (Tilauhun & Zewide, 2021).

Vegetation vulnerability is caused by several environmental factors at different intensities, causing vegetation loss and a reduction in soil quality. Among these factors, human activities are the most harmful (Minale & Gelaye, 2021). Without persistent pressure from human activities, vegetated regions are less subject to degradation. This degradation leads to significant changes in ecosystems and the environment, making it essential to monitor these transformations.

Change detection is a technique for examining and analysing changes in an object or spatial data at several intervals to detect its variations (Rokni & Musa, 2019; Elnaggar et al., 2020). It compares spatial datasets that cover the same areas at different times while limiting variations caused by differences in irrelevant factors and measuring changes caused by those differences in relevance (Asokan & Anitha, 2019). In change detection, remote sensing and geographic information systems are essential tools for assessing vegetation sensitivity because of the broad spatial dimensions in which it is assessed and tracked (Minale & Gelaye, 2021). Loss of vegetation cover is a critical indicator of land degradation. Vegetation indices were used to measure vegetation cover.

Vegetation indices are mathematical combinations of two or more bands associated with the spectral properties of vegetation (Rokni & Musa, 2019). It enhances sensitivity to vegetation traits while reducing elements such as soil background reflectance (Basso et al., 2004; Hatfield et al., 2019). There are several types of vegetation indices, such as the Normalised Difference Vegetation Index (NDVI) and the Soil-Adjusted Vegetation Index (SAVI) (Vani & Mandla, 2017). The SAVI is a vegetation index that accounts for and minimises the influence of soil background conditions (Xue & Su, 2017). NDVI, on the other hand, is a spectral index that is used to quantify vegetation by detecting the spatial change in vegetation cover (Agone & Bhamare, 2012). Tasselled Cap Transformation (TCT) optimises data viewing for vegetation studies, which have three bands describing brightness, wetness, and greenness. Tasselled Cap Brightness (TCB) specifically analyses the variations in soil background reflectance by providing a measurement of soil brightness (Vorovencii, 2007).

Military operations also affect the physical environment. Activities such as live fire training, testing drills, and heavy vehicle movement can intensify soil erosion, and if prolonged, can decrease the soil's nutrient supply and vegetation cover (Henrico et al., 2021; Smit & Henrico, 2021). This results in the loss of vegetation cover and soil erosion on the landscape surface. The intensity of activities, such as those involving armoured vehicles, is linked to land cover changes. Frequent and aggressive exercise inadvertently degrades vegetation and causes significant disturbances (Magagula, 2016; Demarais et al., 1999). Live fire exercises involving various weapon systems and ammunition are highly destructive and significantly contribute to vegetation disturbance and loss (Magagula, 2014; Lawrence et al., 2015; Mokalapa, 2022). Heavy metals and explosive munitions also contaminate vegetation, negatively impacting wildlife and other species that depend on them. Land degradation is almost unavoidable in areas of concentrated movement of heavy vehicles, which is observed as vegetation loss (Haugen et al., 2003). In conclusion, intense military operations contribute to vegetation vulnerability.

This study aimed to determine and map the degradation caused by military training exercises at the General De Re Lay Training Area by utilising remote sensing (RS) data and geographic information system (GIS) methods. This includes using spectral data to calculate the NDVI, SAVI, and TCB to determine how all indices delineate changes in vegetation cover and bare soil of the land cover. These three indices were used because they provide an overview of vegetation health and land surface conditions. The NDVI measures the amount of vegetation on the land surface, the SAVI accounts for soil brightness in areas with sparse vegetation cover, and the TCB index represents the overall brightness of the land surface. Low NDVI and SAVI values indicate poor vegetation cover and soil, while high TCB values indicate surfaces such as bare soil, in which vegetation loss and increased erosion are associated with land degradation. NDVI, SAVI, and TCB can indicate whether vegetation and soil deteriorate over time and how susceptible the area is to erosion, which may intensify land degradation.

Literature review

According to Bajocco et al. (2018), land degradation is one of the world's most pressing environmental problems. Humans and their activities are the primary cause of land degradation, while climate change is another major factor (Tilauhun & Zewide, 2021). Changes will always occur in the physical landscape (Dwivedi et al., 2005). One of the major causes of land degradation is low vegetation cover (Tilauhun & Zewide, 2007, 2021). Soil degradation can reduce soil productivity and prevent essential nutrients from reaching plant roots. Parmar et al. (2021) suggest that to monitor and assess land degradation, it is necessary to use a complex, multidisciplinary approach and integrate datasets from various sources and methods. Remote sensing and GIS have developed into separate spatial data tools with unique approaches to data analysis and depiction. Hutchinson et al. (2015) support the idea that remote sensing data is widely used to monitor degradation and highlight the use of geographical surveys as a complementary approach.

As stated earlier, military activities are a source of land degradation. According to Simmons (2004), there is not enough area accessible for training exercises, and as a result, the same training lands are used repeatedly, causing further degradation. In general, the movement of military vehicles causes minor disruption to the soil and vegetation, whereas aggressive tactical manoeuvres cause severe damage (Haugen et al., 2003). According to Lawrence et al. (2015), vehicle weight is directly proportional to land degradation and is sufficient to exacerbate soil erosion and reduce vegetation cover. However, the most significant negative impact of military vehicle movement on soil is compaction, which can lead to the degradation of land. Military training operations usually entail explosions that can remove significant amounts of vegetation cover (Althoff & Thien, 2005; Broomandi et al., 2020; Zentelis et al., 2017). Additionally, military munitions contain non-biodegradable components that contaminate the land and lead to environmental degradation. A study conducted by Magagula (2016) at the Grahamstown Military Training Area (GMTA) and the South African Army Combat Training Centre (SAACTC) showed that remote sensing techniques can be used to study land cover changes, especially in military training areas, without the need for in-situ analysis. The aggressive nature of military exercises, which severely affects land cover, is thus important.

Vegetation can be used to indicate the type of environmental conditions that exist at different spatial and temporal times (Cilliers et al., 1999; Augusto et al., 2003). A change detection technique can be used to identify and analyse land changes over time. This is possible because the land surface has properties that can be analysed, including vegetation and soil coverage (Polykretis et al., 2020). Vegetation indices are efficient methods for assessing vegetation cover, as they are linked to vegetation parameters derived from various spectral ranges of remotely sensed data (Gilabert et al., 2002). Xue and Su (2017) emphasise that selecting the ideal approach for a given problem depends on the applicability of remote sensing and the various vegetation indices it generates, which are largely influenced by the platforms and instruments used.

Normalised Difference Vegetation Index (NDVI)

The NDVI is a satellite-derived vegetation index that is strongly correlated with surface net primary productivity (Pettorelli et al., 2005). The NDVI, first introduced by Rouse et al. (1973), exhibits a strong correlation between the quantity of green biomass and leaf surface area. Consequently, it is among the most extensively utilised indicators for monitoring changes in agricultural activities (Lv, 2014; Coetzee, 2022). The NDVI is sensitive to background factors such as soil brightness, soil colour, atmosphere, cloud cover, cloud shadow, and leaf canopy shadow (Xue & Su, 2017; Huang et al., 2021). The formula defined by Rouse et al. (1973) for the NDVI is as follows:

$$\mathrm{NDVI} = \frac{\mathrm{NIR} - \mathrm{RED}}{\mathrm{NIR} + \mathrm{RED}}$$

where:

NIR represents the vegetation reflectance in the near-infrared (NIR) band.
RED is the vegetation reflectance in the red band.

NDVI values range from -1 to +1, reflecting the vegetation conditions. Values close to zero indicate little to no green vegetation, whereas positive values signify increasing vegetation density, with higher values representing the greatest potential leaf cover (Shahabi et al., 2012). However, owing to soil influence, the NDVI tends to underestimate sparse vegetation and saturates at very high vegetation densities (Solymosi et al., 2019).

A study by Henrico et al. (2021) conducted at the South African Army Combat Training Centre, Lohathla, suggests that prolonged military exercises intensify the effects of degradation, such as soil erosion and vegetation loss. They further stated that conditions such as extended military exercises, heavy vehicle movement, frequent firepower use, and seasonal variations make it easier to use the NDVI to calculate the change in vegetation cover. However, despite its usefulness, the NDVI is often considered less effective in accurately representing vegetation recovery, particularly in forested areas, because grass and other non-woody vegetation can repopulate a site after a disturbance and revert the NDVI signal to its pre-disturbance condition (Hislop et al., 2018).

According to Ke et al. (2015), surface reflectance measurements are inconsistent because of variations in sensor viewing circumstances, spectral band responses, equipment performance, and atmospheric variables at the time of observation. This makes it more challenging to use NDVI data obtained from multiple sensors. Land degradation, deforestation, and habitat monitoring have all benefited from the NDVI's ability to detect changes in the amount and health of vegetation (Huete et al., 2002). To address the NDVI's limitations in areas with sparse vegetation and soil, the SAVI was introduced to improve accuracy by minimising background soil effects.

Soil-Adjusted Vegetation Index (SAVI)

The SAVI was created to counter the limitations of NDVI in regions with high soil brightness, where the vegetation signal can be concealed by soil reflection (Huete, 1988). The SAVI is a vegetation index similar to the NDVI, except that it additionally considers the reflectance of the soil background (Sashikkumar et al., 2017; Ren et al., 2018). The SAVI uses the spectrum reflectance of solar radiation from the soil and the vegetation, which enables it to measure the density and intensity of green vegetation growth (Nanasaheb & Paswan, 2020). The formula defined by Huete (1988) is as follows:

$$\mathrm{SAVI} = \frac{\mathrm{NIR} - \mathrm{RED}}{\mathrm{NIR} + \mathrm{RED} + \mathrm{L}}(1 + \mathrm{L})$$

where:

NIR is the vegetation reflectance in the NIR spectral bands.
RED is the vegetation reflectance in the red spectral bands.

L is a constant parameter representing the soil adjustment factor, with values ranging from 0 to 1 (Vani & Mandla, 2017; Ren et al., 2018).

Huete (1988) suggested an optimal adjustment factor of L = 0.5 for areas with moderate vegetation cover to reduce soil background variation. The selection of L is determined by the density of vegetation: L = 1 was applied for analysing areas with very low vegetation cover, L = 0.5 for areas with intermediate vegetation cover or when the extent of vegetation cover is unclear, and L = 0.25 for areas with denser vegetation cover (Rondeaux et al., 1996). The SAVI is more suited for dense vegetation because its L parameter can be adjusted, unlike the NDVI. The value of SAVI is equal to NDVI when L is near 0 (Xue & Su, 2017). While SAVI equals NDVI when L is near 0, the flexibility of SAVI allows for more accurate vegetation assessment in areas with different soil conditions and vegetation densities. The introduction of the SAVI marked a substantial advancement in the precision of vegetation assessments, particularly in regions with diverse soil conditions (Bannari et al., 1995).

The SAVI uses the spectrum reflectance of solar radiation to measure the density and vigour of green vegetation growth (Bannari et al., 1995; Nanasaheb & Paswan, 2020). It is an essential tool for analysing vegetation dynamics in various environments because of its capacity to consider soil differences. The use of SAVI has become more widespread owing to developments in remote sensing technology, notably the availability of high-resolution satellite imagery. To better understand environmental changes, the SAVI can be used to monitor vegetation at finer spatial scales (Nagendra, 2001). It is a useful technique for evaluating vegetation in areas with a variety of soil types, as research has shown that it is effective at lowering the impact of soil brightness on vegetation indices. For instance, the SAVI has been successfully used in agricultural studies to track crop health in regions where soil brightness has a major impact on the NDVI (Tucker, 1979; Huete, 1988). This includes studies conducted by Hishe et al. (2017) in Ethiopia and Zhen et al., (2021) in Apiacás in the northernmost of Mato Grosso, Brazil. Thus, the SAVI provided more precise results than the NDVI in regions with varying soil types.

Tasselled Cap Transformation (TCT)

Xue and Su (2017) defined the TCT as a technique that expresses more physical features of the landscape than other vegetation indices by considering the spectral properties of the vegetation. The TCT is widely used in remote sensing because it can link features with physical landscape characteristics (Liu et al., 2014). It considers the linear combinations of the spectral bands and transforms six Landsat bands into three bands, which include brightness, greenness, and wetness (Shahabi et al., 2012; Xue & Su, 2017):

1. Tasselled Cap Brightness (TCB) shows variations in soil background reflectance (Li & Chen, 2014).

2. Tasselled Cap Greenness (TCG) shows variations in the health of green vegetation and responds to the increased absorption of chlorophyll and the high infrared reflectance caused by the interior structure of leaves (Vorovencii, 2007).

3. Tasselled Cap Wetness (TCW) considers the interrelationship between soil and canopy moisture by comparing the differences between the water content and densely vegetated areas (Liu et al., 2014).

TCT indices also offer a thorough representation of many geographical features, including water and bare terrain (Samarawickrama et al., 2017). In a 2014 study conducted by Liu et al. (2014) on the landscapes of the Yellow River Delta in China, the TCT parameters were calculated for the Landsat 8 Operational Land Imager (OLI) and Top of Atmosphere (TOA) reflectance images, and the differences in the parameters for the TCT between spring and autumn were compared. The study concluded that TCT parameters in October were most appropriate for regional remote sensing applications when atmospheric correction is impractical. TCB is an important indicator of the general reflectivity and brightness of land surfaces. According to Crist and Cicone (1984), the TCB formula for Landsat data is a linear combination of the spectral bands. In the current study, the computational model for the TCB was based on Li and Chen's (2014) formula:

$$TCB = [0.3037(OLS_2)] + [0.2786(OLS_3)] + [0.4733(OLS_4)] + [0.5599(OLS_5)] + [0.5082(OLS_6)] + [0.1872(OLS_7)]$$

where:

OLS_i represents the Landsat 8 Surface Reflectance (SR) Digital Number (DN) values of corresponding bands.

Brightness is the weighted average of all bands and is responsible for the greatest amount of image variability. It naturally coexists with bare or partially covered soil, natural or artificial features, and topographical changes (Maity & Mane, 2020). It shows surface brightness, with high values indicating bare soil or disturbed land and low values suggesting dense vegetation. TCB is useful for several tasks, such as classifying land cover, tracking land use, and assessing the environment. This was demonstrated in a study by Dymond et al. (2002), who improved the precision of mapping and land cover classification using tasselled cap indices. TCB is used to convert multispectral data into orthogonal components that capture diverse land cover attributes (Maity & Mane, 2020). The tasselled cap indices were successfully used in a study conducted by Mbow et al. (2004) to reflect and show the distribution of fuel moisture in savanna areas.

In remote sensing applications, the SAVI, NDVI, and TCB have specific functions. The SAVI accounts for soil background effects, the NDVI serves as a common indicator of vegetation health, and the TCB represents the overall brightness of the land surface.

Materials and methods

Study area

The General De La Rey Military Training Area is located in the North West province of South Africa. The military base is situated in Potchefstroom under the J.B. Marks Municipality (Figure 2.1). It is situated between latitudes 26°37'49" S and 26°40'37" S and between longitudes 27°02'39" E and 27°04'20" E. The region has a warm, temperate climate characterised by moderate seasonal and daily temperature changes. The training area experiences a warm, summer-like climate with an average annual rainfall of 660 mm (Daemane et al., 2010). The characteristics of the subsurface drainage networks in the area are influenced by the location of the area at the edge of a dolomitic landscape (Smit, 2020). Its environment is categorised as a grassland biome, with vegetation primarily composed of low grasses and a sizable shrub component (Cilliers et al., 1999; Daemane et al., 2010). Owing to its proximity to the Highfield Nature Reserve, the training area is home to numerous plant species.

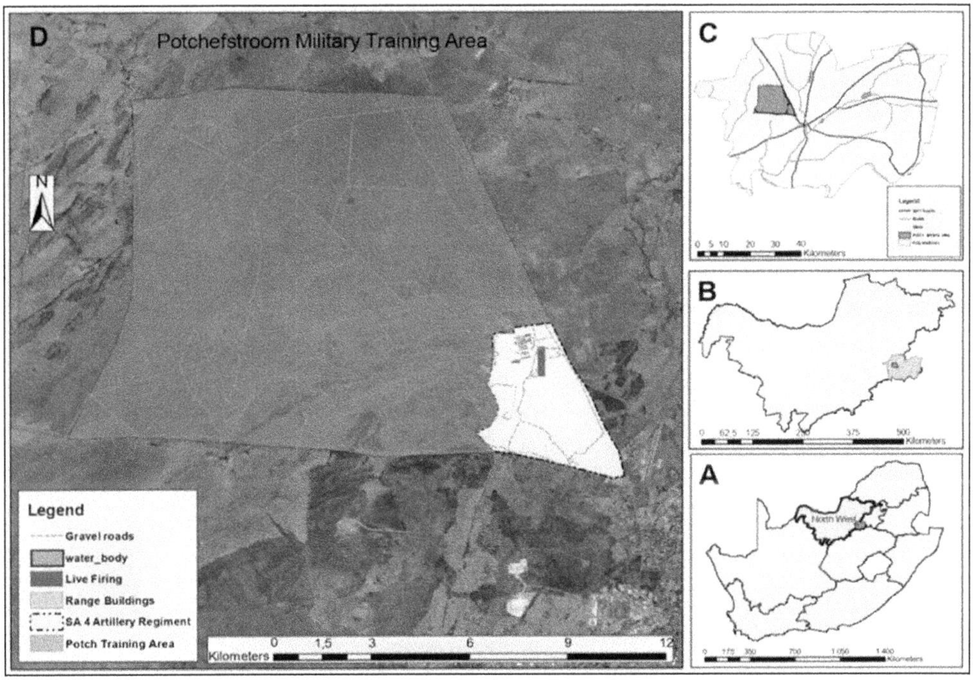

Figure 2.1: Illustration of the study area. The inlet maps indicate (D) the General De La Rey Military Training Area located in (C) Potchefstroom in (B) the North West Province of (A) South Africa.
(*Source:* Authors' own work using ArcGIS/Esri)

The Army Support Base (ASB) Potchefstroom, which provides support services to the local militaries, is responsible for the administrative control of the General De La Rey Military Training Area (Mokalapa, 2022). It is one of the largest military training areas in the North West Province and is used for live firing exercises (Cock & McKenzie, 1998). During the exercises, various weapons are fired, ranging from small arms to infantry, armour, and artillery munitions (Mokalapa, 2022).

Methodology

This desktop study aimed to use geospatial techniques to map and determine land degradation. The study used remotely sensed data from Landsat 8 Operational Land Imager (OLI) and Thermal Infrared Sensor (TIRS) to obtain real-time data and information on current earth observations at the General De La Rey Military Training Area. Landsat 8 OLI provides data in multiple spectral bands, allowing for the detailed characterisation of land cover, vegetation health, water quality, and geological features (USGS, 2016). The TIRS instrument on Landsat 8 provides valuable thermal infrared data that are crucial for applications such as land degradation (USGS, 2016). A total of seven scenes covered the General De La Rey Military Training Area, which were downloaded from the USGS EarthExplorer website.[1] The same months were investigated in 2013 and 2023, as the timeframe proved sufficient to map any significant changes. With the area categorised as a grassland biome, the NDVI assesses the density of the vegetation, and the SAVI improves the results of the NDVI by adjusting the influence of the soil brightness, while the TCB measures the surface brightness by analysing the land cover. Utilising spatio-temporal analysis techniques, including the NDVI, the SAVI, and the TCB, changes in land cover from 2013 to 2023 were analysed.

Radiometrically and geometrically corrected data images were available; hence, Landsat 8 data were used. Landsat 8 OLI/TIRS photos of the study area were downloaded from the Earth Explorer website. The NDVI, SAVI, and TCB were computed using the applicable formulas. ArcGIS 10.8.1. was used to create a spectral profile categorising features into vegetation, buildings, bare soil, and water bodies. For each category, 15 data points were collected from the Landsat 8 SR image, and a signature file representing the spectral characteristics of the identified categories in the study area was generated. The Spatial Analyst tool within ArcGIS 10.8.1 was employed to generate the NDVI, SAVI, and TCB surfaces, applying the formulas outlined in the literature. These indices were standardised to ensure consistent index ranges, facilitating the assessment of the most effective indices. All three indices were normalised to range between 0 and +1 using the following equation (Patro & Sahu, 2015):

$$\mathrm{z} = \frac{\mathrm{x} - \min(\mathrm{x})}{\max(\mathrm{x}) - \min(\mathrm{x})}$$

where:

z is the normalised index.
x is the index to be normalised.
$\min(x)$ is the minimum value of the index.
$\max(x)$ is the maximum value of the index.

1 https://earthexplorer.usgs.gov/

The normalised indices (the NDVI, SAVI, and TCB) were calculated using the raster calculator tool in ArcGIS and compared using the identical classification points utilised for spectral analysis. These points were used to extract pixel values for the normalised indices, enabling a detailed comparison. The information was analysed using interpretive methods to establish the connection between vegetation cover and land degradation. An accuracy assessment was done to quantitatively evaluate the efficacy of the pixels' sampling in assigning the appropriate land cover categories. The spatial analyst tool in ArcGIS 10.8.1 was used to perform the accuracy assessment computation. To successfully complete the accuracy assessment, an error matrix was created to compare the values assigned to each land feature using the classification process. A random set of 100 points was generated and compared with aerial photos from the USGS EarthExplorer website. Cohen's kappa coefficient, a crucial accuracy assessment metric, was also computed using the error matrix.

Results and discussion

Different forms of land cover can be distinguished using spectral analyses. This is possible because different materials reflect and absorb different wavelengths of light. The analysis of spectral bands makes it possible to visualise image features that share similar combinations of spectral reflectance. Figure 2.2 shows the reflectance of the land cover in the General De La Rey Military Training Area in 2013, using spectral pattern analysis.

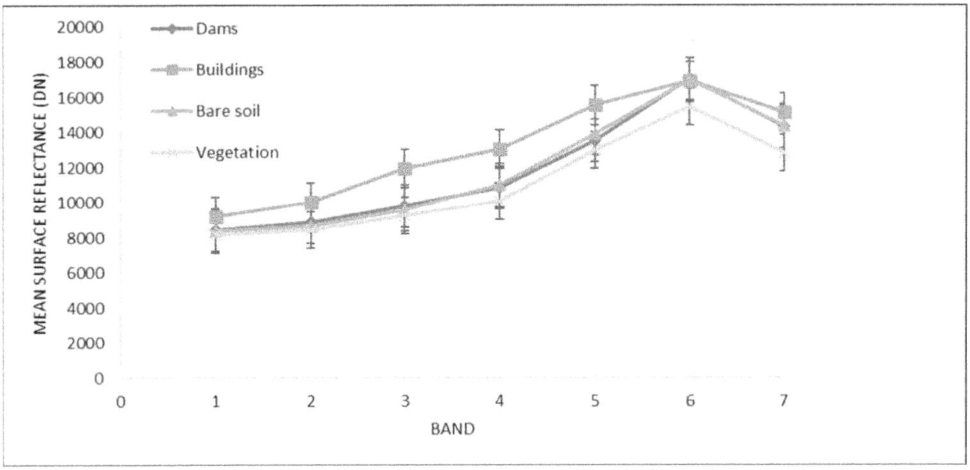

Figure 2.2: The spectral profile of the four categories in the year 2013.

Similarly, Figure 2.3 shows the reflectance of the land cover in the General De La Rey Military Training Area for 2023. The features in Figures 2.2 and 2.3 were divided into four categories: dams, buildings, bare soil, and vegetation.

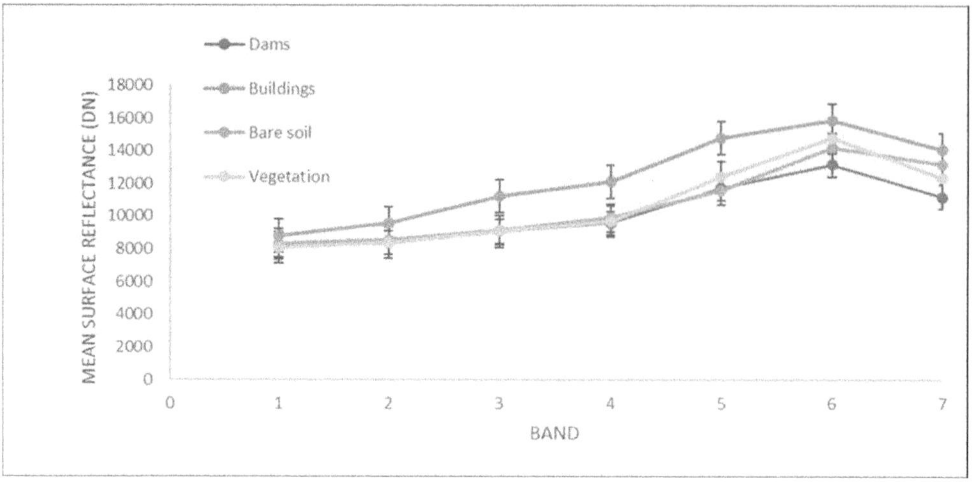

Figure 2.3: The spectral profile of the four categories in the year 2023.

The features show similar tendencies on both graphs from band 1 (coastal aerosol) and band 4 (red). The differences between the features are apparent in band 5 near-infrared (NIR) to band 7 shortwave infrared 2 (SWIR2). These differences indicate changes in land cover conditions between 2013 and 2023. The spectral profile illustrates how Landsat 8 SR data can specifically identify and distinguish vegetation and bare soil by analysing their reflectance values across bands.

When the NDVI, SAVI, and TCB findings were compared visually, all three indices identified and delineated some areas of bare soil. The results were given a green-to-red colour ramp symbology (Figures 2.4, 2.5, and 2.6), which shows high positive values as green and low negative values as red. The NDVI values for the years 2013 and 2023 corresponded to the ranges of -0.1 to 0.1. The NDVI values for 2013, illustrated in Figure 2.4(a), show a high value of 0.571 and a low value of -0.305. High positive NDVI values within the range of 0.2 to 0.5 indicate sparse vegetation, such as shrubs and grasslands. An NDVI value of 0.571 suggests the presence of dense and healthy vegetation. An NDVI value of -0.305 generally signifies non-vegetated surfaces or sparsely vegetated areas. This may indicate water bodies, barren land, urban areas, or areas with limited vegetation cover.

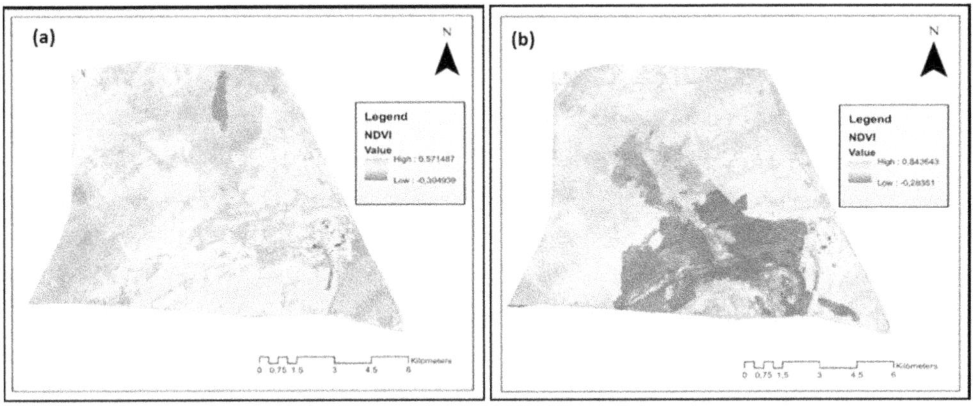

Figure 2.4: NDVI map for 2013 (a) and NDVI map for 2023 (b) calculated for the General De La Rey Military Training Area.
(*Source:* Authors' own work using ArcGIS 10.8.1)

The results indicate that, in 2013, the General De La Rey Military Training Area consisted of sparsely vegetated areas, with some parts consisting of bare soil and water bodies. However, the 2023 results, as illustrated in Figure 2.4(b), indicate a change in the landscape, with NDVI values between -0.283 and 0.844. The NDVI index of 0.844 indicates a very dense and healthy vegetation cover, whereas the low value of -0.283 indicates features like water bodies, barren land, or urban areas where there is little or no live vegetation in some areas.

A comparison of NDVI values between 2013 and 2023 indicated an improvement in vegetation health and stability in non-vegetated areas found in 2013. The noticeable increase in vegetation health (as indicated by the higher NDVI values) is encouraging and could be a result of the good land management techniques adopted between 2013 and 2023. This may also suggest a significant decrease in training exercises in the area or possibly restricted heavy vehicle movements. The upper NDVI value increased significantly between 2013 and 2023 (from 0.571 to 0.844), indicating a notable improvement in the density and health of the vegetation during this period. In regions with little to no vegetation, the lowest NDVI readings (from -0.305 to -0.283) showed minimal change, indicating stability. It is possible that during this time, not much changed in terms of the environment or land use in these locations.

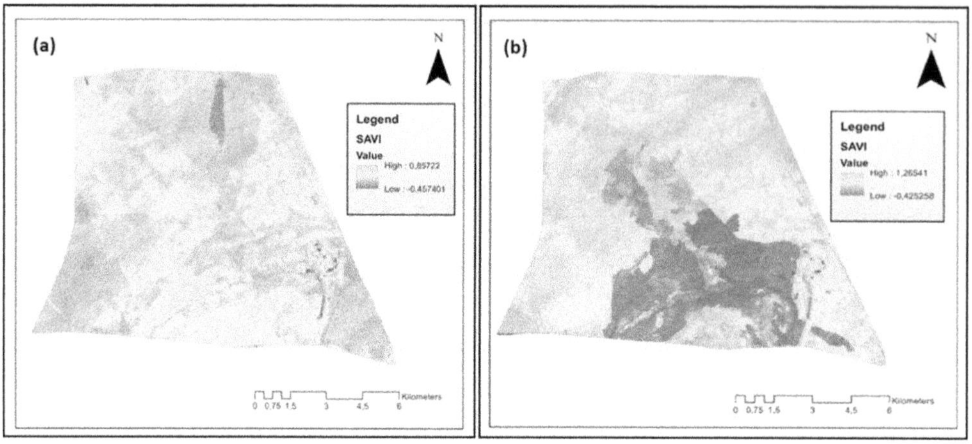

Figure 2.5: SAVI map for 2013 (a) and SAVI map for 2023 (b) calculated for the General De La Rey Military Training Area.
(*Source*: Authors' own work using ArcGIS 10.8.1)

Figure 2.5(a) presents a map of the calculated SAVI values in 2013, with a maximum of 0.857 and a minimum of -0.457. A maximum value of 0.857 indicates an area with very healthy and dense vegetation cover, such as grasslands with abundant green cover. The minimum negative value of -0.457 indicates non-vegetated or sparsely vegetated areas. Figure 2.5(b) shows a map of the calculated SAVI values in 2023, with a maximum of 1.265 and a minimum of -0.425. As mentioned previously, SAVI values close to 1 indicate optimal vegetation health. An index of 1.265 suggests that the area is extremely healthy and densely vegetated. The negative low SAVI index can be attributed to areas with little or no vegetation. An increase was found in the SAVI from 2013 to 2023 (from 0.857 to 1.265), which points to a significant improvement in vegetation cover and health over the period. This can be attributed to favourable climate conditions or fewer military activities and training exercises.

Figures 2.6(a) and (b) show the TCB indices (in greyscale) for 2013 and 2023, respectively. The results showed an increase in the TCB value between 2013 and 2023, which might indicate an increase in the reflectance of certain areas in the General De La Rey Military Training Area. The maximum TCB index's rise from 2013 to 2023 (from 136407.0 to 150869.0) indicates modifications to the surface characteristics that can improve reflectivity. These modifications can be attributed to infrastructural expansion or changes in land-use practices. The minimum TCB index (16580.9 in 2013 and 16725.2 in 2023) suggests relatively stable regions with persistently low reflectance values. This suggests that the region is composed of sections with continuous vegetative cover or permanent natural features, such as bodies of water.

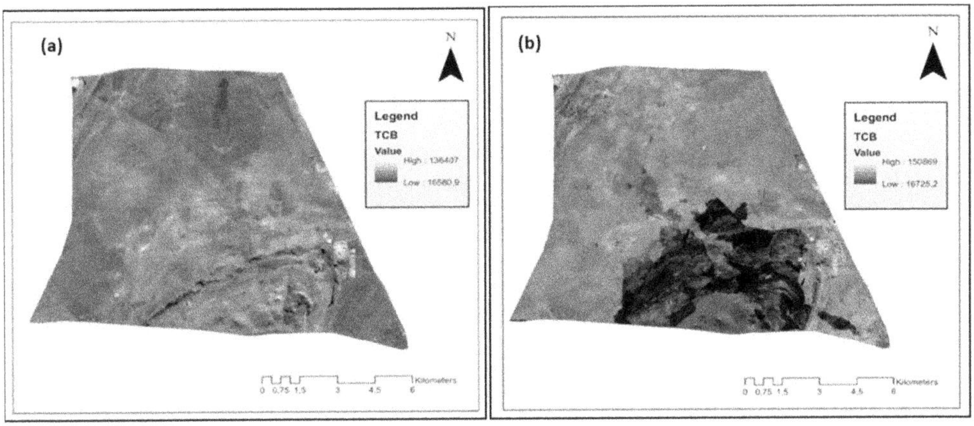

Figure 2.6: TCB map for 2013 (a) and TCB map for 2023 (b) calculated for the General De La Rey Military Training Area.
(*Source:* Authors' own work using ArcGIS 10.8.1)

The NDVI and SAVI results for 2013 and 2023 for the General De Le Ray Military Training Area showed an increase in vegetation, indicating that no land degradation had occurred. The SAVI indicates changes in vegetation cover while minimising the influence of soil background and brightness conditions. The TCB index also showed an increase in the reflectivity of the features on land cover. The absence of land degradation implies that there was no vegetation vulnerability or lack of vegetation due to soil nutrients that could cause soil erosion. Possible factors contributing to the absence of land degradation include a reduction in military activities, such as live ammunition training, training drills, and heavy vehicle movement.

Table 2.1 shows the NDVI, SAVI, and TCB threshold values that represent the high and low values (range). The volume change indicates the change in vegetation cover in terms of the percentage of vegetation that increased or decreased during the ten years. The NDVI results showed an overall increase of 30.51% in the area's vegetation health between 2013 and 2023, but only a modest rise of 7.28% in the sparsely vegetated areas. The SAVI results also showed a 38.46% increase in the health of the vegetation. However, despite the healthy vegetation, there was a decrease of 7.28% in vegetation in certain areas, meaning that some areas of the training site lost vegetation. Finally, the TCB results showed an overall increase of 10.07% in the reflectance and brightness of land surfaces in the areas, which assisted in differentiating between various land cover types. However, the 0.87% growth shows that the land cover in the area has remained largely stable over the past ten years.

Table 2.1: Summary of results. The threshold ranges (rounded to three decimal places) were used to classify the NDVI, SAVI, TCB, and volume change (%) of high and low values

Class	**2013**	**2023**	**Volume change (%)**
NDVI threshold value	High value = 0.571	High value = 0.843	High value = 38.46
	Low value = 0.304	Low value = -0.283	Low value = 7.28
SAVI threshold value	High value = 0.857	High value = 1.265	High value = 38.46
	Low value = 0.457	Low value = -0.425	Low value = -7.28
TCB threshold value	High value = 136407	High value = 150869	High value = 10.07
	Low value = 16580.9	Low value = -16725.2	Low value = 0.87%

In calculating the change in vegetation volume, all indices showed similar characteristics. Consequently, the SAVI is better suited for this study than the NDVI because the study location is a humid subtropical region where the soil background is more prominent. This is because the adjusted L factor allows the SAVI to provide better assessments in areas with significant soil exposure. However, by analysing the fluctuations in soil background reflectance, TCB plays a significant role in determining reflectance. One of the challenges encountered was the use of a small number of data points for each feature, considering the size of the study area.

Accuracy assessment

The following formulas were used for the accuracy assessment, and the results are presented in Table 2.2:

$$\text{Users accuracy} = \frac{\text{(number of correctly classified pixels in each category)}}{\text{(total number of classified pixels in that category)}} \text{x100}$$

$$\text{Producers accuracy} = \frac{\text{(Number of Correctly Classified Pixels in each Category)}}{\text{(Total Number of Reference Pixels in that Category (The Column Total))}} \text{x100}$$

$$\text{Overall accuracy} = \frac{\text{(Total Number of Correctly Classified Pixels (Diagonal))}}{\text{(Total Number of Reference Pixels)}} \text{x100}$$

$$\text{Kappa coefficient} \left(\text{T}\right) = \frac{((\text{TS} \times \text{TCS}) - \sum(\text{Column Total} \times \text{Row Total}))}{(\text{TS}^2 - \sum(\text{Column TotalxRow Total}))} \text{x100}$$

where:

TS is the total sample.
TCS is the total corrected sample.

Table 2.2: Accuracy assessment calculated using the error matrix

	Water body	**Bare soil**	**Built up area**	**Sparse vegetation**	**Moderate vegetation**	**Dense vegetation**	**Total**	**User's accuracy**
Water body	6	0	0	0	0	2	8	75%
Bare soil	3	17	2	0	0	1	23	74%
Built up area	0	1	10	5	4	1	21	76%
Sparse vegetation	0	0	0	12	2	1	15	80%
Moderate vegetation	1	2	0	1	7	3	14	50%
Dense vegetation	0	0	0	0	0	19	19	100%
Total producer	10	20	12	18	13	27	100	
Producer's accuracy	60%	85%	83%	67%	54%	70%		
Kappa = 0.69								

The overall accuracy achieved by the supervised classification of the General De Le Ray Military Training Area was 71%. The overall accuracy was 71% for the correctly classified values, and the total number of referenced pixels was 100, which represents the total classified values. The highlighted numbers in Table 2.2, which represent all correctly classified values in each category, were used to calculate the overall accuracy.

In ArcMap, the overall accuracy was determined by summing the correctly classified values and dividing them by the total number of classifications. The correctly classified values are located along the principal diagonal of the confusion matrix, running from the upper left corner to the lower right corner

- Correctly classified values: 6+17+10+12+7+19 = 71
- Total classified values: 100
- Overall accuracy: 71/100 = 71%

According to the calculated accuracy results, the highest user's accuracy was associated with a percentage of 100% for dense vegetation, while the highest producer accuracy was associated with bare soil. A hundred points were allocated for validation and classification accuracy evaluation to achieve good study outcomes. The kappa index serves as a measure of the overall accuracy of classification. A kappa value of 1 indicates complete agreement, whereas a value of 0 indicates the absence of any agreement (Fenstermaker, 1991). The kappa coefficient was determined using ArcGIS 10.8.1 to be 69% via a confusion matrix, which is substantial. However, given the size of the study area, such a percentage is more likely because the number of points (100) used for the accuracy assessment was rather small (Figure 2.7). This suggests that there were insufficient points in each category to allow for a more accurate classification. Therefore, in future research, the accuracy of classification could be improved by increasing the number of training samples while maintaining a balanced number of points in each category.

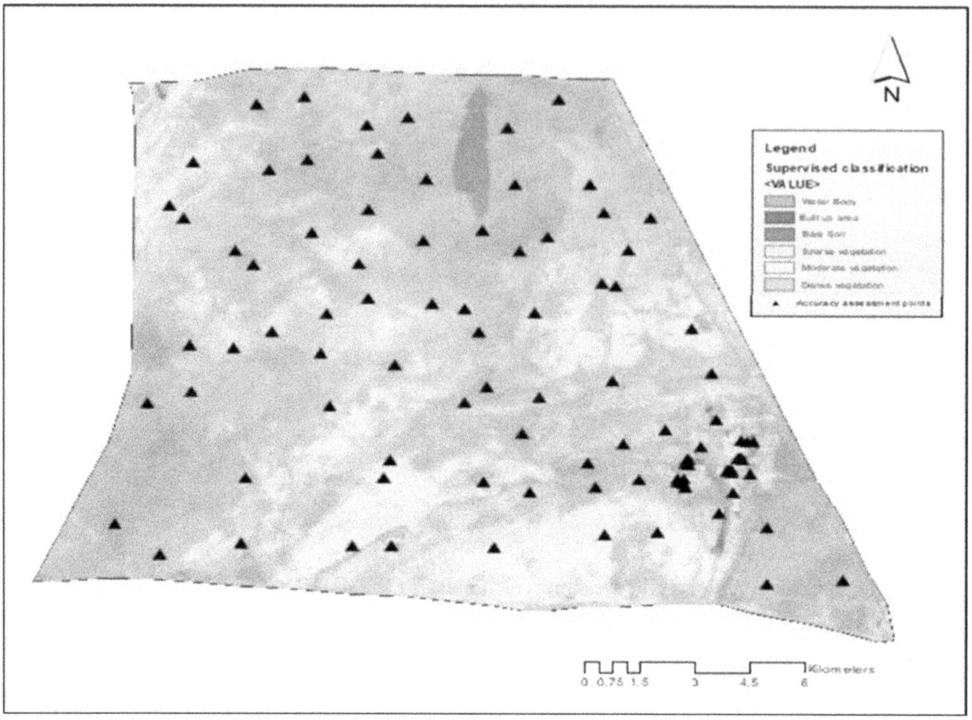

Figure 2.7: Random points used to calculate the error matrix.
(*Source:* Authors' own work using ArcGIS 10.8.1)

Conclusion

This study aimed to map and determine land degradation in the General De Le Ray Military Training Area using three vegetation indices: the NDVI, SAVI, and TCB. Each index delineates changes in vegetation cover, which can be used as an indicator of land degradation. The results showed an increase in vegetation cover in the study area, possibly due to a reduction in military training exercises. No indications of soil erosion that could lead to land degradation were found. For land degradation to occur, vegetation cover must be absent or diminished, leading to increased soil erosion, which accelerates the degradation process. The study found that land cover has improved in the General De Le Ray Military Training Area according to the calculated vegetation indices.

The findings of this study are important because they indicate that the area has shown improvement in its vegetation cover, indicating little or no soil erosion in the area over ten years, from 2013 to 2023. This is possible if there was a decrease in the number of activities carried out each year, a decline in the availability of live ammunition, and the implementation of secure and sound environmental practices over the years. One of the factors that possibly contributed to the increase in vegetation cover is the decrease in military activities post-COVID.

The effectiveness of the suggested indices for detecting vegetation change was assessed using the NDVI, SAVI, and TCB differencing techniques. Each index played a role, with the SAVI improving the vegetation classification in areas dominated by soil, the NDVI providing the general state of the vegetation, and the TCB detecting changes in land brightness. All three indices detected changes in vegetation cover between 2013 and 2023. The integration of these indices' techniques was found to be successful in mapping and determining land degradation in the study area.

The accuracy assessment indicated a reasonable percentage of agreement, with an overall accuracy of 71% and a kappa coefficient of 0.69. Although the classification was performed, particularly for dense vegetation and bare soil, the small number of validation points may have influenced the results. Increasing the number of data points and ensuring that the points are distributed more evenly among the categories could improve the classification accuracy in future research.

For future research, it is recommended that a larger number of training samples be used for areas that occupy a large area. This study recommends that annual vegetation data should be used instead of specific seasonal data to validate or obtain more accurate results. To improve the quality of the classification, it is advisable that future research increase the number of points, allocate the same number of points to each category, and create training samples (data points) using a suitable pixel number. If the training area had a significant lull in military activities, it was safe to prescribe a certain number of years for land rehabilitation for military environmental management. To further reduce the influence of military exercises on land cover, rotational training at different sites can be implemented to avoid the overuse of land.

References

Agone, V. & Bhamare, S.M. 2012. Change detection of vegetation cover using remote sensing and GIS. *Journal of Research and Development*, 2(4).

Althoff, P.S. & Thien, S.J. 2005. Impact of M1A1 main battle tank disturbance on soil quality, invertebrates, and vegetation characteristics. *Journal of Terramechanics*, 42(3–4): 159–176. https://doi.org/10.1016/j.jterra.2004.10.014

Asokan, A. & Anitha, J.J.E.S.I. 2019. Change detection techniques for remote sensing applications: A survey. *Earth Science Informatics*, 12: 143–160. https://doi.org/10.1007/s12145-019-00380-5

Augusto, L., Dupouey, J.L. & Ranger, J. 2003. Effects of tree species on understory vegetation and environmental conditions in temperate forests. *Annals of Forest Science*, 60(8): 823–831. https://doi.org/10.1051/forest:2003077

Bajocco, S., Smiraglia, D., Scaglione, M., Raparelli, E. & Salvati, L. 2018. Exploring the role of land degradation on agricultural land use change dynamics. *Science of the Total Environment*, 636: 1373–1381. https://doi.org/10.1016/j.scitotenv.2018.04.412

Bannari, A., Morin, D., Bonn, F. & Huete, A. 1995. A review of vegetation indices. *Remote Sensing Reviews*, 13(1–2): 95–120. https://doi.org/10.1080/02757259509532298

Basso, B., Cammarano, D. & De Vita, P. 2004. Remotely sensed vegetation indices: Theory and applications for crop management. *Rivista Italiana di Agrometeorologia*, 1(5): 36–53.

Broomandi, P., Guney, M., Kim, J.R. & Karaca, F. 2020. Soil contamination in areas impacted by military activities: A critical review. *Sustainability*, 12(21). https://doi.org/10.3390/su12219002

Cilliers, S.S., Van Wyk, E. & Bredenkamp, G.J. 1999. Urban nature conservation: Vegetation of natural areas in the Potchefstroom municipal area, North West province, South Africa. *Koedoe: African Protected Area Conservation and Science*, 42(1): 1–30. https://doi.org/10.4102/koedoe.v42i1.218

Cock, J. & McKenzie, P. (eds.). 1998. *From Defence to Development: Redirecting Military Resources in South Africa*. Ottawa: IDRC.

Coetzee, C. 2022. Change detection of vegetation cover using remote sensing and GIS: A case study of the West Coast region of South Africa. *Geography, Environment, Sustainability*, 15(2): 91–102. https://doi.org/10.24057/2071-9388-2021-067

Crist, E.P. & Cicone, R.C. 1984. A physically-based transformation of Thematic Mapper data – The TM Tasselled Cap. *IEEE Transactions on Geoscience and Remote Sensing*, (3): 256–263. https://doi.org/10.1109/TGRS.1984.350619

Daemane, M.E., Bezuidenhout, H. & Cilliers, S.S. 2010. An ecological study of the plant communities in the proposed Highveld National Park, in the peri-urban area of Potchefstroom, South Africa. *Koedoe: African Protected Area Conservation and Science*, 52(1): 1–8. https://doi.org/10.4102/koedoe.v52i1.708

Demarais, S., Tazik, D.J., Guertin, P.J. & Jorgensen, E.E. 1999. Disturbance associated with military exercises. In: L.R. Walker (ed.). *Ecosystems of Disturbed Ground. Ecosystems of the World 16*. Amsterdam: Elsevier. 385–396.

Dwivedi, R.S., Sreenivas, K. & Ramana, K.V. 2005. Land-use/land-cover change analysis in part of Ethiopia using Landsat Thematic Mapper data. *International Journal of Remote Sensing*, 26(7): 1285–1287. https://doi.org/10.1080/01431160512331337763

Elnaggar, A.A., Azeez, A.A. & Mowafy, M. 2020. Monitoring spatial and temporal changes of urban growth in Dakahlia Governorate, Egypt, by using remote sensing and GIS techniques. *Mansoura Engineering Journal*, 39(4): 1–14. https://doi.org/10.21608/bfemu.2020.102842

Fenstermaker, L. 1991. A proposed approach for national to global scale error assessments. Proceedings of the GIS/LIS '91, ASPRS, ACSM, AAG, AM/FM International and URISA, 1: 293–300.

Gilabert, M.A., González-Piqueras, J., García-Haro, F.J. & Meliá, J. 2002. A generalized soil-adjusted vegetation index. *Remote Sensing of Environment*, 82(2–3): 303–310. https://doi.org/10.1016/S0034-4257(02)00048-2

Hatfield, J.L., Prueger, J.H., Sauer, T.J., Dold, C., O'Brien, P. & Wacha, K. 2019. Applications of vegetative indices from remote sensing to agriculture: Past and future. *Inventions*, 4(4). https://doi.org/10.3390/inventions4040071

Haugen, L.B., Ayers, P.D. & Anderson, A.B. 2003. Vehicle movement patterns and vegetative impacts during military training exercises. *Journal of Terramechanics*, 40(2): 83–95. https://doi.org/10.1016/j.jterra.2003.09.004

Henrico, I., Vukea, M. & Smit, H. 2021. Examining the impact of military training exercises on vegetation: Case of the South African Army Combat Training Centre Lohatla. *Spatial Information Research*, 29: 557–568. https://doi.org/10.1007/s41324-020-00370-6

Hishe, S., Lyimo, J. & Bewket, W. 2017. Effects of soil and water conservation on vegetation cover: A remote sensing-based study in the Middle Suluh River Basin, Northern Ethiopia. *Environmental Systems Research*, 6(1): 1–16. https://doi.org/10.1186/s40068-017-0103-8

Hislop, S., Jones, S., Soto-Berelov, M., Skidmore, A., Haywood, A. & Nguyen, T.H. 2018. Using Landsat spectral indices in time series to assess wildfire disturbance and recovery. *Remote Sensing*, 10(3). https://doi.org/10.3390/rs10030460

Hoffman, M.T. & Todd, S. 2000. A national review of land degradation in South Africa: The influence of biophysical and socio-economic factors. *Journal of Southern African Studies*, 26(4): 743–758. https://doi.org/10.1080/713683611

Huang, S., Tang, L., Hupy, J.P., Wang, Y. & Shao, G. 2021. A commentary review on the use of normalized difference vegetation index (NDVI) in the era of popular remote sensing. *Journal of Forestry Research*, 32(1): 1–6. https://doi.org/10.1007/s11676-020-01155-1

Huete, A.R. 1988. A soil-adjusted vegetation index (SAVI). *Remote Sensing of Environment*, 25(3): 295–309. https://doi.org/10.1016/0034-4257(88)90106-X

Huete, A., Didan, K., Miura, T., Rodriguez, E.P., Gao, X. & Ferreira, L.G. 2002. Overview of the radiometric and biophysical performance of the MODIS vegetation indices. *Remote Sensing of Environment*, 83(1–2): 195–213. https://doi.org/10.1016/S0034-4257(02)00096-2

Hutchinson, J.M.S., Jacquin, A., Hutchinson, S.L. & Verbesselt, J. 2015. Monitoring vegetation change and dynamics on US Army training lands using satellite image time series analysis. *Journal of Environmental Management*, 150: 355–366. https://doi.org/10.1016/j.jenvman.2014.08.002

Ke, Y., Im, J., Lee, J., Gong, H. & Ryu, Y. 2015. Characteristics of Landsat 8 OLI-derived NDVI by comparison with multiple satellite sensors and in-situ observations. *Remote Sensing of Environment*, 164: 298–313. https://doi.org/10.1016/j.rse.2015.04.004

Lawrence, M.J., Stemberger, H.L., Zolderdo, A.J., Struthers, D.P. & Cooke, S.J. 2015. The effects of modern war and military activities on biodiversity and the environment. *Environmental Reviews*, 23(4): 443–460. https://doi.org/10.1139/er-2015-0039

Li, S. & Chen, X. 2014. A new bare-soil index for rapid mapping developing areas using Landsat 8 data. *International Archives of the Photogrammetry, Remote Sensing and Spatial Information Sciences – ISPRS Archives*, 40(4): 139–144. https://doi.org/10.5194/isprsarchives-XL-4-139-2014

Liu, Q., Liu, G., Huang, C., Liu, S. & Zhao, J. 2014. A tasselled cap transformation for Landsat 8 OLI TOA reflectance images. 2014 IEEE Geoscience and Remote Sensing Symposium. 541–544. https://doi.org/10.1109/IGARSS.2014.6946479

Lv, X. 2014. Remote sensing, normalized difference vegetation index (NDVI), and crop yield forecasting. Doctoral dissertation. Champaign, IL: University of Illinois at Urbana-Champaign.

Magagula, B. 2014. The environmental management system of the South African National Defence Force at the Grahamstown military installation. *Scientia Militaria: South African Journal of Military Studies*, 42(2): 143–163. https://doi.org/10.5787/42-2-1097

Magagula, H.B. 2016. Land cover change induced by military action: A case study of two South African military training areas. Proceedings of the Centenary Conference of the Society of South African Geographers. 301–314.

Maity, S. & Mane, S.P. 2020. Assessment of Tasselled Cap transformation of Mumbai Urban Agglomeration, India. *Aayushi International Interdisciplinary Research Journal*, (80): 158–164.

Mbow, C., Goïta, K. & Bénié, G.B. 2004. Spectral indices and fire behaviour simulation for fire risk assessment in savanna ecosystems. *Remote Sensing of Environment*, 91(1): 1–13. https://doi.org/10.1016/j.rse.2003.10.019

Minale, A.S. & Gelaye, S. 2021. Vegetation vulnerability analysis by using GIS and remote sensing techniques: A case study of Sensawuha Watershed, Ethiopia. *GeoJournal*, 86(1): 475–488. https://doi.org/10.1007/s10708-019-10082-6

Mokalapa, L.K. 2022. Environmental impact of military explosive munitions at General De La Rey Training Area, Potchefstroom. Doctoral dissertation. Potchefstroom: North-West University.

Nagendra, H. 2001. Using remote sensing to assess biodiversity. *International Journal of Remote Sensing*, 22(12): 2377–2400. https://doi.org/10.1080/01431160117096

Nanasaheb, A.A.V.P.H. & Paswan, P.K. 2020. Soil adjusted vegetation index (SAVI), in the Uben river basin of Gujarat. *International Journal of Chemical Studies*, 8(5): 693–698. https://doi.org/10.22271/chemi.2020.v8.i5j.10381

Parmar, M., Bhawsar, Z., Kotecha, M., Shukla, A. & Rajawat, A.S. 2021. Assessment of land degradation vulnerability using geospatial technique: A case study of Kachchh District of Gujarat, India. *Journal of the Indian Society of Remote Sensing*, 49: 1661–1675. https://doi.org/10.1007/s12524-021-01349-y

Patro, S.G.K. & Sahu, K.K. 2015. Normalization: A preprocessing stage. *International Advanced Research Journal in Science, Engineering and Technology*, 2(3): 20–22. https://doi.org/10.17148/IARJSET.2015.2305

Pettorelli, N., Vik, J.O., Mysterud, A., Gaillard, J.M., Tucker, C.J. & Stenseth, N.C. 2005. Using the satellite-derived NDVI to assess ecological responses to environmental change. *Trends in Ecology & Evolution*, 20(9): 503–510. https://doi.org/10.1016/j.tree.2005.05.011

Polykretis, C., Grillakis, M.G. & Alexakis, D.D. 2020. Exploring the impact of various spectral indices on land cover change detection using change vector analysis: A case study of Crete Island, Greece. *Remote Sensing*, 12(2). https://doi.org/10.3390/rs12020319

Ren, H., Zhou, G. & Zhang, F. 2018. Using negative soil adjustment factor in soil-adjusted vegetation index (SAVI) for aboveground living biomass estimation in arid grasslands. *Remote Sensing of Environment*, 209: 439–445. https://doi.org/10.1016/j.rse.2018.02.068

Rokni, K. & Musa, T.A. 2019. Normalized difference vegetation change index: A technique for detecting vegetation changes using Landsat imagery. *Catena*, 178: 59–63. https://doi.org/10.1016/j.catena.2019.03.007

Rondeaux, G., Steven, M. & Baret, F. 1996. Optimization of soil-adjusted vegetation indices. *Remote Sensing of Environment*, 55(2): 95–107. https://doi.org/10.1016/0034-4257(95)00186-7

Rouse, J.W., Haas, R.H., Schell, J.A. & Deering, D.W. 1973. Monitoring vegetation systems in the Great Plains with ERTS. Proceedings of Third Earth Resources Technology Satellite Symposium, Washington, DC: NASA. Goddart Space Flight Center, 1: 309–317. (NASA SP-351).

Samarawickrama, U., Piyaratne, D. & Ranagalage, M. 2017. Relationship between NDVI with Tasselled cap indices: A remote sensing-based analysis. *International Journal of Innovative Research in Technology*, 3(12): 13–19.

Sashikkumar, M.C., Selvam, S., Karthikeyan, N., Ramanamurthy, J., Venkatramanan, S. & Singaraja, C. 2017. Remote sensing for recognition and monitoring of vegetation affected by soil properties. *Journal of the Geological Society of India*, 90: 609–615. https://doi.org/10.1007/s12594-017-0759-8

Shahabi, H., Allahvirdiasl, H. & Zabihian, H. 2012. Monitoring vegetation conditions of catchments using multispectral remote sensing data (Case study: Central Zab basin). *International Journal of Engineering Research & Technology*, 1(7).

Simmons, K.J. 2004. Vegetative recovery of military vehicle impacts at Fort Lewis, WA. Master's dissertation. Knoxville, TN: University of Tennessee.

Smit, H.A.P. 2020. Developing military environmental literacy in the South African Army through a dedicated military environmental management course. *Scientia Militaria: South African Journal of Military Studies*, 48(1): 75–96. https://doi.org/10.5787/48-1-1247

Smit, H. & Henrico, I. 2021. "Green" militaries: The military-environment conundrum. Society of South African Geographers and the Southern African Association of Geomorphologists Joint Online Conference. 6–8 September 2021.

Solymosi, K., Kövér, G. & Romvári, R. 2019. The development of vegetation indices: A short overview. *Acta Agraria Kaposváriensis*, 23(1): 75–90. https://doi.org/10.31914/aak.2264

Tilauhun, E. & Zewide, I. 2021. Causes and impacts of land degradation. *British Journal of Earth Sciences Research*, 9(2): 28–37.

Tucker, C.J. 1979. Red and photographic infrared linear combinations for monitoring vegetation. *Remote Sensing of Environment*, 8(2): 127–150. https://doi.org/10.1016/0034-4257(79)90013-0

USGS [United States Geological Survey]. 2016. *Landsat 8 Data Users Handbook*. https://www.usgs.gov/land-resources/nli/landsat/landsat-8-data-users-handbook [23 January 2025].

Vani, V. & Mandla, V.R. 2017. Comparative study of NDVI and SAVI vegetation indices in Anantapur district semi-arid areas. *International Journal of Civil Engineering Technology*, 8(4): 559–566.

Vorovencii, I. 2007. Use of the "Tasselled Cap" transformation for the interpretation of satellite images. *RevCad – Journal of Geodesy and Cadastre*, 7: 75–82.

Wessels, K.J., Prince, S.D., Malherbe, J., Small, J., Frost, P.E. & Van Zyl, D. 2007. Can human-induced land degradation be distinguished from the effects of rainfall variability? A case study in South Africa. *Journal of Arid Environments*, 68(2): 271–297. https://doi.org/10.1016/j.jaridenv.2006.05.015

Xue, J. & Su, B. 2017. Significant remote sensing vegetation indices: A review of developments and applications. *Journal of Sensors*, 2017. https://doi.org/10.1155/2017/1353691

Zentelis, R., Banks, S., Roberts, J.D., Dovers, S. & Lindenmayer, D. 2017. Managing military training-related environmental disturbance. *Journal of Environmental Management*, 204: 486–493. https://doi.org/10.1016/j.jenvman.2017.09.029

Zhen, Z., Chen, S., Yin, T., Chavanon, E., Lauret, N., Guilleux, J., Henke, M., Qin, W., Cao, L., Li, J. & Lu, P. 2021. Using the negative soil adjustment factor of soil-adjusted vegetation index (SAVI) to resist saturation effects and estimate leaf area index (LAI) in dense vegetation areas. *Sensors*, 21(6). https://doi.org/10.3390/s21062115

CHAPTER 3

Temporal analysis of the effects of warfare on vegetation in Gaza

T. Henrico, D. Naidoo, Z. Lourens, and Z. Münch

Abstract

The Gaza Strip, an area perpetually affected by conflict, provides critical yet underexplored context for assessing the environmental consequences of warfare, particularly on vegetation cover. Vegetation is central to ecosystem stability, supporting biodiversity, agriculture, and water management; however, it is vulnerable to direct destruction and indirect stressors, such as soil degradation and pollution. This study used Sentinel-2A satellite imagery and advanced geospatial methodologies to analyse the temporal effects of warfare on vegetation in the Gaza Strip between March 2023 and March 2024. By employing three vegetation indices – Normalised Difference Vegetation Index (NDVI), Soil-Adjusted Vegetation Index (SAVI), and Modified Soil-Adjusted Vegetation Index (MSAVI) – the research quantifies vegetation dynamics and categorises them into negative, no, and positive changes. Statistical analysis revealed no significant temporal differences across the indices at the 95% confidence level. However, a strong positive correlation ($r = 0.985$, $p = 0.015$) between the SAVI and MSAVI confirmed their reliability in detecting vegetation changes in sparse arid environments. In contrast, the NDVI showed weaker correlations, reflecting its focus on vegetation density rather than soil conditions. These findings highlight the environmental toll of conflict on vegetation and underscore the importance of using a multi-index approach for ecological monitoring in conflict zones. This study also demonstrates the potential of geospatial technologies to inform environmental recovery efforts, which are essential for supporting broader peacebuilding and reconstruction strategies in post-conflict regions.

Keywords: conflict zones, environmental impacts of warfare, Gaza Strip, statistical analysis, remote sensing applications, vegetation change detection

Introduction

The Gaza Strip, a region perpetually embroiled in conflict, offers a unique yet critical opportunity to study the environmental consequences of warfare. While much attention is rightly devoted to the human toll and infrastructural damage caused by prolonged conflict, ecological impacts, particularly on vegetation, are often overlooked. Vegetation is a cornerstone of ecological stability and plays a vital role in biodiversity, agriculture, and water management. Its degradation, whether due to direct destruction or indirect stressors such as pollution and soil erosion, has far-reaching implications for the environment and human livelihoods. Despite the global significance of understanding the environmental effects of warfare, there is limited research on vegetation dynamics in active conflict zones such as Gaza. Remote sensing technologies, specifically the use of satellite-derived vegetation indices, provide an effective way to quantify the impacts of these fires (Al-Doski et al., 2013). These indices, NDVI (Normalised Difference Vegetation Index), SAVI (Soil-Adjusted Vegetation Index), and MSAVI (Modified Soil-Adjusted Vegetation Index), enable the monitoring of vegetation health, density, and changes over time. However, studies applying these indices to conflict zones remain sparse, leaving a critical gap in the academic literature and practical environmental monitoring. This study addresses this gap by employing Sentinel-2A satellite imagery acquired from the Copernicus browser to analyse the temporal effects of warfare on vegetation in the Gaza Strip between March 2023 and March 2024.

Environmental impacts of warfare

The environmental impacts of warfare extend beyond immediate human suffering and infrastructural damage and significantly affect ecosystems and vegetation. Vegetation, which serves as the foundation for biodiversity and human livelihoods, is particularly vulnerable in conflict zones owing to direct destruction, land degradation, and pollution. Remote sensing technologies have emerged as indispensable tools for understanding and quantifying these impacts (Kaplan et al., 2022). Al-Doski et al. (2013) used remote sensing data (Landsat, SPOT, and IKONOS) to investigate the environmental degradation caused by the Gulf War between Iraq and Kuwait, while Garzón and Valánszki (2020) identified deforestation, land-use change, and abandonment of arable land as some of the negative impacts of warfare identified using remote sensing.

An important aspect of vegetation heavily affected by conflict environments is soil health, which is crucial to overall vegetation health. According to Certini et al. (2013), there are three types of disturbances caused by warfare on soil health: physical, chemical, and biological. This study focused on the physical aspect, as this is the aspect that can be monitored using vegetation indices. It is important to monitor the impacts of warfare on vegetation because, according to Certini et al. (2013), if the appropriate techniques are not followed to reconstitute the soil, some of the soil functions may be lost forever, thus negatively affecting the vegetation.

Vegetation monitoring using remote sensing

Vegetation indices are mathematical formulations derived from satellite imagery that highlight vegetation health, density, and coverage. Each index used in this study has unique characteristics that make it suitable for monitoring vegetation in conflict-affected areas. NDVI provides a broad overview of vegetation health and degradation (Yengoh et al., 2015), whereas SAVI accounts for soil brightness (Huete, 1988). To refine these insights in sparsely vegetated areas, the MSAVI further enhances the analysis by reducing the soil influence, enabling the detection of early-stage vegetation regrowth (Xue & Su, 2017). By integrating multiple vegetation indices (NDVI, SAVI, and MSAVI) to effectively capture the complex dynamics of vegetation in conflict-affected regions, these indices offer a robust and nuanced framework for monitoring the environmental consequences of warfare.

Normalised Difference Vegetation Index (NDVI)

NDVI is calculated (1) as follows (Huang et al., 2021):

$$\mathrm{NDVI} = \frac{(\mathrm{NIR} - \mathrm{RED})}{(\mathrm{NIR} + \mathrm{RED})}$$

where:

NIR is the near-infrared band reflectance.
RED is the red band reflectance value.

NDVI values range from -1 to 1, with the following thresholds commonly used to interpret vegetation cover (Mehta et al., 2021): -1 to 0: water bodies, 0 to 0.2: no vegetation, 0.2 to 0.4: sparse vegetation, 0.4 to 0.6: moderate vegetation, and > 0.6 as dense vegetation. The NDVI is sensitive to chlorophyll and is widely used to assess vegetation health and cover extent, making it a key indicator of ecological conditions in conflict zones.

Soil Adjusted Vegetation Index (SAVI)

SAVI is an improvement over NDVI, designed to minimise the influence of soil brightness in areas with sparse vegetation. This index ranges from -1 to 1. The thresholds are similar to those of the NDVI but provide better accuracy in low-vegetation environments (Sadiq & Saleh, 2020). SAVI is calculated (2) as follows (Vani & Mandla, 2017):

$$\mathrm{SAVI} = \frac{(\mathrm{NIR} - \mathrm{RED})}{(\mathrm{NIR} + \mathrm{RED} + \mathrm{L})} \times \left(1 + \mathrm{L}\right)$$

where:

NIR is the near-infrared band reflectance.
RED is the red band reflectance value.
L is the soil brightness correction factor, which was set to 0.5 for moderate vegetation conditions.

Modified Soil-Adjusted Vegetation Index (MSAVI)

MSAVI refines the SAVI by dynamically adjusting the soil correction factor to further reduce soil influence, particularly during early-stage vegetation growth. The MSAVI values range from -1 to 1 and offer detailed insights into vegetation at different growth stages, such as bare soil, germination, leaf development, and dense vegetation (EOS Data Analytics, 2024). The formula for calculating (3) MSAVI is (Voitik et al., 2023):

$$\mathrm{MSAVI} = \frac{2 \times \mathrm{NIR} + 1 - \sqrt{(2 \times \mathrm{NIR} + 1)^2 - 8 \times \left(\mathrm{NIR} - \mathrm{RED}\right)}}{2}$$

where:

NIR is the near-infrared band reflectance.
RED is the red band reflectance value.

Change detection methods

Change detection is used to note the changes in a study area using satellite images. There are many types of change detection techniques, including algebra-based (e.g. image differencing), transformation-based (e.g. Principal Component Analysis), and classification-based approaches (e.g. post-classification comparison), GIS (for integration of source information), and advanced models (e.g. spectral mixture models) (Asokan & Anitha, 2019). Object-based change detection (OBCD) groups pixels into meaningful objects using segmentation techniques before analysing changes over time. Unlike pixel-based methods, which can be sensitive to noise and minor radiometric variations, OBCD reduces false positives by considering the spatial context, shape, and texture, in addition to spectral information (Zhang et al., 2017). Despite their advantages in feature generalisation and reduced misclassification, object-based methods require parameter tuning and may be computationally intensive compared to pixel-based approaches.

The method chosen for this study is a pixel-based image differencing approach (Nordberg & Evertson, 2005), which allows for a direct comparison of a study area between two time periods. This method emphasises the differences in the spectral response of different features while reducing the impact of topographic effects and illumination (Minu & Shetty, 2015). Using a higher spatial resolution, such as Sentinel-2A, allows for a relatively easy quantitative analysis of the results of change detection (Mas, 1999). The decision to use a pixel-based approach in this study was influenced by the need for a straightforward quantitative assessment of spectral variations without requiring extensive segmentation and classification. However, an object-based approach may be more suitable for applications in which contextual information is crucial, such as urban expansion studies.

Aim and objectives

This study used Sentinel-2A satellite imagery and advanced geospatial methodologies to analyse the temporal effects of warfare on vegetation in the Gaza Strip between March 2023 and March 2024. By employing change detection on three vegetation indices (NDVI, SAVI, and MSAVI), the research quantified vegetation dynamics and categorised them into negative, no, and positive changes. While the primary focus of this research was to assess the spatio-temporal effects of warfare on vegetation in the Gaza Strip, the reliability of the quantitative analysis was assessed using statistical analysis.

The significant contribution of this study lies in its application to the Gaza Strip, a region where environmental monitoring is often hampered by security constraints. The use of freely available Sentinel-2A satellite imagery and accessible geospatial tools demonstrates the feasibility of conducting detailed vegetation analyses in resource-limited settings. This approach provides a replicable model for researchers and policymakers to assess vegetation dynamics in other conflict zones where traditional ground-based data collection may not be viable.

Furthermore, this study addresses a critical gap in the literature by focusing on the environmental toll of warfare in arid and semi-arid regions. While much attention has been given to the human and infrastructural impacts of conflict, this study highlights the long-term ecological consequences, including vegetation degradation and disruption of ecosystem services.

By combining advanced geospatial techniques with a focus on conflict zones, this study aims to provide valuable insights into the ecological costs of warfare. The findings aim to inform strategies for environmental recovery and sustainable land management while contributing to a growing body of literature on the intersection of conflict and environmental science.

Study area

The Gaza Strip, a narrow coastal enclave along the eastern Mediterranean, spans approximately 365 km², with predominantly flat terrain transitioning from sandy coastal areas to loess plains inland (Figure 3.1). The temperate Mediterranean climate features mild, rainy winters and hot, dry summers, with annual precipitation ranging from 200 to 300 mm, mostly between October and March, contributing to semi-arid conditions (Zaineldeen & Aish, 2012). These climatic challenges, compounded by groundwater over-extraction, have led to salinisation and declining agricultural productivity (Selby, 2003).

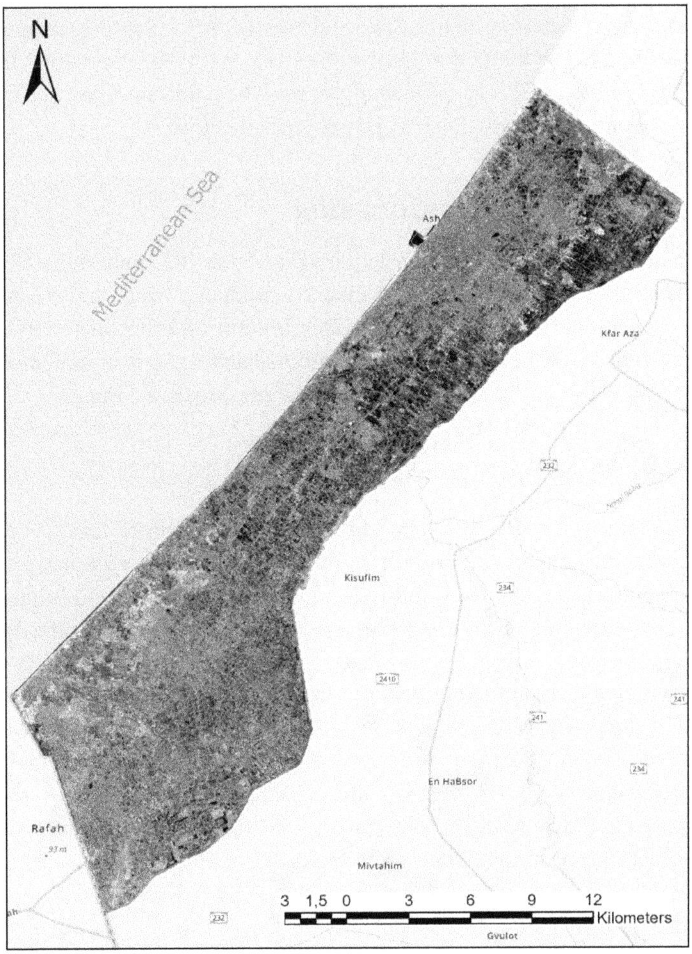

Figure 3.1: Location of the Gaza Strip with Sentinel-2A satellite imagery overlay on the Esri ArcGIS Topographic Basemap, illustrating the urban footprint and surrounding geography (2024).

Natural vegetation reflects a transitional zone between Mediterranean and desert biomes, with coastal dunes supporting flora such as Salsola kali, as well as cultivated citrus and olive groves (Dardona, 2016). However, urbanisation, intensive agriculture, and prolonged conflict have degraded vegetation cover, biodiversity, and natural habitats (Abuelaish & Olmedo, 2016). Military activities have further exacerbated these issues, causing widespread vegetation destruction, soil contamination, and erosion, while debris from conflict inhibits natural regeneration (Biswas, 2000; UNEP, 2009).

In this context, vegetation indices such as the NDVI, SAVI, and MSAVI are critical tools for assessing ecological impacts, enabling the monitoring of vegetation health, and identifying areas of degradation or recovery. These indices are essential for quantifying the environmental toll of warfare and informing sustainable land management strategies.

Methodology

The analysis focused on data from March 2023 and March 2024 to capture changes in vegetation over a one-year period. The methodological framework involved calculating three vegetation indices (NDVI, SAVI, and MSAVI), performing change detection analyses, and classifying land cover based on vegetation density. Each step is described below.

Data acquisition and preprocessing

Sentinel-2A imagery of the Gaza Strip was acquired in March 2023 and March 2024 to maintain seasonal consistency. The data used were Level-2A products, which underwent atmospheric correction using the Sen2Cor processor, providing bottom-of-atmosphere (BOA) reflectance values (Sentinel Hub, 2023). Consequently, no additional atmospheric or radiometric corrections were necessary, allowing for the direct analysis of the pre-processed imagery.

Vegetation indices calculation

Three vegetation indices (NDVI, SAVI, and MSAVI) were calculated for 2023 and 2024 using ArcGIS Pro 3.2.2. These indices were selected due to their effectiveness in assessing different aspects of vegetation health and their ability to reduce the impact of soil brightness, which is a critical consideration in regions with sparse vegetation, such as Gaza (Sadiq & Saleh, 2020).

The NDVI was computed to highlight variations in vegetation density and health. Higher NDVI values indicate healthier and denser vegetation. To account for the influence of exposed soil, the SAVI was applied. This index incorporates a correction factor of $L = 0.5$, making it suitable for areas with limited vegetation cover. Additionally, the MSAVI was calculated to further minimise the effects of soil brightness. MSAVI is particularly effective in detecting early-stage vegetation growth, providing a more refined analysis than NDVI in environments where vegetation is sparse.

Change detection techniques

Change detection was conducted for each vegetation index to assess the dynamics of the vegetation cover over the study period. This analysis followed a per-pixel approach, comparing the individual pixel values for the NDVI, SAVI, and MSAVI between 2023 and 2024, for a detailed assessment of spatial variations in vegetation health across the region.

For this study, the per-pixel change detection tool in ArcGIS Pro 3.2.2, which compared individual pixel values between March 2023 and March 2024 for each index. This method ensures a high level of spatial precision in the detection of vegetation dynamics. Vegetation dynamics were categorised into negative, none, and positive changes. The innovative combination of these indices with per-pixel change detection followed by raster-to-polygon conversions demonstrates the utility of advanced geospatial techniques in analysing fine-scale vegetation changes.

To categorise these changes, histogram slopes were analysed. These histogram slopes were used to categorise the observed changes in vegetation into three distinct classes. A negative slope indicates a decline in vegetation, signifying vegetation loss or degradation. In contrast, a flat slope suggests no significant change, reflecting stability in vegetation cover. A positive slope represented vegetation recovery or growth (Nordberg & Evertson, 2005). This classification method enabled a nuanced understanding of the spatial extent and temporal patterns of vegetation changes within the Gaza Strip.

Classification of vegetation and land cover

The analysis also included the spatial classification of vegetation into categories such as bare land, sparse vegetation, and dense vegetation within the indices, enabling detailed spatial comparisons. These categories were based on predefined index thresholds, as shown in Table 3.1.

Table 3.1: Classification of vegetation and land cover based on the NDVI, SAVI, and MSAVI

Index	Range	Category	Reference
NDVI	-1 to 0.2	Water/Bare soil	Mehta et al. (2021)
	0.2 to 0.5	Sparse vegetation	
	0.5 to 0.7	Moderate vegetation	
	0.7 to 1.0	Dense vegetation	
SAVI	-0.4 to 0.2	Water/Bare soil	Sadiq & Saleh (2020)
	0.2 to 0.4	Low vegetation	
	0.4 to 0.6	Moderate vegetation	
	0.6 to 1.0	Dense vegetation	
MSAVI	-1 to 0.2	Bare soil	EOS Data Analytics (2024)
	0.2 to 0.4	Seed germination stage	
	0.4 to 0.6	Leaf development stage	
	0.6 to 1.0	Vegetation dense enough to cover the soil	

Each classified raster cell was then converted to a polygon based on its grid value. This allowed for the spatial quantification of each vegetation category and enabled direct comparisons between the indices.

Spatial analysis

The total area covered by each vegetation category was calculated for each index in both years. This provided a clear measure of vegetation change over time and allowed cross-index comparisons. The change detection results for each index were classified into three categories: negative change, no change, and positive change, based on the histogram slope analysis. The total area for each change detection category was calculated and compared across the indices. This step was critical for understanding the magnitude, spatial distribution, stability, and recovery of vegetation loss.

The results from the index calculations, raster-to-polygon conversion, and change detection analysis were synthesised to provide a comprehensive picture of vegetation dynamics in the Gaza Strip. The integration of multiple indices ensured that the nuances of vegetation change were captured, considering the limitations and strengths of each index.

Statistical validation

To evaluate the reliability of the vegetation indices (NDVI, SAVI, and MSAVI), a comparative analysis was conducted using IBM SPSS Statistics 30.0, using three key statistical tests. Firstly, a paired samples t-test assessed differences in vegetation index values between March 2023 and March 2024. Secondly, a repeated measures ANOVA examined overall differences across the indices and tested whether the interaction between year and index type was significant (Hais et al., 2009). Finally, a bivariate correlation analysis evaluated whether the changes detected by one index were correlated with the changes detected by the other indices.

Results and findings

The temporal changes between March 2023 and March 2024 revealed the complex dynamics of vegetation degradation and recovery across the region, which are described in the following subsections.

Normalised Difference Vegetation Index (NDVI)

The NDVI analysis of the Gaza Strip revealed notable changes in vegetation health between March 2023 and March 2024. The changes in spatial distribution are visualised in Figure 3.2, which highlights the temporal shifts in vegetation cover across the study area.

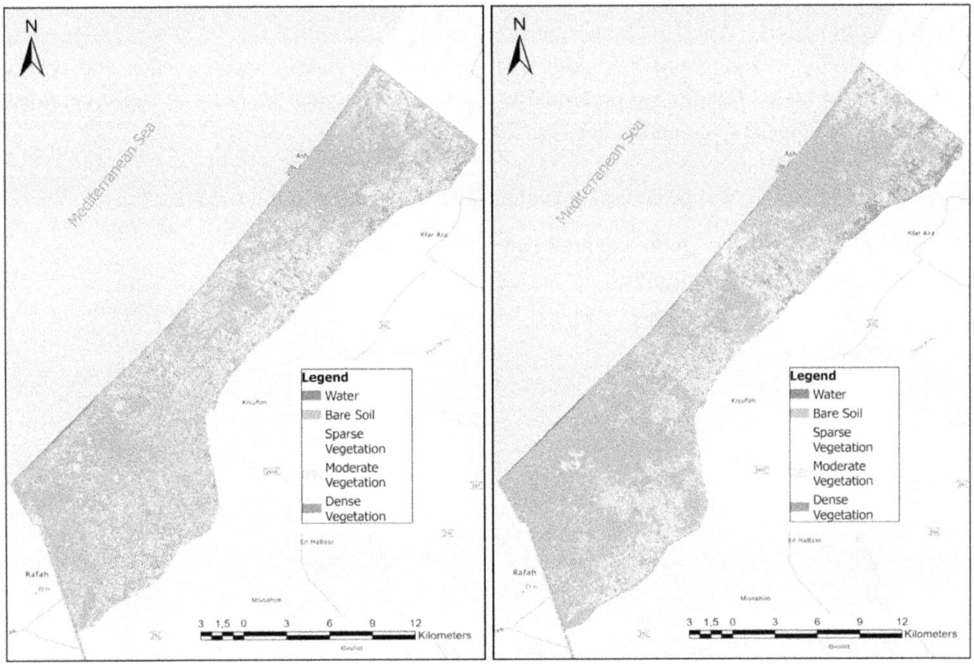

Figure 3.2: Comparison of the NDVI values for the Gaza Strip between March 2023 (left) and March 2024 (right).

In 2023, the NDVI indicated that bare soil was the dominant land cover in the Gaza Strip, encompassing approximately 214.79 km², which highlights the extensive lack of vegetation in the region. Sparse vegetation accounted for 70.09 km², whereas areas of moderate vegetation covered only 62.16 km². By 2024, the extent of bare soil increased significantly to 238.72 km², representing a growth of 23.93 km². Concurrently, sparse vegetation exhibited a substantial decrease, covering only 55.36 km², and moderate vegetation declined to 49.26 km². These shifts indicate a continued trend of vegetation loss, likely driven by the compounded effects of conflict, land degradation, and environmental stressors. The distribution of the NDVI categories for 2023 and 2024 based on the corresponding areas is presented in Table 3.2. This table provides a quantitative overview of the observed changes in land cover during the study period.

Table 3.2: Land cover classification based on the NDVI values for 2023 and 2024 in the Gaza Strip

	Area covered km²	
Category	**NDVI 2023**	**NDVI 2024**
Water	1.00	1.39
Bare soil	214.79	238.72
Sparse vegetation	70.09	55.36
Moderate vegetation	62.16	49.26

The change detection analysis identified that negative changes in vegetation, covering approximately 167.77 km², were the most prominent, reflecting widespread loss likely caused by conflict-related environmental stress. Conversely, positive changes, spanning 158.51 km², represented localised areas of recovery or regrowth, potentially influenced by natural or anthropogenic efforts. Areas with no significant change accounted for 36.79 km², suggesting limited stability within specific pockets of the region. The quantitative results and spatial distribution of these changes are presented in Table 3.3, showing the areas of negative, none, and positive changes in vegetation between 2023 and 2024.

Table 3.3: The NDVI change detection summary for vegetation in the Gaza Strip

Category		Area covered (km²)	Illustration
Negative change		167.77	
No change		36.79	
Positive change		158.51	
Total		**363.07**	

Soil-Adjusted Vegetation Index (SAVI)

The SAVI analysis of the Gaza Strip accounted better for the more prominent arid landscape. The temporal shifts in vegetation and bare soil are clearly illustrated in Figure 3.3, which visualises the extent of degradation and localised recovery during the study period.

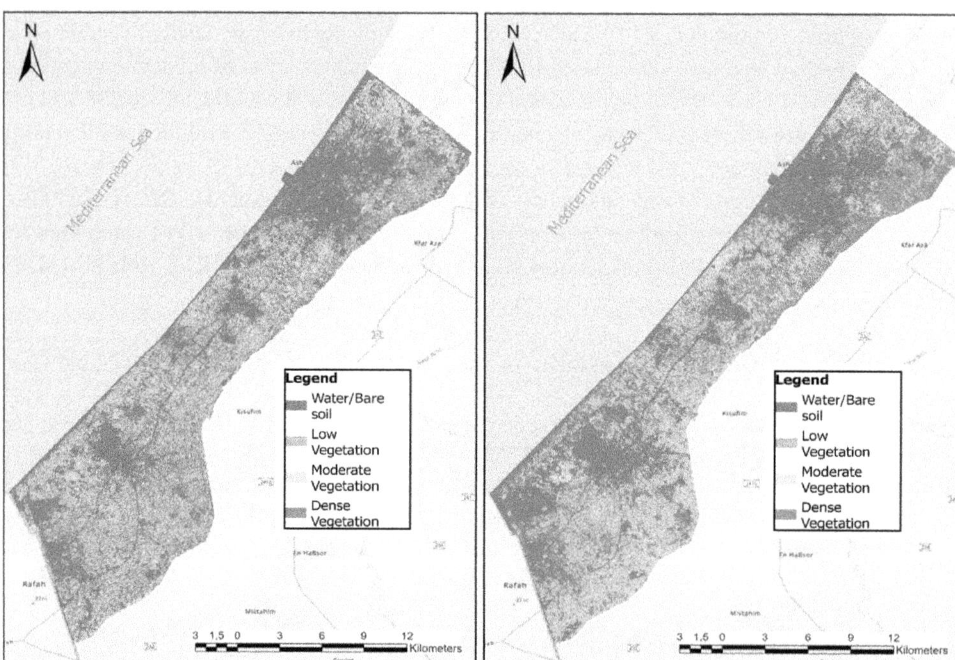

Figure 3.3: Comparison of SAVI values for the Gaza Strip in March 2023 (left) and March 2024 (right).

In 2023, bare soil/water covered approximately 108.16 km², and low vegetation dominated the landscape at 118.37 km². Moderate vegetation accounted for 84.86 km², and dense vegetation covered 51.68 km², reflecting a fragmented vegetative structure under the pressures of arid conditions and anthropogenic stressors. By 2024, a notable expansion in bare soil/water was observed, increasing to 137.89 km², whereas low vegetation decreased to 111.14 km². Similarly, moderate vegetation declined sharply to 65.40 km² and dense vegetation contracted to 48.64 km². These shifts highlight the sensitivity of the SAVI to soil-vegetation interactions, capturing the effects of ongoing conflict and environmental degradation. The detailed distribution of land cover based on the SAVI values is summarised in Table 3.4.

Table 3.4: Land cover classification based on SAVI values for 2023 and 2024 in the Gaza Strip

	Area covered km²	
Category	**SAVI 2023**	**SAVI 2024**
Water/Bare soil	108.16	137.89
Low vegetation	118.37	111.14
Moderate vegetation	84.86	65.40
Dense vegetation	51.68	48.64

Change detection using the SAVI indicated a larger area of negative change than the NDVI, with 178.92 km² showing vegetation loss. Positive changes were recorded over 160.35 km², suggesting localised recovery, whereas 22.07 km² exhibited no significant change, reflecting areas of relative stability. The results are shown in Table 3.5, illustrating the areas of negative, none, and positive changes in vegetation between 2023 and 2024, with corresponding spatial distribution.

Table 3.5: The SAVI change detection summary for vegetation in the Gaza Strip

Category	**Area covered (km²)**	**Illustration**
Negative change	178.92	
No change	22.07	
Positive change	160.35	
Total	361.33	

Modified Soil-Adjusted Vegetation Index (MSAVI)

The MSAVI figures for the Gaza Strip illustrated temporal shifts in vegetation and bare soil that are shown in Figure 3.4.

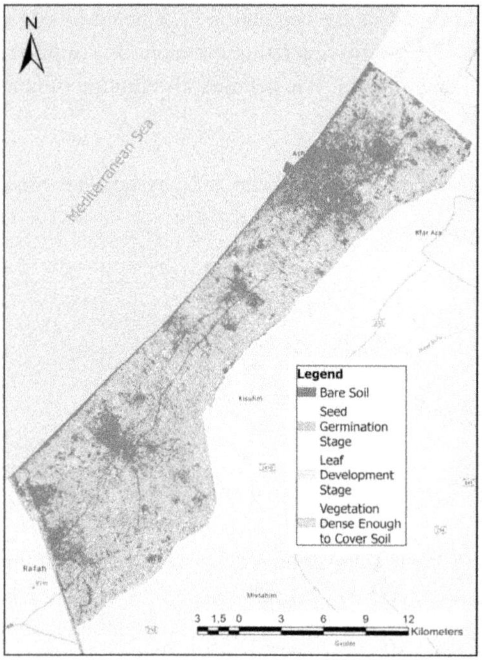

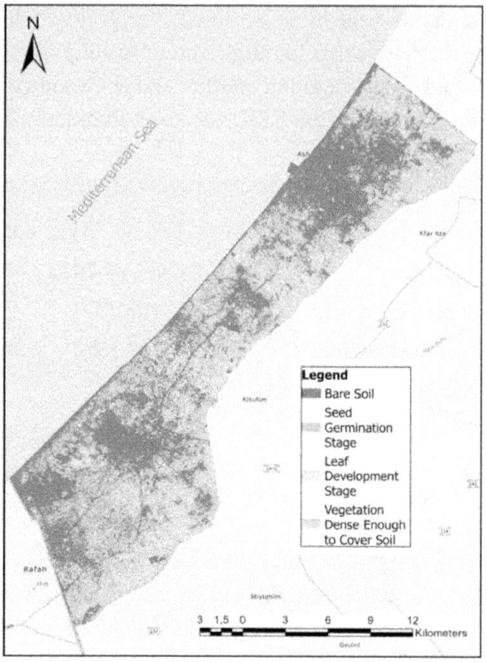

Figure 3.4: Comparison of the MSAVI values for the Gaza Strip in March 2023 (left) and March 2024 (right).

In 2023, bare soil accounted for 85.66 km², whereas areas in the seed germination stage dominated the landscape, covering 128.37 km². Vegetation in the leaf development stage covered 112.15 km² and dense vegetation occupied 36.89 km². By 2024, the area of bare soil expanded to 115.19 km², indicating a significant increase in the extent of exposed land. The area in the seed germination stage decreased slightly to 123.48 km², whereas the vegetation in the leaf development stage showed a notable decline to 87.01 km². Dense vegetation, however, experienced a minor increase to 37.38 km², suggesting localised vegetation recovery despite widespread degradation. These results reflect the MSAVI's sensitivity to subtle changes in vegetation and soil conditions, particularly in arid areas. The land cover classification based on the MSAVI values for 2023 and 2024 is summarised in Table 3.6, providing a clear quantitative breakdown of the vegetation and soil categories over the study period.

Table 3.6: Land cover classification based on the MSAVI values for 2023 and 2024 in the Gaza Strip

	Area covered km^2	
Category	**MSAVI 2023**	**MSAVI 2024**
Bare soil	85.66	115.19
Seed germination stage	128.37	123.48
Leaf development stage	112.15	87.015
Vegetation dense enough to cover soil	36.89	37.38

Negative changes, covering 170.88 km^2, were the most widespread, indicating extensive vegetation loss, likely exacerbated by soil erosion and environmental stress. Positive changes, spanning 168.34 km^2, highlighted areas of recovery, predominantly in agricultural zones and regions less affected by human activity. Approximately 23.53 km^2 exhibited no significant change, reflecting stability in certain areas. The detailed change detection results are listed in Table 3.7, showing the areas of negative, none, and positive changes in vegetation between 2023 and 2024, with corresponding spatial distribution.

Table 3.7: The MSAVI change detection summary for vegetation in the Gaza Strip

Category	**Area covered (km^2)**	**Illustration**
Negative change	170.88	
No change	23.53	
Positive change	168.34	
Total	**362.75**	

Comparative analysis of vegetation indices

Paired samples t-test

The analysis included a comparison of the mean values (Table 3.8), the correlation between the two time periods, and the paired differences (Ross 2017). The results show that the mean NDVI values decreased slightly from 87.01 in 2023 to 86.18 in 2024, whereas the SAVI and MSAVI values remained consistent at 90.77 across both time periods. The correlation between the 2023 and 2024 values was strong for the NDVI ($r = 0.993$) and relatively lower for the SAVI ($r = 0.872$) and the MSAVI ($r = 0.837$). Despite these observable differences, the paired t-test results revealed no statistically significant differences in vegetation index values across the two time periods. The p-values for all three indices were greater than 0.05 (NDVI = 0.932, SAVI = 1.000, and MSAVI = 1.000), indicating that the temporal changes observed may not be substantial enough to rule out random variations.

Table 3.8: Paired samples t-test summary of vegetation indices

Pair	Mean		Correlation	Mean difference	Std. deviation	p-value (two-tailed)
	2023	2024				
NDVI	87.01	86.18	0.993	0.828	17.83	0.932
SAVI	90.77	90.77	0.872	0.001	21.01	1.000
MSAVI	90.77	90.77	0.837	0.001	22.57	1.000

Repeated measures ANOVA

The results (Table 3.9) indicate that there were no significant main effects for the year ($F = 0.001$, $p = 0.976$), suggesting that there were no substantial temporal changes in the vegetation indices between the two years. Similarly, the main effect of the indices was not significant ($F = 0.008$, $p = 0.996$), indicating that the indices did not differ significantly in their ability to detect vegetation changes across the study period. Furthermore, the interaction effect between the year and index type was not significant ($F = 0.004$, $p = 0.999$), indicating that the vegetation indices performed consistently across both periods.

Table 3.9: Summary of repeated measures ANOVA for vegetation indices

Source	Type III sum of squares	df	Mean square	F	p-value
Year	0.459	1	0.459	0.001	0.976
Indices	92.765	2	46.382	0.008	0.996
Year × indices	0.912	2	0.456	0.004	0.999

Bivariate correlation analysis

The results (Table 3.10) indicate a strong and significant positive correlation between SAVI and MSAVI differences ($r = 0.985$, $p = 0.015$). This suggests that both indices capture similar patterns of vegetation change, particularly in environments where the effects of soil brightness are prominent. The 95% confidence interval for this correlation ranged from 0.447 to 1.000, reinforcing the reliability of this association. In contrast, the correlations between NDVI and SAVI differences ($r = 0.146$, $p = 0.854$) and NDVI and MSAVI differences ($r = 0.214$, $p = 0.786$) were weak and not significant. These findings suggest that the NDVI captures different aspects of vegetation dynamics compared to the SAVI and the MSAVI, likely because of its primary sensitivity to vegetation density rather than soil-vegetation interactions.

Table 3.10: Bivariate correlation analysis of vegetation index differences

Correlation pair	Pearson correlation (r)	p-value (two-tailed)	95% confidence interval
NDVI_Diff – SAVI_Diff	0.146	0.854	-0.948 to 0.971
NDVI_Diff – MSAVI_Diff	0.214	0.786	-0.941 to 0.975
SAVI_Diff – MSAVI_Diff	0.985*	0.015	0.447 to 1.000

* Significant at the 0.05 level (two-tailed)

The comparative analysis highlighted that the SAVI and the MSAVI are more reliable in arid regions with sparse vegetation, as their ability to adjust for soil brightness effects makes them more suitable for detecting subtle changes in vegetation health. The NDVI, on the other hand, is effective in assessing vegetation density but may not perform as well in environments where soil exposure dominates.

These findings underscore the importance of employing a multi-index approach to vegetation monitoring, especially in regions with complex environmental conditions. They also highlighted the need for further research incorporating ground-based validation and more advanced geospatial techniques to improve the accuracy and reliability of vegetation assessments in conflict zones.

Discussion

This study examined the environmental consequences of the ongoing conflict in the Gaza Strip, focusing specifically on its impact on vegetation. Using Sentinel-2A satellite imagery and key vegetation indices (NDVI, SAVI, and MSAVI), the study tracked temporal changes in vegetation health from March 2023 to March 2024. The analysis revealed widespread vegetation loss, underscoring the severe ecological toll of prolonged regional warfare.

The NDVI analysis captured broad patterns of vegetation degradation, particularly the significant expansion of bare soil in urban and peri-urban areas, such as in Gaza City. These findings highlight the direct destruction caused by military activities and the indirect effects of environmental stressors. The decline in sparse and moderate vegetation further reflects a diminished capacity for natural regeneration in conflict-affected zones, which compounds the ecological challenges faced by the region (UNEP, 2009).

The SAVI offered additional insights by accounting for soil brightness, which is a critical factor in arid environments such as Gaza. This index provides a more refined interpretation of vegetation health, particularly in areas with sparse vegetation cover. The results confirmed the expansion of bare soil and a decline in vegetation across degraded areas, reflecting the interplay between environmental stressors and the soil conditions. The sensitivity of the SAVI to soil-vegetation dynamics is valuable for capturing the extent of degradation in regions where traditional vegetation indices might overlook subtle changes.

The MSAVI, which dynamically adjusts for soil influence, provided nuanced insights into the early stages of vegetation recovery. Although the overall trend showed continued vegetation decline, isolated signs of regrowth were detected, particularly in areas classified as dense vegetation. These pockets of recovery may be attributed to natural processes or targeted interventions aimed at restoring the vegetation cover. However, the persistence of large expanses of bare soil highlights the ongoing barriers to ecological restoration in conflict-affected areas (Solokha et al. 2023).

The change detection analysis further emphasised the extent of vegetation degradation. Negative changes dominated the landscape, particularly in urban and conflict-intensive zones, confirming significant vegetation loss (Holail et al., 2024). Areas with no significant change were limited, suggesting that stability was restricted to regions with minimal human activity or more resilient vegetation types. Positive changes, though present, were relatively small and largely concentrated in agricultural zones on the periphery of the region, indicating localised recovery efforts or a temporary reduction in conflict intensity.

The results have broader implications for environmental monitoring and management in conflict-affected areas. By demonstrating the effectiveness of multi-index analyses, this study advances the field of remote sensing as a tool for ecological assessments. The methodology presented here can be adapted to other regions, providing a scalable framework for monitoring vegetation dynamics and informing targeted interventions, such as reforestation and soil conservation.

While these findings provide valuable insights into the environmental impact of warfare, certain limitations should be acknowledged. Comparative and statistical analyses were conducted using aggregated area values rather than pixel-level data. This approach, while useful for broad-scale analyses, may reduce sensitivity to localised changes in vegetation cover. Consequently, statistical tests may not fully capture the nuanced variations in vegetation dynamics across the study area. Future research should incorporate per-pixel comparisons to enhance the precision of the analysis and validate observed patterns of degradation and recovery. Additionally, this study relied solely on satellite-derived indices without ground-based validation, which could further improve the accuracy and contextual understanding of vegetation changes.

Overall, the findings paint a stark picture of environmental degradation in the Gaza Strip, driven by prolonged conflict. However, the detection of isolated areas of recovery offers a glimmer of hope for future ecological restoration. These results underscore the need to integrate reforestation, soil conservation, and sustainable land management strategies into post-conflict recovery plans. By addressing the long-term ecological impacts of warfare, such initiatives can enhance the resilience of affected ecosystems and contribute to broader peacebuilding efforts.

Conclusion

This study highlights the profound environmental consequences of conflict in the Gaza Strip by examining vegetation dynamics over a one-year period using Sentinel-2 satellite imagery and advanced geospatial techniques. The results revealed extensive vegetation loss across conflict-affected areas, with a significant expansion of bare soil and a corresponding decline in sparse and moderate vegetation. Despite these concerning trends, isolated pockets of positive change suggest that some recovery is possible, particularly in less conflict-intense zones, which provides a foundation for future restoration initiatives.

The findings emphasise that the environmental impacts of warfare extend far beyond immediate human and infrastructural damage, leaving behind lasting scars on ecosystems. Vegetation loss disrupts biodiversity, undermines soil stability, reduces agricultural productivity, and hampers water regulation, all of which threaten long-term sustainability in the region. Addressing these impacts through targeted interventions is essential to support the recovery and resilience of the affected communities.

The multi-index approach employed in this study provided a comprehensive assessment of vegetation health and recovery patterns in the study area. Each index provided unique insights: the NDVI captured broad trends in vegetation loss, the SAVI refined interpretations in areas with sparse vegetation cover, and the MSAVI detected early signs of regrowth. This combination of indices proved valuable for understanding the complex vegetation dynamics within the Gaza Strip's arid environment.

However, this study has limitations, particularly the reliance on satellite-derived indices without field-based validation and the use of aggregated area values for statistical analyses. Future research should address these limitations by incorporating ground truth data and performing per-pixel analyses to achieve greater accuracy and precision. Additionally, expanding the temporal scope of the analysis would provide insights into longer-term vegetation trends and recovery trajectories, further enhancing the understanding of post-conflict environmental recovery.

This study illustrates the far-reaching ecological costs of conflict, providing critical insights into the patterns of vegetation degradation and recovery in the Gaza Strip. By identifying these patterns, this research highlights the importance of integrating environmental recovery strategies into broader peacebuilding efforts. As conflicts continue to affect ecosystems globally, restoring vegetation and promoting sustainable land management should remain central to post-conflict rehabilitation efforts. The findings of this study emphasise the need for post-conflict recovery strategies that prioritise ecological restoration alongside social and infrastructural rebuilding. This study not only informs restoration efforts in the Gaza Strip but also offers a scalable framework for assessing and mitigating environmental damage in other conflict-affected regions.

In conclusion, this study contributes to the growing body of literature on the environmental impacts of conflict by combining innovative geospatial methodologies with accessible data sources. The findings provide valuable insights for researchers, policymakers, and practitioners seeking to understand and mitigate the ecological consequences of warfare while offering a scalable and adaptable model for future research in similar contexts.

References

Abuelaish, B. & Olmedo, M.T.C. 2016. Scenario of land use and land cover change in the Gaza Strip using remote sensing and GIS models. *Arabian Journal of Geosciences*, 9(274). https://doi.org/10.1007/s12517-015-2292-7

Al-Doski, J., Mansor, S.B. & Shafri, H.Z.M. 2013. War impacts studies using remote sensing. *IOSR Journal of Applied Geology and Geophysics*, 1(2): 11–15. https://doi.org/10.9790/0990-0121115

Asokan, A. & Anitha, J. 2019. Change detection techniques for remote sensing applications: A survey. *Earth Science Informatics*, 12: 143–160. https://doi.org/10.1007/s12145-019-00380-5

Biswas, A.K. 2000. Scientific assessment of the long-term environmental consequences of war. In: J.E. Austin & C.E. Bruch (eds.). *The Environmental Consequences of War: Legal, Economic, and Scientific Perspectives*. Cambridge: Cambridge University Press. 303–315. https://doi.org/10.1017/CBO9780511522321.017

Certini, G., Scalenghe, R., & Woods, W.I. 2013. The impact of warfare on the soil environment. *Earth-Science Reviews*, 127: 1–15. https://doi.org/10.1016/j.earscirev.2013.08.009

Dardona, A.W.Y. 2016. Floristic biodiversity and phytogeographical study in few sites of Gaza Strip, Palestine. *International Journal of Current Science*, 19: 165–182.

EOS Data Analytics. 2024. *MSAVI: Modified Soil Adjusted Vegetation Index*. https://eos.com/make-an-analysis/msavi/ [19 November 2024].

Garzón, F.A.M. & Valánszki, I. 2020. Remote sensing tendencies in the assessment of areas damaged by armed conflicts. *Land Reclamation, Earth Observation & Surveying, Environmental Engineering*, IX: 223–234.

Hais, M., Jonášová, M., Langhammer, J. & Kučera, T. 2009. Comparison of two types of forest disturbance using multitemporal Landsat TM/ETM+ imagery and field vegetation data. *Remote Sensing of Environment*, 113(4): 835–845. https://doi.org/10.1016/j.rse.2008.12.012

Holail, S., Saleh, T., Xiao, X., Xiao, J., Xia, G.S., Shao, Z., Wang, M., Gong, J. & Li, D. 2024. Time-series satellite remote sensing reveals gradually increasing war damage in the Gaza Strip. *National Science Review*, 11(9). https://doi.org/10.1093/nsr/nwae304

Huang, S., Tang, L., Hupy, J.P., Wang, Y. & Shao, G. 2021. A commentary review on the use of normalized difference vegetation index (NDVI) in the era of popular remote sensing. *Journal of Forestry Research*, 32: 1–6. https://doi.org/10.1007/s11676-020-01155-1

Huete, A.R. 1988. A soil-adjusted vegetation index (SAVI). *Remote Sensing of Environment*, 25(3): 295–309. https://doi.org/10.1016/0034-4257(88)90106-X

Kaplan, G., Rashid, T., Gasparovic, M., Pietrelli, A., & Ferrara, V. 2022. Monitoring war-generated environmental security using remote sensing: A review. *Land Degradation & Development*, 33(10): 1513–1526. https://doi.org/10.1002/ldr.4249

Mas, J.-F. 1999. Monitoring land-cover changes: A comparison of change detection techniques. *International Journal of Remote Sensing*, 20: 139–152. https://doi.org/10.1080/014311699213659

Mehta, A., Shukla, S. & Rakholia, S. 2021. Vegetation change analysis using normalized difference vegetation index and land surface temperature in greater Gir landscape, *Journal of Scientific Research*, 65(3): 1–6. https://doi.org/10.37398/JSR.2021.650301

Minu, S. & Shetty, A. 2015. A comparative study of image change detection algorithms in MATLAB. *Aquatic Procedia*, 4: 1366–1373. https://doi.org/10.1016/j.aqpro.2015.02.177

Nordberg, M.L. & Evertson, J. 2005. Vegetation index differencing and linear regression for change detection in a Swedish mountain range using Landsat TM® and ETM+® imagery. *Land Degradation & Development*, 16(2): 139–149. https://doi.org/10.1002/ldr.660

Ross, S.M. 2017. *Introductory Statistics*. Fourth edition. London: Academic Press.

Sadiq, M.R. & Saleh, A.M. 2020. Spatial distribution of vegetative covers in Baquba district, Diyala Governorate using the means of remote sensing. *Plant Archives*, 20(2): 7161–7166.

Selby, J. 2003. *Water, Power and Politics in the Middle East*. London: I.B. Tauris. https://doi.org/10.5040/9780755620968

Sentinel Hub. 2023. Sentinel-2 L2A. https://docs.sentinel-hub.com/api/latest/data/sentinel-2-l2a/ [27 March 2025].

Solokha, M., Pereira, P., Symochko, L., Vynokurova, N., Demyanyuk, O., Sementsova, K., Inacio, M. & Barcelo, D. 2023. Russian-Ukrainian war impacts on the environment. Evidence from the field on soil properties and remote sensing. *Science of The Total Environment*, 902. https://doi.org/10.1016/j.scitotenv.2023.166122

UNEP [United Nations Environment Programme]. 2009. *Environmental Assessment of the Gaza Strip following the escalation of hostilities in December 2008–January 2009*. https://www.unep.org/resources/assessment/environmental-assessment-gaza-strip [27 March 2025].

Vani, V. & Mandla, V.R. 2017. Comparative study of NDVI and SAVI vegetation indices in Anantapur district semi-arid areas. *International Journal of Civil Engineering and Technology*, 8: 559–566.

Voitik, A., Kravchenko, V., Pushka, O., Kutkovetska, T., Shchur, T. & Kocira, S. 2023. Comparison of NDVI, NDRE, MSAVI and NDSI indices for early diagnosis of crop problems. *Agricultural Engineering*, 27: 47–57. https://doi.org/10.2478/agriceng-2023-0004

Xue, J. & Su, B. 2017. Significant remote sensing vegetation indices: A review of developments and applications. *Journal of Sensors*, 2017(1). https://doi.org/10.1155/2017/1353691

Yengoh, G.T., Dent, D., Olsson, L., Tengberg, A.E., & Tucker III, C.J. 2015. *Use of the Normalized Difference Vegetation Index (NDVI) to Assess Land Degradation at Multiple Scales: Current Status, Future Trends, and Practical Considerations*. Cham: Springer. https://doi.org/10.1007/978-3-319-24112-8

Zaineldeen, U. & Aish, A. 2012. Geology, geomorphology and hydrology of the Wadi Gaza catchment, Gaza Strip, Palestine. *Journal of African Earth Sciences*, 76: 1–7. https://doi.org/10.1016/j.jafrearsci.2012.07.005

Zhang, X., Xiao, P., Feng, X. & Yuan, M. 2017. Separate segmentation of multi-temporal high-resolution remote sensing images for object-based change detection in urban areas. *Remote Sensing of Environment*, 201: 243–255. https://doi.org/10.1016/j.rse.2017.09.022

CHAPTER 4

Investigating the distribution of sediments using natural radionuclides through in-situ measurements on the Kilindini Harbour, Mombasa, Kenya

R.K. Botlholo, J. Bezuidenhout, and B. Mtshawu

Abstract

A survey and mapping of the amounts of naturally occurring radionuclides in the Port Reitz–Kilindini hydraulic system were conducted. In-situ measurements of uranium, thorium, and potassium were collected using a Delta Underwater Gamma System (DUGS). Radionuclides were used as tracers to identify and model the presence of sediment in aquatic systems. The data were gathered within a period of three days and then merged for final use. The collected radionuclide data were interpolated using the kriging tool on Esri's ArcMap. This tool can be understood as a two-step process: first, the spatial covariance structure of the sampled points is determined by fitting a variogram, and second, weights derived from this covariance structure are used to interpolate values for unsampled points across the spatial field. In this study, kriging was used as a visualisation tool to show the distribution of naturally occurring radionuclides. Potassium concentrations were higher than those of uranium and thorium, and heightened areas of potassium concentrations were associated with mud over the entire region. The Mwache River has been recognised as the primary contributor of sediment to the harbour. Mangrove trees and waves promote sediment deposition, and thorium and uranium are used as tracers to determine the origin of silt. This study enhances the existing knowledge of tracers and offers useful insights into the application of natural radionuclides for sediment tracing. Subsequent investigations should include additional examinations and sample evaluations of hydraulic systems in controlled laboratory environments.

Keywords: geostatistical analysis (kriging), in-situ gamma-ray spectroscopy, Kilindini Harbour, natural radionuclides, sediment tracing

Introduction

Sediments are particles or soil deposits found in the bottom layers of various water bodies, including oceans, lakes, and rivers (Kamp-Nielsen, 2008). The increase in fluvial sediment in aquatic systems causes environmental deterioration and loss of ecological habitat. The increase in sediment in water bodies impacts the effectiveness of fish spawning, migrating, and feeding (Kamp-Nielsen, 2008). Sedimentation build-up is a major challenge in harbours because it impacts the navigation channels of vessels. This necessitates high-cost maintenance procedures, such as periodic dredging operations, to maintain safe and usable harbour depths. It also enables the construction of new wharves and piers and the enhancement of the existing harbour. Apart from the financial implications, which account for approximately 75% of all maintenance costs in small harbours, dredging operations have detrimental effects on the aquatic environment, such as elevated turbidity in the water and the dispersion of contaminants because of the disturbance of fine-grained sediment (Ferreira & Santos, 2018).

All sediments contain varying concentrations of naturally occurring radionuclides (Kamp-Nielsen, 2008). The major naturally occurring radionuclides are primordial and mostly consist of uranium (^{238}U), thorium (^{232}Th), and potassium (^{40}K). These radionuclides are typically found in small quantities in sediments. Natural radionuclides may serve as tracers for modelling sediment movement (Bezuidenhout, 2020). Gamma-ray spectroscopy is used to quantify these radionuclides because gamma rays are the most penetrating type of ionising radiation released by natural and artificial sources. Each radionuclide uniquely emits gamma rays of specific energies recorded as total count measurements (IAEA, 2003). A radio fingerprint can be generated by the collective concentrations of naturally occurring nuclides in sediment (Bezuidenhout, 2020). Potassium emits a single gamma ray, and it is used to determine its concentration in an area, whereas uranium and thorium undergo a series of decays with some radionuclide daughters also emitting gamma rays (Caciolli et al., 2012). The unique energies and gamma-ray intensities were converted into activity concentrations. Specific locations have distinctive activity concentrations of these radionuclides corresponding to geological features (Bezuidenhout, 2020). Thus, a radio fingerprint can be created using the combined concentrations of natural nuclides in sediments. As a result, it is possible to track the migration of a specific sediment at a location using its radio fingerprint. Consequently, the sediment can also be related to its source (Bezuidenhout, 2020). The presence of radionuclides in nature is strongly correlated with geology. This is because radioactive materials can be found in the Earth's rocks. Radioactive material is released from these rocks when they weather and encounter the environment. For instance, because igneous rocks are created deep inside the earth, radionuclide concentrations are lower in sedimentary rocks than in igneous rocks (Missimer et al., 2019). This study endeavours to identify regions in Kilindini Harbour with low and high sedimentation and investigate their sources.

Materials and methods

Study area

Kilindini Harbour is a natural harbour with a deep inlet leading directly to the ocean, allowing easy access for large seafaring vessels. Consequently, the northeastern region of the harbour has seen the development of several ports and shipyards. River Mwache, the main source of fresh water in Kilindini Harbour, contributes around 3 m^3/s, while Mteza Creek, situated approximately 10 km from the main navigation channel of the harbour, contributes 1.4 m^3/s of water and sediment to the harbour basin. The deposition of silt gives rise to the formation of mudflats and mangrove forests in the vicinity of the river (Rees et al., 1996). The objective of this study was to identify areas with significant sedimentation and trace the source of the sediment using natural radionuclides as indicators in the Port Reitz–Kilindini hydraulic system. The study area (Figure 4.1) upon which in-situ measurements were carried out is the Port Reitz–Kilindini hydraulic system (**39°38' E, 4°04'** S). Sedimentation characteristics were inferred from the patterns in the distribution of the three natural radionuclides at this port.

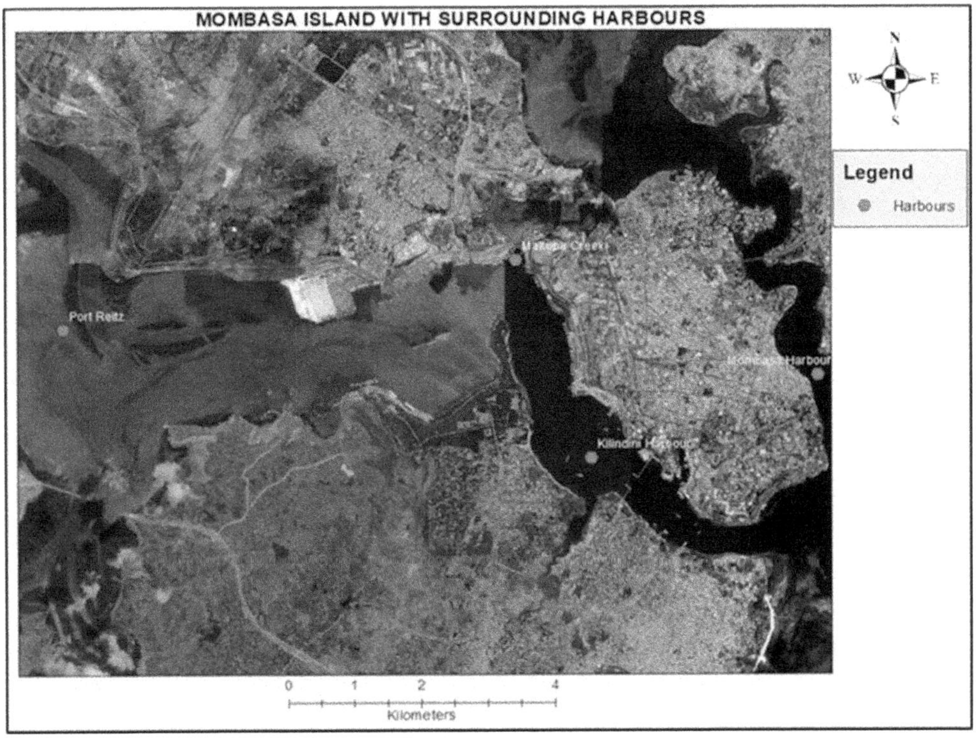

Figure 4.1: Map of the study area showing Mombasa Island and the surrounding harbours, including Port Reitz, Kilindini Harbour, Mombasa Harbour, and Mtwapa Creek. The locations of the harbours are marked with red dots. This area forms the focus of the in-situ gamma-ray spectrometry measurements conducted to investigate sediment distribution using natural radionuclides.
(*Source:* Authors' own work using ArcGIS 10.8.1; Google Earth Pro – imagery date: 25 January 2025; https://earth.google.com/web/)

Measuring systems

Gamma-ray spectra can be obtained using in-situ and laboratory measurements. Because of the spatial characteristics required, in-situ measurements are more efficient than sampling followed by laboratory measurements, which collect samples for analysis in a controlled environment. The advantage of laboratory measurements is that they yield highly accurate results with low uncertainty when quantifying radionuclide concentrations (Povinec et al., 2006). However, with in-situ measurements, large areas can be covered, enabling the acquisition of data in real-time (Kilel et al., 2023). In-situ measurements of naturally occurring radionuclides were performed using a Delta Underwater Gamma System (DUGS). The DUGS is a tailor-made underwater gamma detection system (Figure 4.2) designed and built by a research team at Stellenbosch University in South Africa. The DUGS is made of a 3"×3" NaI(Tl) scintillation detector, a photomultiplier tube, an Amptek® TB-5 multichannel analyser (MCA), a DC-DC converter (step-down), an external computer, and all other associated electronics in waterproof housing. The external computer features a built-in GPS that logs in-situ measurements and geographic coordinates.

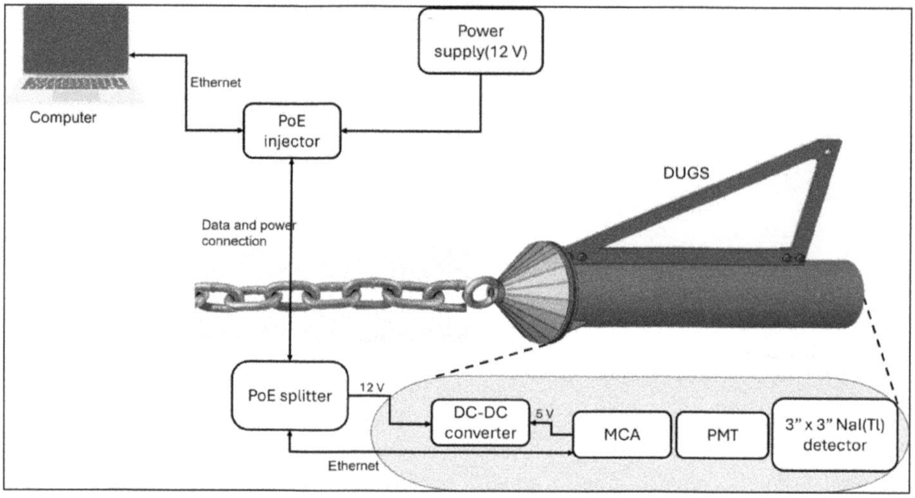

Figure 4.2: Isometric representation of the DUGS with a block system of its components. (*Source:* Bezuidenhout et al., 2023)

The low cost, durability, and high efficiency of the components led to their selection. The DUGS is designed to be dragged underwater continuously; thus, it can withstand hydrostatic pressure as well as impacts from rocks and all other loose natural materials. To counter the drag and other impacts, the front of the system was fitted with a 4 mm-thick-walled stainless steel cone. To acquire this data effectively, a software was developed that allowed the extraction of activity concentrations of the naturally occurring radionuclides. Ethernet was used to establish communication with the detection system using a Cat5e connection. Power-over-Ethernet (PoE) was used to supply power to the system. A boat was used to tow the DUGS along the area of the Port Reitz–Kilindini hydraulic system, while a computer system with the installed software was on board to acquire the data. To maintain contact with the sea floor, the DUGS is attached to a chain and a righting buoy (Bezuidenhout et al., 2023) (Figure 4.3). The data were acquired for three days and combined for final use. The system was calibrated before and after every measurement using either a point radium source or a thorium source, and the calibration parameters were recorded for analysis.

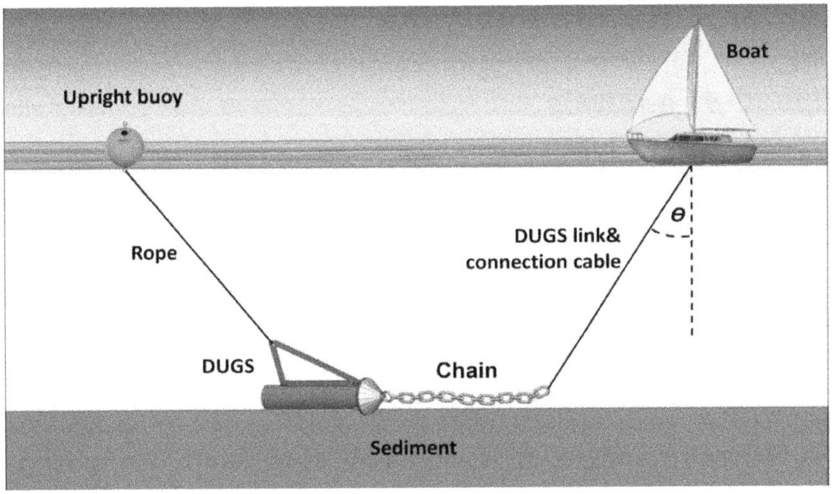

Figure 4.3: Measuring configuration of the DUGS connected to a boat during in-situ measurements.
(*Source:* Mtshawu et al., 2023)

Figure 4.4 shows the potassium peak and most of the daughter nuclides of uranium and thorium. According to Caciolli et al. (2012), ^{232}Th and ^{238}U do not directly emit gamma rays; however, their activity concentrations are determined by detecting the gamma rays emitted by their daughter nuclides, ^{208}TI and ^{214}Bi, respectively. To use this approach, the assumption of secular equilibrium of the decay chains is required, implying that all the emitted nuclides are in equilibrium with one another. In this study, ^{208}Tl (2615 keV) and ^{214}Bi (1764 keV) were used to conduct a full-spectrum analysis. The potassium concentration was evaluated based on a single peak at ^{40}K (1460 keV). For post-processing and analysis, the gamma-ray spectra, location, and time were recorded every 5 s.

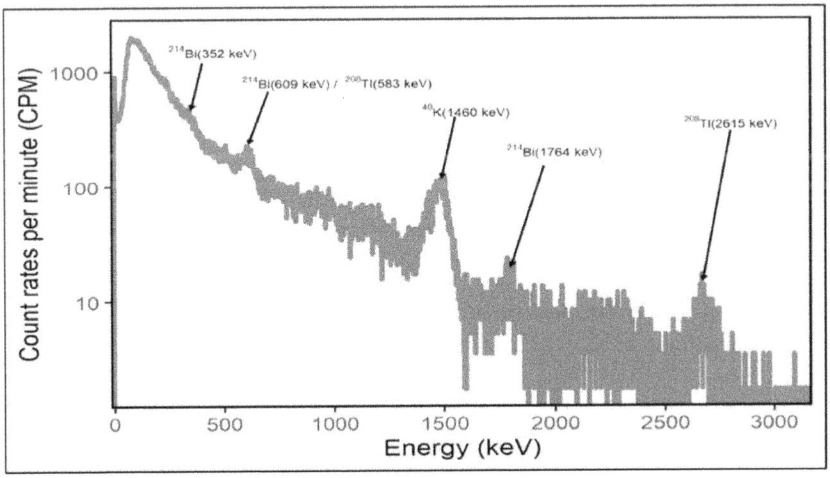

Figure 4.4: Typical spectrum obtained from the DUGS during in situ measurements, showing the natural radionuclides and their daughter nuclides.
(*Source:* Mtshawu et al., 2023)

Measurements and analyses

Full spectrum analysis

Upon the acquisition of data from the onboard software, the data were cleaned, and nuclide concentrations were extracted using a code developed by a team from Stellenbosch University (Bezuidenhout et al., 2023). Full-spectrum analysis was performed to conduct a quantitative assessment and extract nuclide concentrations from the in-situ spectra by binning the channels in the in-situ spectrum and the three available standards into 1 keV bins, starting at 300 keV to 3000 keV, while anything less than 300 keV and more than 3000 keV was neglected. This technique uses all the counts in the obtained in-situ spectra, resulting in minimal variance that yields accurate results. All gamma interactions with matter, such as Compton scattering and pair production, were considered, which is one of the reasons a full-spectrum analysis is preferred. The Nuclear Energy Corporation of South Africa (NECSA) in Pretoria provided standard radiation reference pads, where the three calibration spectra were recorded. The obtained calibration spectra were then utilised as standard spectra after being normalised to one Bq/kg (Bezuidenhout et al., 2023). According to Salmon (1961), full-spectrum analysis is an ideal method for minimising statistical errors when working with low count rates through the method of least squares.

The workings of the full-spectrum analysis are as follows: An in-situ spectrum containing m standard nuclides is obtained using a gamma-ray spectrometer with an n-channel multichannel analyser. Standard calibration spectra of uranium, thorium, and potassium are available, and these are fitted into the in-situ spectra to extract the number of activity concentrations of these radionuclides in the in-situ spectra. This is expressed by the following equation:

$$b_i = \sum_{j=1}^{m} a_{ij} x_j + Z_i$$

where:

b_i is the count rate from the in-situ spectra.

a_{ij} is the count rate in channel i from standard j.

x_j is the ratio of the unknown activity to its corresponding standard.

Z_i is a random error.

The sum of the squares of the random error (2) is optimised with respect to xj and equated to zero to obtain the values of xj.

$$R = \sum_{i=1}^{n} Z^2 = \sum_{i=1}^{n} \left[b_i - \sum_{j=1}^{m} a_i x_j \right]^2$$

where R is the sum of the squares of the random error.

This optimisation of R in (2) leads to a sequence of *m* equations (3) that form a 3 × 3 matrix (7) where the three standard spectra are combined into 2700 ×3(standard spectra matrix) and 2700 × 1(in-situ spectra matrix), as shown in equations (4), (5), and (6). The matrix is over-determined and contains the concentrations of ^{40}K, ^{232}Th, and ^{238}U, whereas matrix *x* gives the total activity concentrations. Solving the matrix leads to obtaining the unknown activity (8) corresponding to a standard in the in-situ spectrum by multiplying the inverse of the known standards matrix with the in-situ spectra. Thus, the amount of each standard in the in-situ spectra is determined, that is, the activity concentrations (x).

$$\sum_{j=1}^{m} x_j \sum_{i=1}^{n} a_{ik} a_{ij} = \sum_{i=1}^{n} a_{ik} b_i$$

Equation (3) can be simplified as follows:

$$x_1 \sum a_{i1}^2 + x_2 \sum a_{i1} a_{i2} + x_3 \sum a_{i1} a_{i3} = \sum a_{i1} b_i$$

$$x_1 \sum a_{i1} a_{i2} + x_2 \sum a_{i2}^2 + x_3 \sum a_{i2} a_{i3} = \sum a_{i2} b_i$$

$$x_1 \sum a_{i1} a_{i3} + x_2 \sum a_{i2} a_{i3} + x_3 \sum a_{i3}^2 = \sum a_{i3} b_i$$

The summations are over 2700 bins, each row of the matrix represents the bins, and the columns are the activity concentrations of the three standard nuclides.

In matrix form, the three equations become

$$\sum_{i=1}^{3} A_i x_i (n) = y\left(n\right)$$

where:

y and A are the in-situ and standard matrix spectra respectively.
$i=1$ *to* 3 represent the standard spectra matrices of ^{40}K, ^{232}Th, and ^{238}U, respectively.

Matrix equation (7) can be rearranged to solve the activity concentrations.

$$x = A^{-1} y$$

The variance in the activity concentration can be obtained using the following equation:

$$VAR(x_j) = \frac{R}{n-m} [e_{jj}]^{-1}$$

where:

$[e_{jj}]^{-1}$ is the diagonal element corresponding to the inverse matrix .

The shortfall of the full-spectrum technique is that it yields imprecise results when applied over short times; likewise, it experiences the NaI (TI) detector's low energy resolution (Caciolli et al., 2012).

Geographical analysis

Further analysis of the data was performed using geographic information systems (GIS) 10.8.1 ArcMap software to display and develop maps representing the spatial variation of the nuclide concentrations. The geostatistical analysis tool is a package in GIS software that enables users with spatially continuous data to interpolate optimum surfaces and employ statistical methods for data exploration and analysis from measured values. The interpolation method used was kriging. It assumes that the spatial change in the activity concentrations is a function of the distance between points. This allows semi-variogram calculations which compare successive point measurements. The weights for interpolation are then determined using the semi-variogram (Huisman & De By, 2009). The probabilistic nature of radioactive decay causes gamma-ray data to inherently contain a noise component. Kriging considers this noise component and offers minimal variance (IAEA, 2003).

Cross-validation techniques

The two most widely used techniques for determining the accuracy of spatial interpolation are validation and cross-validation. Validation uses the subset data to create a model without including all the data. Cross-validation removes a point in the dataset, interpolates a value for that point using the remaining data, and compares the measured and predicted values using all the data (Mtshawu, 2014). In this study, cross-validation was used to indicate how well the interpolation predicted unknown values.

Prediction vs. measured

The geostatistical analysis compares the measured and predicted values, which can be seen in a scatter plot, to show whether the interpolators predicted the data well. The scatter plots are presumed to be on the line of approximately 1:1 (grey line). However, owing to the slope, which is usually less than 1, this may not always be the case. The blue line indicates the fitted line through the scatter plot (Figures 4.5–4.7). If the blue line is horizontal, this indicates that there is no autocorrelation; otherwise, the blue line should be closer to the grey line to indicate autocorrelation. The difference between the measured and predicted values yielded a prediction error. The calculated statistics of the prediction errors serve as an indication of whether the interpolator is a good tool for producing maps.

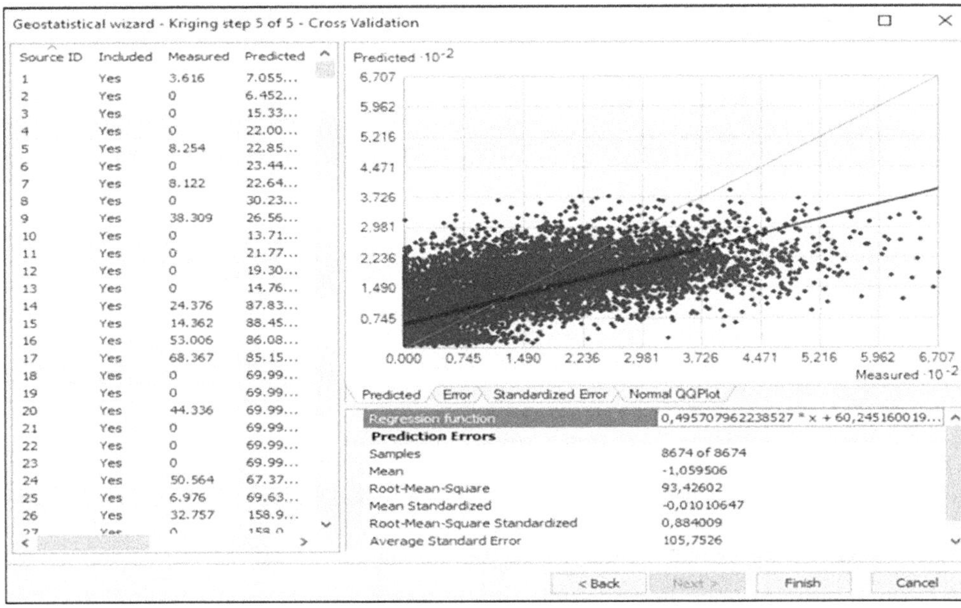

Figure 4.5: Graph showing the predicted values versus the measured values of ^{40}K.

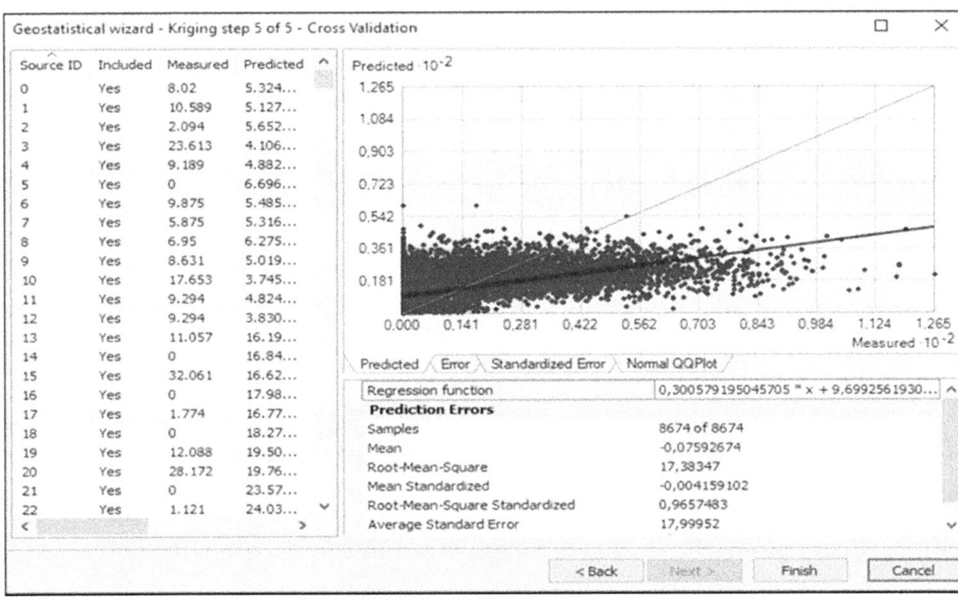

Figure 4.6: Graph showing the predicted values versus the measured values of ^{232}Th.

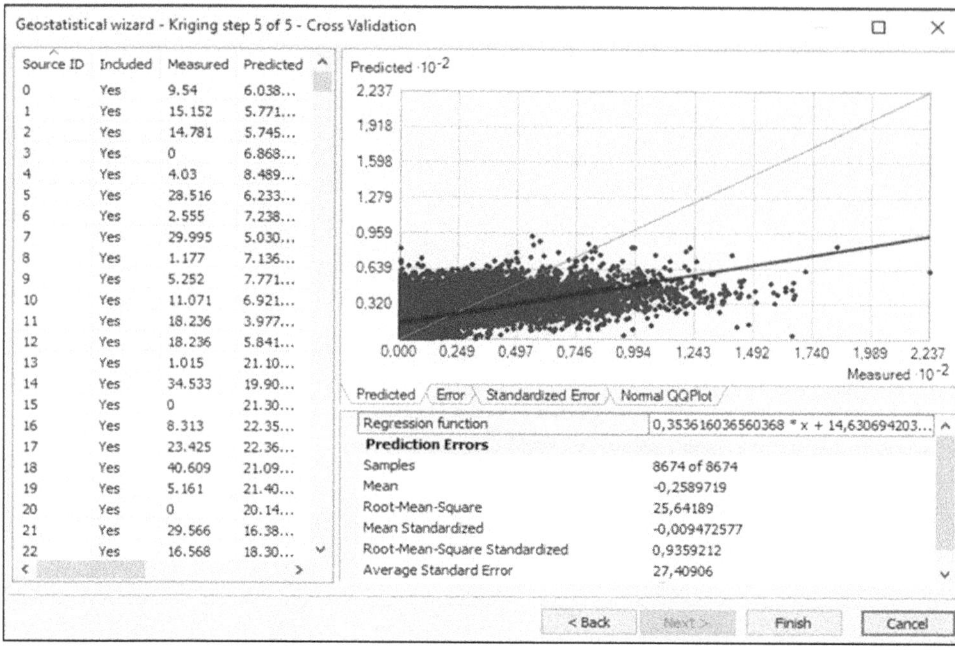

Figure 4.7: Graph showing the predicted values versus the measured values of and ^{238}U.

Table 4.1 indicates the evaluation performance, that is, the mean prediction errors (MPE) which measures biasness of the model. Values near zero mean predictions are not systematically too high or too low. The obtained MPE was less than 0, indicating an unbiased prediction of the results. Root mean squared error (RMSE) measures the average size of the prediction errors; the smaller the RMSE, the better the model. The average standard error (ASE) computes the model's average estimated uncertainty of predictions. The ideal ASE should be closer to the RMSE. This was the case with the study's prediction. The root mean square standardised error (RMSSE) indicates whether the model's estimated errors are realistic; values near 1 are ideal (Bouhout et al., 2024).

Table 4.1: Summary of the error predictions

Evaluation of the performance of predictions on Ordinary Kriging				
Radionuclide	*Ordinary MPE*	*Ordinary RMSSE*	*Ordinary RMSE*	*Ordinary ASE*
^{40}K	-1,059506	0,884009	93,42602	105,7526
^{232}Th	-0,07592674	0,9657483	17,38347	17,99952
^{238}U	-0,2589719	-0,009472577	25,64169	27,40906
Comparing RMS values of the measured points versus predicted points				
Radionuclide	*RMS Measured*		*RMS Predicted*	
^{40}K	8,812-670,674		93,42602	
^{232}Th	0,645-126,462		17,38347	
^{238}U	0,967-223,446		25,64189	

Root-mean-square standardised prediction error	
Radionuclide	*Root-mean-square standardised prediction error*
^{40}K	0,884009
^{232}Th	0,96574
^{238}U	0,9359212

Quantile-Quantile plots

One of the few ways to explore the data is by looking at the Quantile-Quantile plot (QQP), which indicates the normality in the data by comparing the distribution of the data to a normal distribution (refer to Figures 4.8–4.10). The sample data will most probably follow a normal distribution if the data points lie as close as possible to the $45°$ grey line (Johnston et al., 2003). The three graphs show normality in the data.

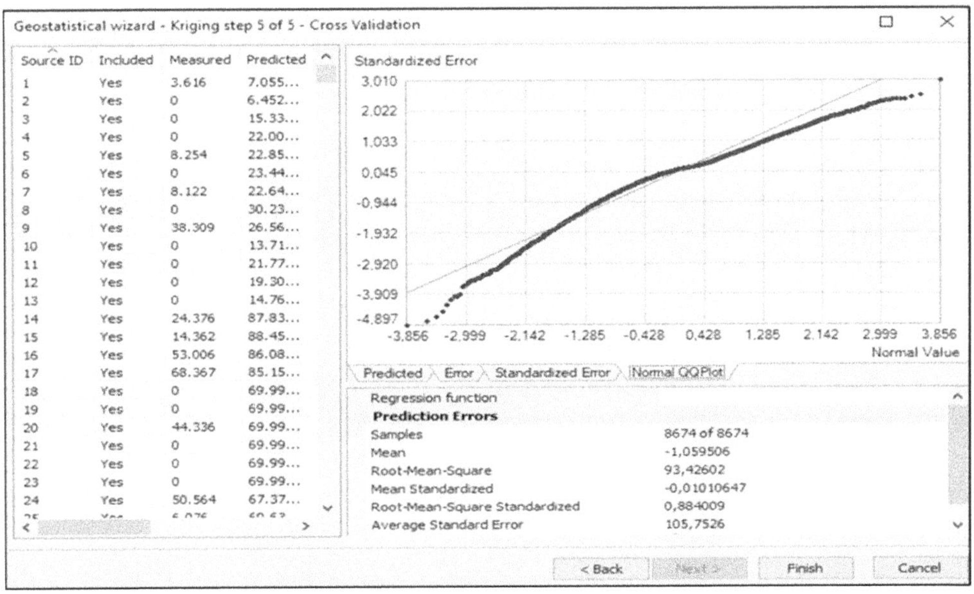

Figure 4.8: Cross-validation (QQP) graph of ^{40}K.

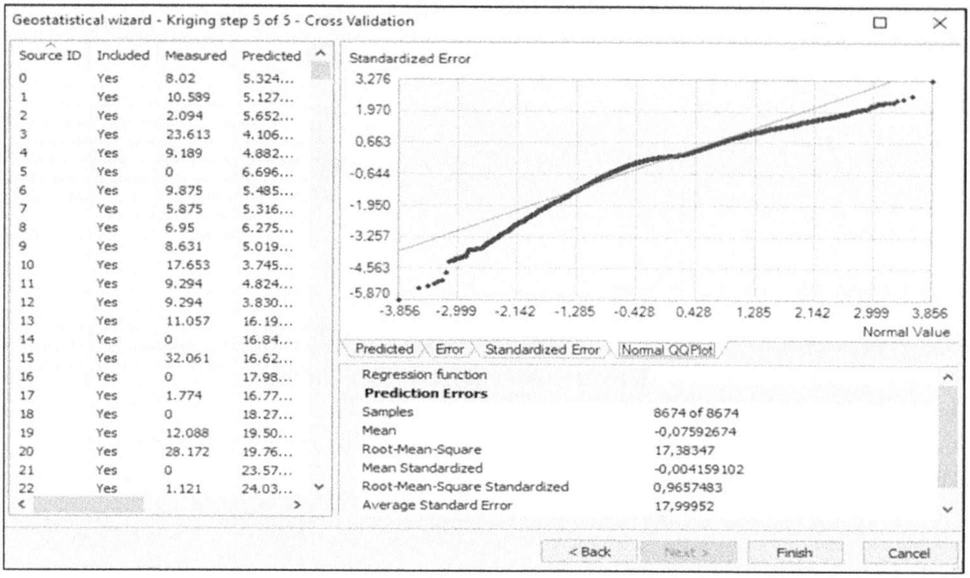

Figure 4.9: Cross-validation (QQP) graph of ^{232}Th.

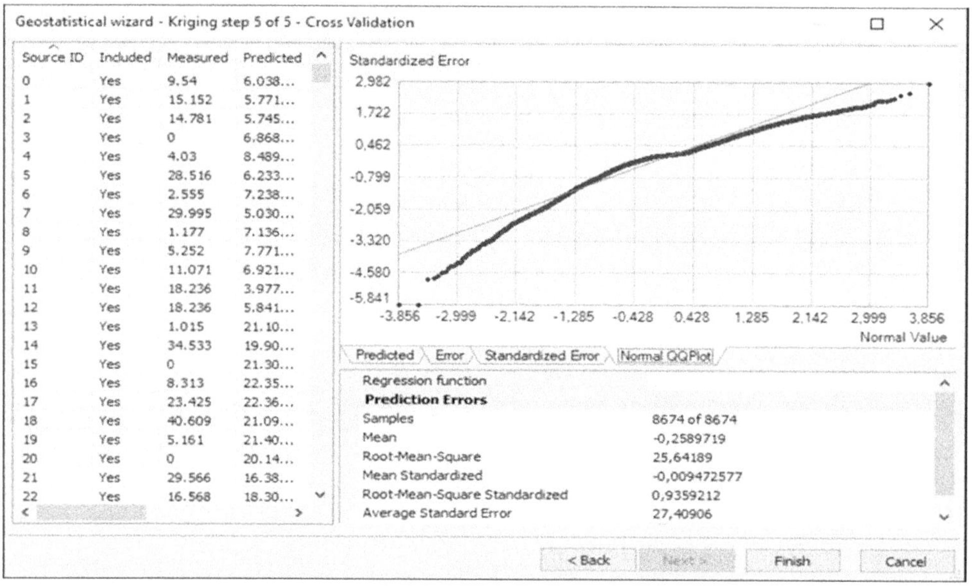

Figure 4.10: Cross-validation (QQP) graph of ^{238}U.

Semi-variance

Semi-variogram plots the semi-variance against the lag distance. Pairs of points that are near each other in distance will have more similar values than pairs that are far apart if there is spatial dependence among the points. To indicate the presence of spatial dependence, the semi-variance should increase as the distance increases, as shown in Figures 4.11–4.13. The lag size indicates the size of the lag distance into which points are grouped to reduce the number of larger combinations of points.

This is a process known as binning, which averages the semi-variance data by distance (Chang, 2019). The modelling of a semi-variogram requires three parameters: nugget, sill, and range. The nugget intercepts the abscissa and shows the non-spatial variation in the data generated by noise factors. The sill, where the levelling occurs, is the amplitude reached at a particular distance known as the range, where the semi-variance starts to level off. Any observations beyond this range are not spatially dependent (IAEA, 2003). The blue line indicates the best-fit line through the semi-variance points. The lag size in the semi-variance graphs (refer to Figures 4.11–4.13) is 12 m, meaning that pairs of points separated by less than 12 m are grouped into the lag class of 0-12. The red points are grouped according to the lag size they correspond with (Johnston et al., 2003).

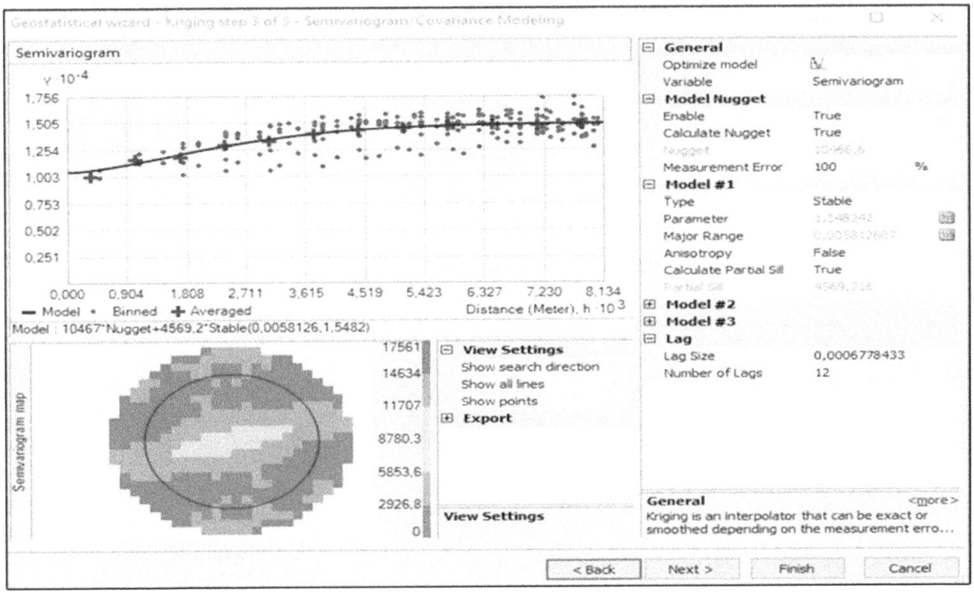

Figure 4.11: Semi-variance plot of ^{40}K.

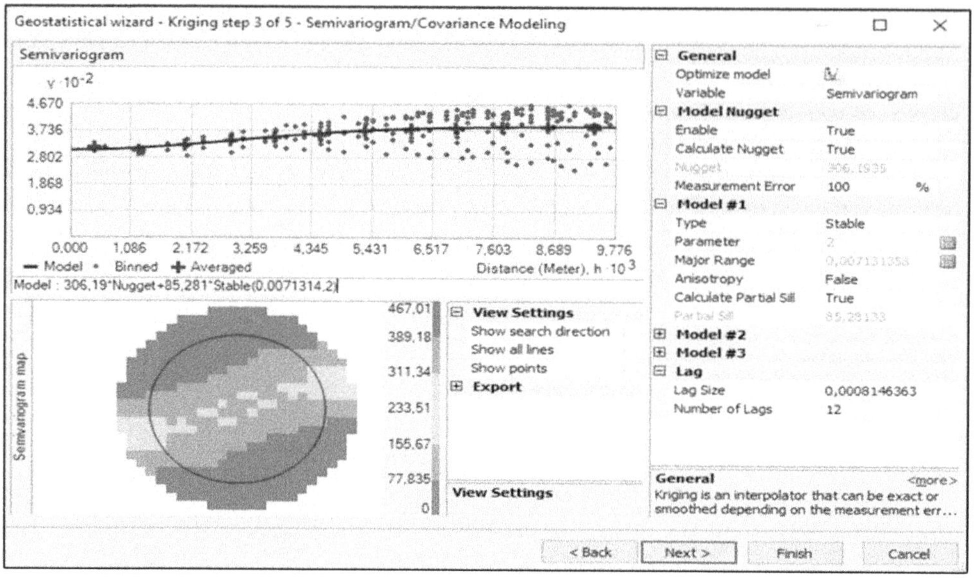

Figure 4.12: Semi-variance plot of ^{232}Th.

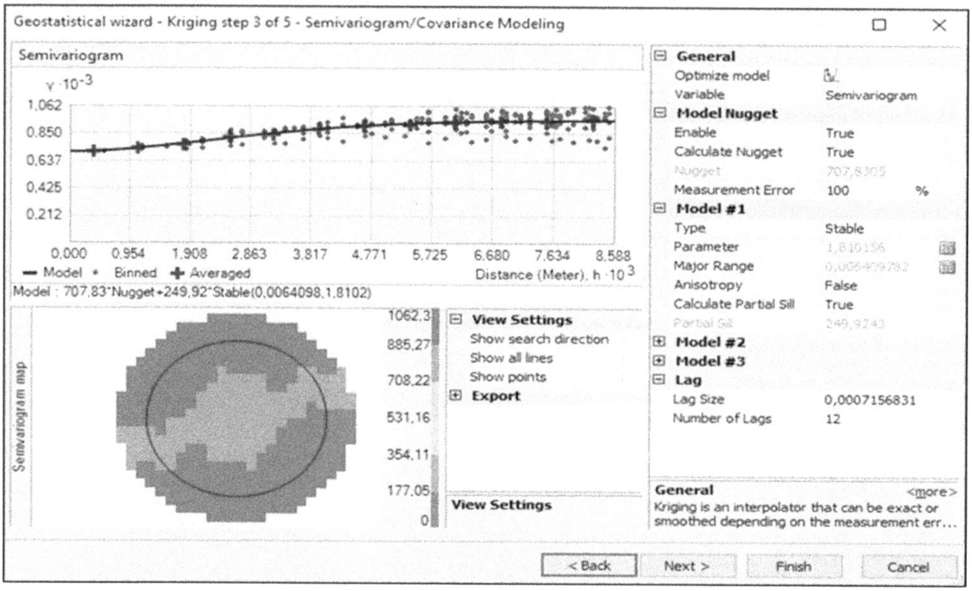

Figure 4.13: Semi-variance plot of ^{238}U.

Discussion and results

Figures 4.14–4.16 show the activity concentrations of the three radionuclides in the Port Reitz–Kilindini hydraulic system. Potassium demonstrated higher concentrations than uranium and thorium, which is anticipated because of its vast presence in the Earth's crust. Small amounts were observed at the easternmost part of the study area inside the harbour. This location was where the dredging occurred. The presence of mangrove trees in the surrounding region enhances sediment deposition on the riverbanks. Furthermore, waves and tides facilitate substantial sediment dispersion across the harbour, resulting in widespread distribution. Thorium and uranium showed a strong correlation and may be used as indicators for tracing the movement of sediment in the Mwache River. The analysis revealed that most of the sediment originated from the Mwache River and was transported downstream to Mteza Creek. Bezuidenhout (2020) established a positive correlation between potassium and mud content. Thus, the elevated potassium concentrations may serve to track mud throughout the research region. Despite dredging efforts in the harbour, sediment may reappear in the channel in the form of muddy regions.

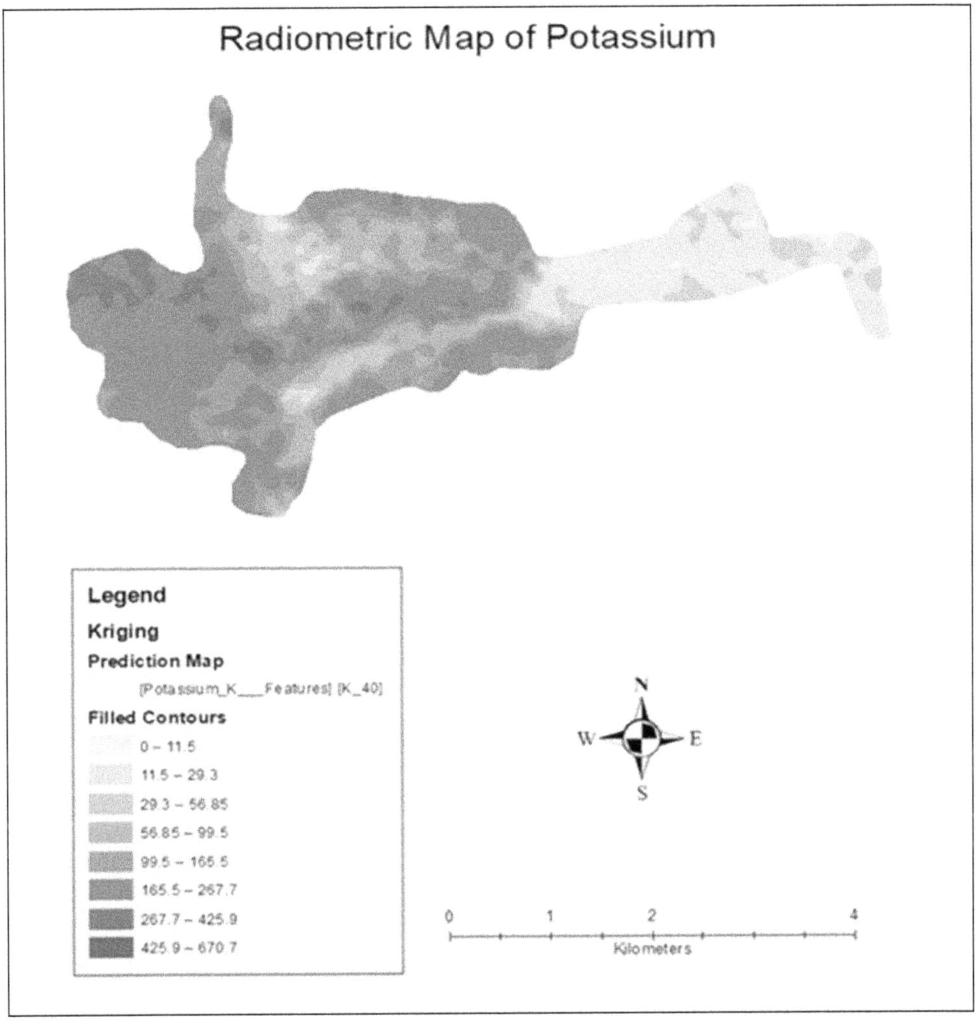

Figure 4.14: ArcMap showing the activity concentrations of ^{40}K (red) in the study area. The dark hue indicates a high concentration, and the light hue indicates a low concentration. (*Source:* Authors' own work using ArcGIS 10.8.1)

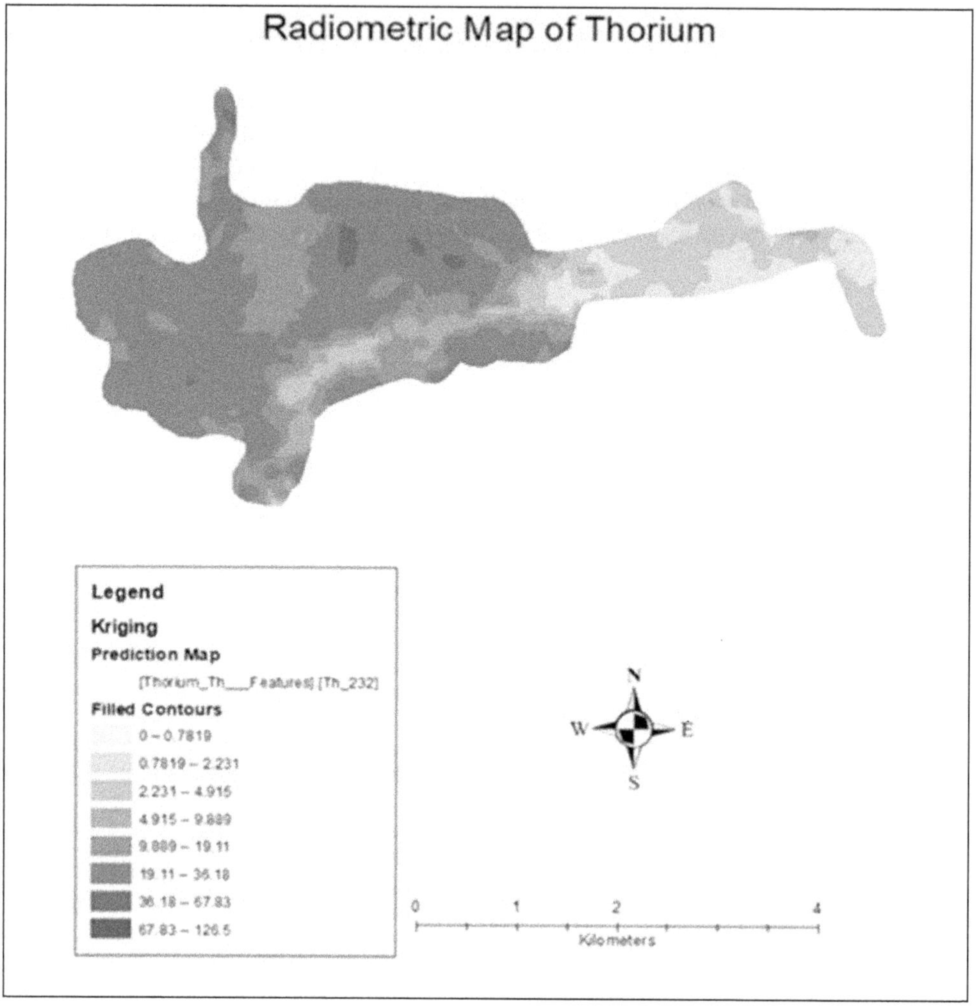

Figure 4.15: ArcMap showing the activity concentrations of ^{232}Th (green) in the study area. The dark hue indicates a high concentration, and the light hue indicates a low concentration. (*Source:* Authors' own work using ArcGIS 10.8.1)

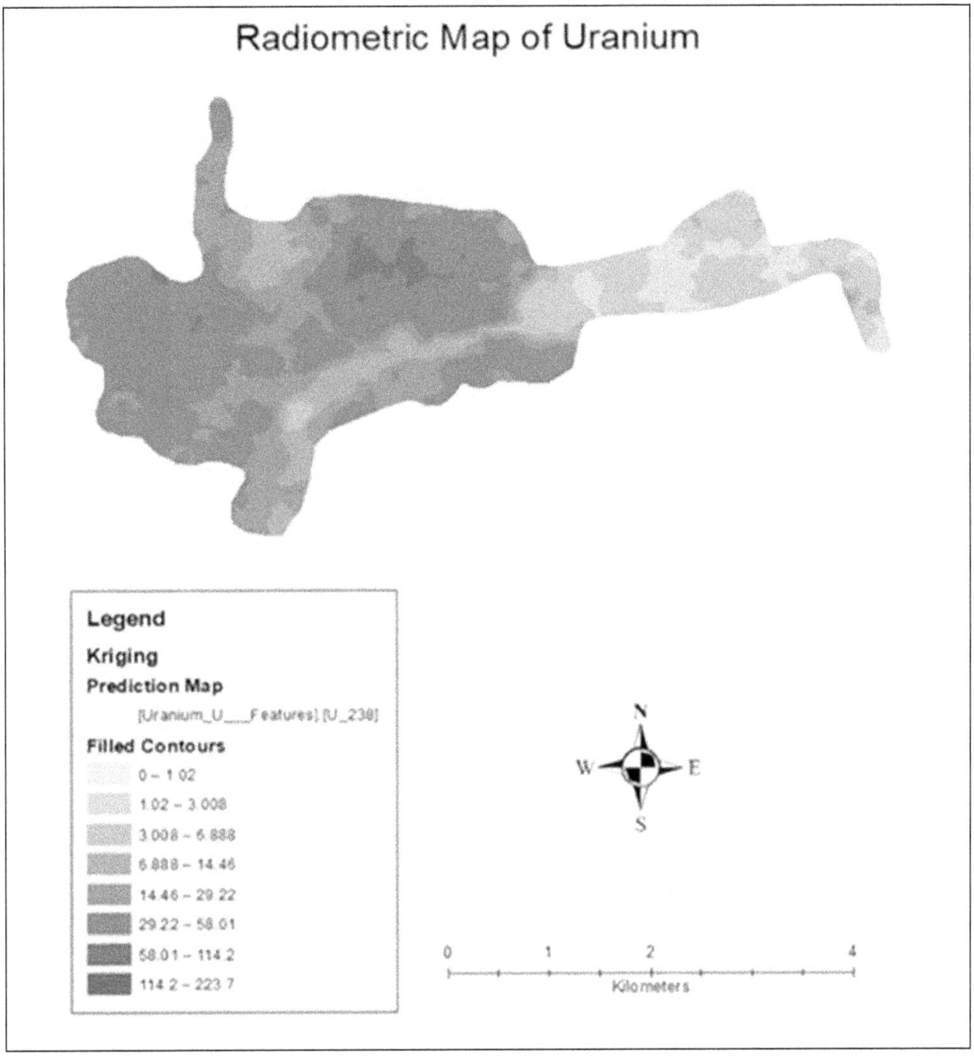

Figure 4.16: ArcMap showing the activity concentrations of ^{238}U (blue) in the study area. The dark hue indicates a high concentration, and the light hue indicates a low concentration. (*Source:* Authors' own work using ArcGIS 10.8.1)

Conclusion

The deployment of the DUGS and recording of measurements on the seafloor Port Reitz–Kilindini hydraulic system were successful. Natural radiation was used to trace the sediments and yielded successful results. The full-spectrum analysis was accurate, and the activity concentrations of ^{40}K, ^{232}Th, and ^{238}U were successfully extracted with minimal errors. Spatial and statistical analyses were conducted using geostatistical tools, resulting in minimal statistical errors, allowing the developed layers of radionuclides to be utilised to draw conclusions about the study area. In-situ measurements revealed the flow of sediments in and around the harbour, and GIS software was used to show the spatial variation of the radionuclide concentration on the seafloor of the harbour.

The primary origin of the sediment was determined by analysing the uranium and thorium concentrations in the rivers. River Mwache was identified as the major source. A previous study conducted by Mtshawu et al. (2023) used the same measuring system of the DUGS in the Berg River to determine the activity concentrations of the naturally occurring radionuclides. The study estimated the effective use of the DUGS and confirmed that naturally occurring radionuclides can be used to trace sediments. To enhance future investigations, further studies and extensive sample analyses should be conducted on hydraulic systems in laboratory settings.

References

Bezuidenhout, J. 2020. The investigation of natural radionuclides as tracers for monitoring sediment processes. *Journal of Applied Geophysics*, 181(104135): 1–16. https://doi.org/10.1016/j.jappgeo.2020.104135

Bezuidenhout, J., le Roux, R. & Kilel, K. 2023. The characterization and optimization of an Underwater Gamma-Ray Detection System. *Journal of Marine Science and Engineering*, 11(171): 1–13. https://doi.org/10.3390/jmse11010171

Bouhout, S., Haboubi, C., Haboubi, K., Elyoubi, M.S., Elabdouni, A., El Bastrioui, M. & El Alaoui, H. 2024. Spatial variability of nitrate leaching and risk assessment of nitrate contamination in the Ghiss-Nekor alluvial aquifer system (Northeastern Morocco) through Disjunctive Kriging. *Scientific African*, 23. https://doi.org/10.1016/j.sciaf.2023.e02009

Caciolli, A., Baldoncini, M., Bezzon, G.P., Broggini, C., Buso, G.P., Callegari, I., Colonna, T., Fiorentini, G., Guastaldi, E., Mantovani, F., Massa, G., Menegazzo, R., Mou, L., Rossi Alvarez, C., Shyti, M., Zanon, A. & Xhixha, G. 2012. A new FSA approach for in situ gamma-ray spectroscopy. *Science of The Total Environment*, 414: 639–645. https://doi.org/10.1016/j.scitotenv.2011.10.071

Chang, K-T. 2019. *Introduction to Geographic Information Systems.* Ninth edition. New York, NY: McGraw-Hill.

Ferreira, A.M.G.S. & Santos, C.S. 2018. Sedimentation and erosion in harbour estuaries. In: A. Amini (ed.). *Sedimentation Engineering*. London: IntechOpen. 31–56. https://doi.org/10.5772/intechopen.74049

Huisman, O. & de By, R. 2009. *Principles of Geographic Information Systems: An Introductory System.* Fourth edition. Netherlands: ITC.

IAEA [International Atomic Energy Agency]. 2003. Guidelines for radioelement mapping using gamma ray spectroscopy data. https://www-pub.iaea.org/mtcd/publications/pdf/te_1363_web.pdf [1 May 2023].

Johnston, K., Ver Hoef, J.M., Krivoruchko, K. & Lucas, N. 2003. *Using ArcGIS geostatistical analyst.* https://content.esri.com/support/documentation/ao_/geostatistical_analyst_tutorial.pdf [2 December 2025].

Kamp-Nielsen, L. 2008. Sediments: Setting, transport, mineralization, and modeling. In: S.E. Jørgensen & B.D. Fath (eds.). *Encyclopedia of Ecology*. Amsterdam: Elsevier. 3181–3188. https://doi.org/10.1016/B978-008045405-4.00429-8

Kilel, K.K., Bezuidenhout, J., Gatari, M.J., le Roux, R.R. & Kaniu, M.I. 2023. A low-cost Delta Underwater Gamma System (DUGS) for in-situ measurement of natural radionuclides in aquatic sediments. *Journal of Radioanalytical and Nuclear Chemistry*, 332(3): 659–667. https://doi.org/10.1007/s10967-022-08701-7

Missimer, T., Teaf, C. & Maliva, R. 2019. Natural radiation in the rocks, soils and groundwater of Southern Florida with a discussion on potential health impacts. *Environmental Research and Public Health*, 16(1793): 1–3. https://doi.org/10.3390/ijerph16101793

Mtshawu, B. 2014. Optimizing interpolation as a tool for the use in soil property mapping. Master's dissertation. Bloemfontein: University of the Free State.

Mtshawu, B., Bezuidenhout, J. & Kilel, K. 2023. Spatial autocorrelation and hotspot analysis of natural radionuclides to study sediment transport. *Journal of Environmental Radioactivity*, 264(107207): 1–9. https://doi.org/10.1016/j.jenvrad.2023.107207

Povinec, P.P., Levy-Palomo, I., Comanducci, J.F., de Oliveira, J., Oregioni, B. & Privitera, A.M. 2006. Submarine groundwater discharge investigations in Sicilian and Brazilian coastal waters using an underwater gamma-ray spectrometer. *Radioactivity in the Environment*, 8: 373–381. https://doi.org/10.1016/S1569-4860(05)08030-7

Rees, J.G., Williams, T.M., Nguli, M.M., Kairu, K.K. & Yobe, A.C. 1996. *Contaminant Transport and Storage in the Estuarine Creek Systems of Mombasa, Kenya*. Nottingham: Natural Environment Research Council.

Salmon, L. 1961. Analysis of gamma-ray scintillation spectra by the method of least squares. *Nuclear Instruments and Methods*, 14: 193–199. https://doi.org/10.1016/0029-554X(61)90071-4

SECTION 2

Urban Dynamics, Crime, and Security (Human-Environmental Interactions)

This section focuses on the spatial dynamics of human settlements, crime patterns, and human security concerns. Each chapter employs GIS and spatial analytics to uncover the underlying geographic processes that shape security environments and urban development. From informal settlement growth to organised criminal activity, the studies highlight the role of spatial analysis in supporting evidence-based planning and response strategies for crime prevention.

Chapter 5 investigates residential expansion in the Middelpos settlement of Saldanha over a 20-year period. Using multi-temporal remote sensing data and post-classification comparison, the study revealed a 25.53% increase in exposed land, indicating unchecked informal development. The results support informed urban planning and highlight the implications of housing policy, land use, and infrastructure provision in semi-urban military-adjacent areas.

Chapter 6 applies hotspot analysis using the Getis-Ord Gi* statistic to visualise patterns of criminal activity in the Saldanha Military Area from 2017 to 2023. Despite low crime volumes, spatial clustering of minor offences, such as trespassing and vehicle theft, was revealed. The study shows how GIS can inform localised military security measures and serve as a practical tool for visual crime intelligence.

Chapter 7 analyses crime statistics from Mamelodi East over an 11-year period, using spatial mapping to highlight the area's persistent ranking among South Africa's most violent police precincts. This study explores the potential of geographic crime analysis to guide the strategic deployment of police services and reflects on policing models, such as community-oriented policing, to curb burglary trends.

Chapter 8 examines evolving human security frameworks and how geospatial technologies are increasingly used to address transnational risks, such as climate migration, environmental degradation, and armed conflict. This chapter situates GIS and remote sensing within broader international security agendas and explores their roles in disaster preparedness, monitoring, and strategic planning.

Collectively, these contributions demonstrate how spatial thinking can inform responses to urbanisation, criminal activity, and complex human security threats by integrating local, national, and global scales of analysis.

CHAPTER 5

Spatio-temporal analysis for monitoring residential expansion: A case of the Middelpos settlement in Saldanha from 2004–2024

N.L. Monyamane and L.P. Welman

Abstract

Rural expansion necessitates effective planning to mitigate adverse environmental and socio-economic consequences. Unregulated informal settlement growth, particularly in developing urban areas, exacerbates challenges such as resource scarcity, environmental degradation, unemployment, and increasing crime rates. Urban sprawl, often characterised by uncoordinated and haphazard development, compromises sustainable urban growth and degrades overall urban quality. To address the risks associated with urban expansion, continuous monitoring is essential for informed planning and decision-making. This study used remote sensing and geographic information system (GIS) techniques to analyse settlement growth over a defined period. Remote sensing data facilitated the efficient mapping and monitoring of settlement expansion across multiple time intervals, providing critical insights into urban development trends. Change detection analysis using a post-classification comparison method was conducted to assess shifts in land use, utilising settlement distribution maps for 2004, 2014, and 2024. The findings indicate that exposed land in the Middelpos settlement of Saldanha Bay decreased by 25.53% over the study period, with informal housing accounting for most of the land conversion. The expansion of low-income housing has further contributed to the reduction of exposed land, intensifying the spatial pressures. This study underscores the value of remote sensing and GIS methodologies in spatio-temporal analysis for tracking settlement expansion and informing sustainable urban planning strategies.

Keywords: Geographic information system, remote sensing, spatio-temporal analysis

Introduction

Becker (2019) states that under the apartheid regime, Black people were restricted in terms of their choice of location. Many Black people were forced to live in their homelands with fewer resources and limited job opportunities. South African History Online (SAHO, 2019) asserts that during the apartheid era, only men were allowed to migrate to cities in search of job opportunities, whereas women remained in their homelands.[1] The situation changed after the

1 Homelands are "places where people have bonded with their landscape in an uncommon manner" (Hurt, 2003).

abolishment of apartheid in 1994, as there were no restrictions. This freedom of movement has led to an influx of people from rural to urban areas. Ramachandra and Kumar (2009) state that urbanisation plays a huge role in the establishment of official, unplanned, unlawful occupation (informal settlements) of land for residential purposes, as it allows people to move freely and quickly, which is referred to as urban sprawl.[2]

Urban sprawl is characterised by the extension of low-density residential areas at the fringes of a township. The expansion of these unlawful residents has become an ongoing nationwide phenomenon. The expansion of informal settlements is fundamentally linked to urban sprawl. As cities grow outwards, they often encroach upon informal settlements, leading to their expansion of informal settlements. In the Middelpos settlement, several key drivers of settlement expansion have been identified. Economic factors, particularly the development of the Saldanha Bay Industrial Development Zone, have played a crucial role in attracting investment, creating jobs, and consequently, an influx of residents seeking employment. However, the rapid pace of expansion has sometimes outstripped infrastructure development, posing challenges to service delivery (Saldanha Bay Municipality, 2025).

Horn (2019) claims that due to the increase in informal settlements, the government was forced to transform and shift its focus to the basic needs approach and economic growth and development of towns and cities. The ongoing mobility and expanding human growth permit a greater mass of migrants worldwide compared to any other time in the past. The spatial patterns of settlement growth observed in Middelpos demonstrate a predominance of outwards peripheral expansion rather than infill development. This pattern is indicative of urban sprawl imitation, characterised by the extension of low-density residential areas at the township's fringes. The analysis showed a significant transformation of agricultural and undeveloped land into residential and commercial zones. This outwards growth pattern is typical of many rapidly urbanising regions but raises concerns about sustainability, resource allocation, and infrastructure provision. This movement is dominated by an influx of people from rural to urban areas (Adger et al., 2015).

Meanwhile, Carling et al. (2020) state that the drivers of migration are diverse, but a common push factor is seeking solutions to ensure a brighter and prosperous future. Most migrants are pushed away by poverty, violence, and hardship. The desire to eliminate poverty. Cummings et al. (2015) argue that factors that influence people's decisions to migrate include personal reasons, secured livelihood opportunities, and political security. Horn and Van Eeden (2018) argue that after apartheid, the Western Cape province has experienced an influx of migrants from other provinces, especially the Eastern Cape. This influx is caused by the continuous need to reduce transport costs by being closer to workplaces.

In addition, Aymes (2021) contended that in the past decade, more effort has been made to create job opportunities, especially in the field of artisans. This was achieved through the establishment of the Saldanha Bay Industrial Development Zone (SBIDZ) (Moses, 2017). The Industrial Development Zone was formally established in Saldanha Bay in 2012 as one of the spatial interventions launched in South Africa (Department of Trade and Industry, 2013; Saldanha Bay Municipality. 2025). The temporal analysis highlights distinct phases of accelerated growth,

2 Urban sprawl is "the spilling over of urban-type buildings into the suburban and agrarian areas, and on the other hand, the disorganized growth of sporadic beginnings of settlements in agrarian regions (separate farms, houses of farm workers, secondary occupation settlements) as well as in early industrialised or commercially permeated areas where ironworks, foundries and mines served as starting points of such sprawling" (Jaeger et al., 2010: 399).

particularly between 2004 to 2014. The Saldanha Bay Municipality (2018) asserts that this period coincides with key economic and policy developments, such as the official launch of the Industrial Development Zone and subsequent infrastructure projects. However, the analysis also indicates periods of slowed growth, possibly attributable to economic downturns, delays in policy implementation, and infrastructure projects.

The establishment of the Saldanha steel plant (1998) led to the coexistence of various businesses and firms with the Saldanha Municipality. The prominent firms within the proximity of the plant include Westarcor, Vaalmac, Profab, Dagbreek, Saldanha Rewinding, and Panda. The firms and the plant provided 650 permanent job opportunities through the outsourcing and insourcing of labour. Although many job opportunities must be absorbed locally, this was not the case. Moreover, the policy stipulates that 30% of skilled workers and 40% of semi-skilled workers should be residents of the West Coast region. However, due to a lack of appropriate skills and qualifications, firms are forced to outsource service provision to subcontractors, causing people to migrate to Saldanha (Welman & Ferreira, 2014). This continuous movement of people in and out of Saldanha influences the spatial distribution and concentration of the Saldanha residential areas, such as Middelpos, White City, Diazville, and the surrounding areas. This study seeks to use spatio-temporal data to determine urban expansion in the Middelpos settlement in Saldanha from 2004 to 2024.

Study area

This study focuses on Middelpos, a residential settlement located within the Saldanha Bay local municipality in the Western Cape (Figure 5.1). The settlement is located 140 km north of Cape Town (Saldanha Bay Municipality, 2014). The settlement has experienced considerable expansion over the past decades, mainly driven by economic activities linked with the SBIDZ and the Saldanha steel plant. The population of the Middelpos area has grown significantly over the last 20 years. In 2004, the population was approximately 12,500 residents (Statistics South Africa, 2001). However, in 2014, the population increased to approximately 17,800, indicating the area's growth due to the establishment of industries. By 2023, the total population had increased to approximately 22,300, indicating continuous growth (Saldanha Bay Local Municipality, 2023).

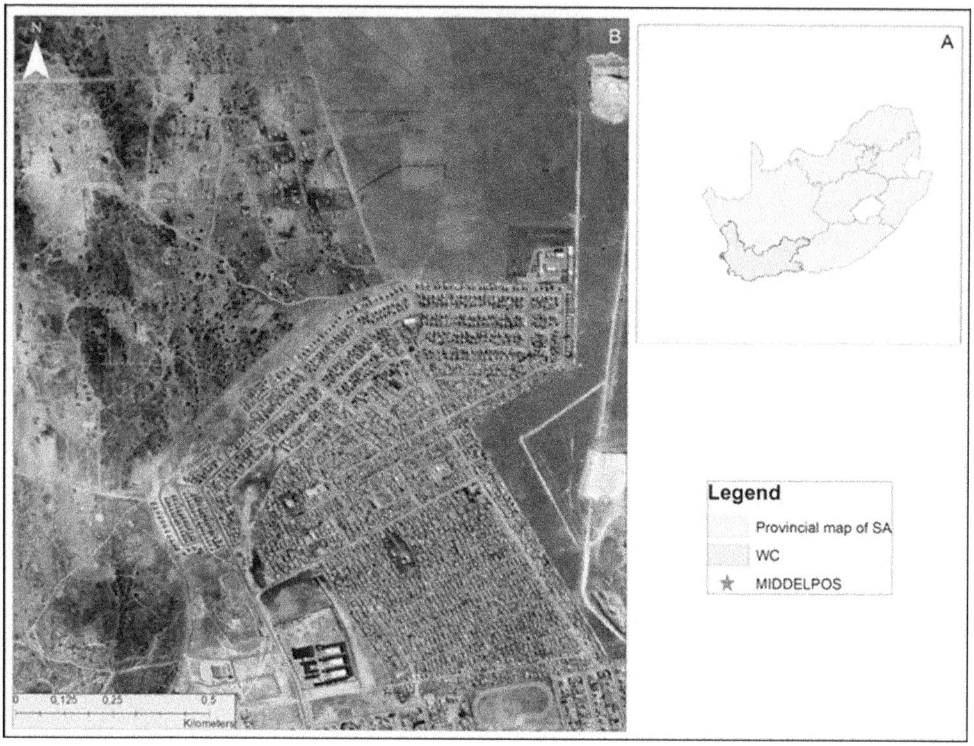

Figure 5.1: Map showing the extent of Middelpos (B) located in the Western Cape province of South Africa (A).
(*Source:* Authors' own work using ArcGIS 10.8.1)

The area is characterised by a combination of formal and informal settlements. However, the informal settle is expanding rapidly because of the ongoing influx of people seeking employment within the municipality. This rapid influx has led to challenges in terms of infrastructure provision, housing, and service delivery (Saldanha Bay Local Municipality, 2023). The settlement is marked by a relatively flat terrain with sandy soil, which influences the expansion of the settlement and infrastructure provision.

Methodology

In this study, satellite imagery for 2004, 2014, and 2024 was acquired from Google Earth Pro. Google Earth Pro was selected as the source because it is accessible and can provide historical imagery at a high spatial resolution. This allowed for a detailed visual interpretation and mapping of small-scale features. GIS have various techniques used for data analysis, including spatial analysis, spatio-temporal analysis, network analysis, and geostatistical analysis. For this study, spatio-temporal analysis was used because the study focused on analysing how the study area evolved. Geometric and radiometric corrections are essential preprocessing steps in remote sensing. In this study, the preprocessing technique used was radiometric correction to standardise image values so that they could correspond to real-world surface features. Radiometric correction was applied using ArcGIS 10.8.1 to adjust the brightness of the image for simplified classification and to improve the visual appearance of the maps. The display function in the image analysis tool was used to enhance the contrast, brightness, and dynamic range. The false colour composition (FSC) technique was used to refine the capability for observing distinct features within the study area, to become familiar with the study area, and to discover land-use categories.

In GIS and remote sensing, two main classification methods are commonly used: unsupervised and supervised classification. Jog and Dixit (2016) describe supervised classification as a method where the analyst manually selects training pixels, while unsupervised classification automatically groups pixels into classes based on spectral similarities without prior knowledge (Duda & Canty, 2002; Nkeki, 2016). Although unsupervised methods are generally simpler and faster, this study applied supervised classification due to the rugged nature of the area and the need for an accurate representation of land use and land cover types. Following radiometric corrections, supervised classification was performed to categorise pixels based on their spectral signatures, using predefined training samples.

Different land-use categories, including formal housing, informal houses, tar roads, schools, and exposed land, were investigated. The number of samples was 12 which was sufficient considering the size of the study area. To validate the effectiveness of the proposed classification method, land use classification was evaluated. A false colour composition was applied to enhance the ability to differentiate land use. The display function in ArcGIS was used to adjust the contrast and brightness to improve the clarity of the image. Lu and Weng (2005) stated that this classification is not error-free; hence, refinements were conducted on the image to improve the accuracy of the results. The most frequently used method to detect changes in the settlement landscape is post-classification image-to-image comparison because it offers quantitative analysis between classified images by class (Nkeki, 2016). An accuracy assessment was conducted using ArcMap and Google Earth Pro. The accuracy assessment of the classified images was compared with Google Earth Pro satellite data to produce ground truth data.

Kantakumar et al. (2016) argue that the important factor in understanding the patterns of urban land use and the dynamics of urban growth is the need to assess the land cover changes. Remote sensing-based change detection methods are mostly utilised because of their cost-effectiveness, high temporal frequency, and extensibility. After classifying the images into various categories, change detection was used to evaluate land use and land cover change over 20 years from 2004 to 2024. However, the accuracy of the results depends on the accuracy of the individual during classification. For analysis, the different classified images were compared to observe and identify similarities and differences in land use patterns.

Results

In 2004, the Middelpos settlement exhibited limited formal development, with extensive tracts of exposed land interspersed with sparsely distributed informal settlements (Figure 5.2). The predominant land use comprises low-density informal housing with large expanses of unoccupied land. The spatial configuration of the settlement was characterised by a dispersed distribution of dwellings, with considerable open spaces between the structures. By 2014, significant urban expansion had transformed the township landscape. The northern section experienced formalisation through the construction of low-income housing, while the central area saw a marked densification of informal settlements. However, the southern portion remained largely unchanged, with substantial areas of exposed land and minimal vegetation cover. This shift indicates a transition from predominantly undeveloped land to an increasingly built environment. Figure 5.2 illustrates the continued spatial transformation of Middelpos until 2024. The proliferation of low-income housing extended into areas previously occupied by informal dwellings; however, the overall number of informal structures also increased. This has resulted in a highly compacted settlement pattern with a significant reduction in open land. Concurrently, vegetated land cover declined as formal and informal housing expanded in the area. Notably, the construction of educational infrastructure, such as schools, further underscores the township's developmental trajectory, reflecting broader trends in urbanisation and spatial restructuring.

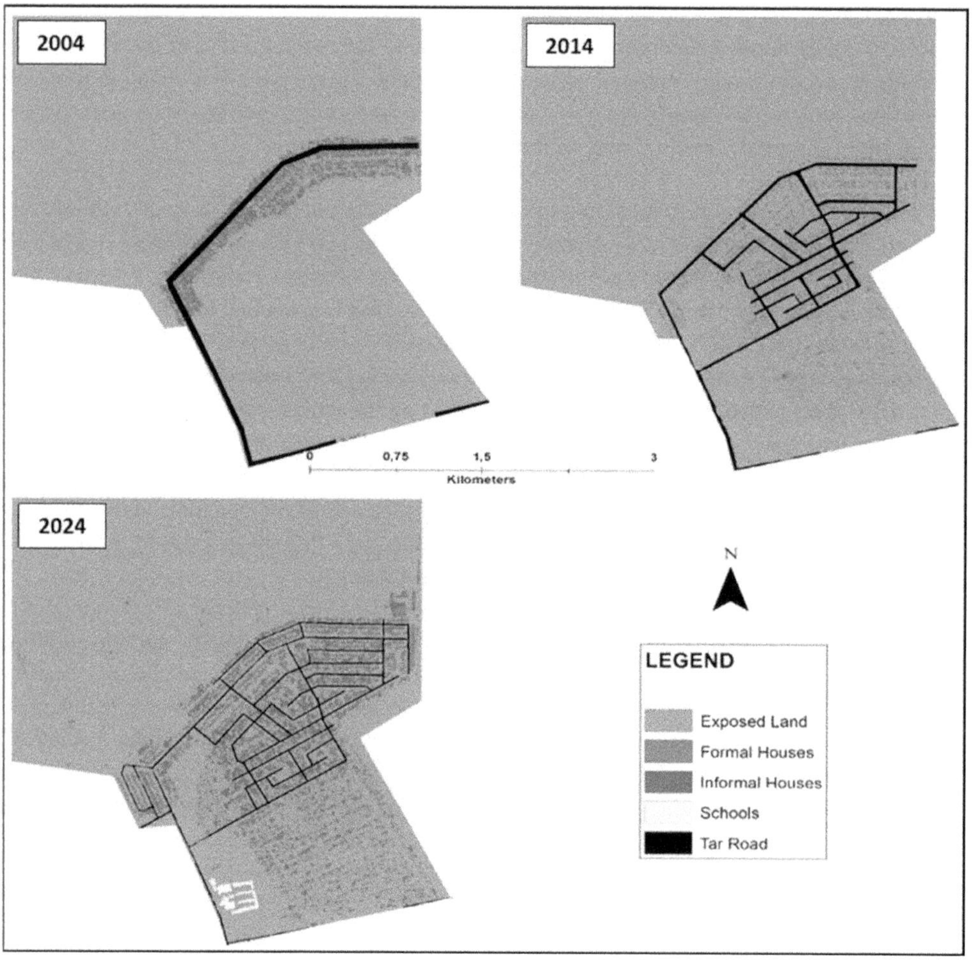

Figure 5.2: Land cover and land use maps of Middelpos in 2004, 2014, and 2024. (*Source:* Authors' own work using ArcGIS 10.8.1)

In 2004, the land cover and land use in Middelpos were relatively homogeneous, with limited anthropogenic features, as indicated in Table 5.1. The predominant land use comprised informal housing and the construction of tar roads. The extent of exposed land was approximately 1,864,294 m², while the combined area occupied by tar roads and informal housing was 135,704 m². This suggests that, at the time, the area experienced relatively low levels of human-induced transformations. Table 5.1 further highlights that by 2014, formal construction in the northern part of the township remained minimal compared to the extensive area occupied by informal settlements. The primary land use consisted of informal settlements, which expanded significantly from 118,403 m² in 2004 to 274,485 m² in 2014, followed by formal housing developments that emerged for the first time, covering 94,500 m². Tar roads also expanded, increasing from 17,301 m² in 2004 to 49,264 m² in 2014. Despite spatial constraints, the number of informal houses continues to rise, leading to increasingly dense and compact settlement patterns. The reason for this densification is unclear. Additionally, a small portion of land was classified as being covered by running water, potentially due to inadequate drainage infrastructure.

Table 5.1: Township land cover and land use in 2004, 2014, and 2024

Land cover / land use	**Area size (m^2)**			**% Change**		
	2004	**2014**	**2024**	**2004–2014**	**2014–2024**	**2004–2024**
Exposed land	1 864 294	1 581 750	1 388 400	-15.16	-12.22	-25.53
Informal houses	118 403	274 485	390 085	131.82	42.12	229.46
Formal houses	Not developed	94 500	144 297	-	52.69	-
Schools	Not developed	Not developed	5 018	-	-	-
Tar road	17 301	49 264	49 264	184.75	-	184.75

By 2024, a further decline in exposed land occurred as the expansion of informal settlements, tar roads, and formal housing continued (Table 5.1). The area of exposed land decreased to 1,388,400 m², while the areas of informal settlements and formal housing expanded to 390,085 m² and 144,297 m², respectively. The construction of a school occupying 5,018 m² contributed to the increased strain on available land resources. Informal settlements expanded northward and southward, reflecting broader urbanisation trends. The introduction of new developments resulted in the emergence of additional land-use classifications, distinguishing the landscape in 2024 from that in 2014. The presence of a newly constructed school is particularly notable, indicating the township's evolving infrastructure.

Figure 5.3 illustrates the increase in land utilisation within the Middelpos settlement over 20 years. A comparative analysis of land use and land cover (LULC) maps, along with pie charts for 2004 and 2014, confirmed a significant expansion of settlement areas. The introduction of formal housing and the concurrent rise in informal dwellings are the primary drivers of this transformation. By 2014, newly constructed formal houses accounted for approximately 3% of the total area, while informal settlements expanded from 6% in 2004 to 13% in 2014, reflecting a 7% increase in the same period. This expansion has contributed to a reduction in vegetated and exposed land, thereby limiting the available space for future formal development from 93% in 2004 to 83% in 2014. This confirms the continued intensification of land use by 2024. The settlement has experienced increased pressure from human activities, with informal dwellings now occupying 18% of the total land area, a 5% increase from 2014. However, due to spatial constraints, these informal houses have become increasingly clustered within the remaining undeveloped areas of the city. This trend underscores the ongoing expansion of informal settlements, which may be attributed to socio-economic factors, population growth, or housing shortages. Figure 5.3 highlights a notable pattern in land occupation, wherein most of the available space has been appropriated for informal housing.

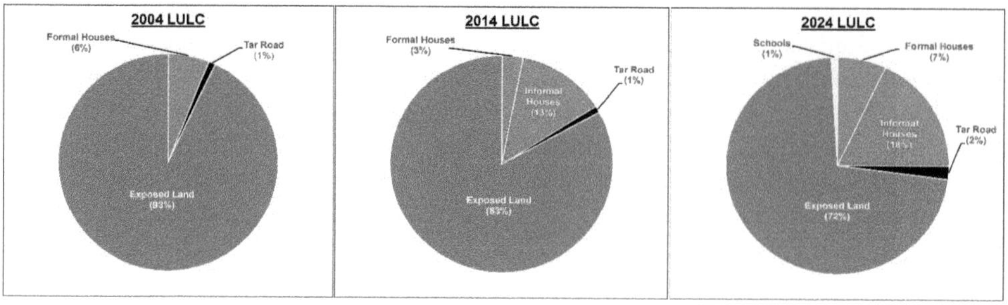

Figure 5.3: Temporal comparison of land use and land cover (LULC) in the Middelpos settlement across 2004, 2014, and 2024. The maps highlight key land cover classes, where brown indicates exposed land, red represents informal houses, grey denotes formal houses, and black corresponds to tar roads.

Over the past two decades, Middelpos has undergone substantial land use transformation, marked by the expansion of formal and informal settlements. The integration of previously informal areas into formal housing developments has contributed to a denser settlement pattern, which is exacerbated by spatial constraints. Informal housing now occupies 390,085 m², whereas exposed land has diminished by 470,000 m². This shift reflects broader urbanisation trends, with increasing infrastructure development and land use intensification shaping the evolving landscape of the township.

Discussion

The findings revealed a significant shift in the settlement pattern and environmental conditions in the Middelpos residential area over 20 years. In 2004, the area was characterised by limited development, with informal settlements dispersed across larger areas of exposed land. The dominant land use was informal settlement with low density and significantly exposed land, suggesting minimal human impact at that time.

By 2014, there was a notable increase in development, particularly in the northern section of the settlements. The introduction of formal housing, alongside a rise in informal dwellings, marked a shift from predominantly exposed land to an increasingly structured environment in the area. The growth of man-made features, as shown in Table 5.1, demonstrates an increase in land use, with informal housing more than doubling in size. However, the southern section remained largely unchanged, with exposed land and sparse vegetation.

By 2024, formal and informal settlements continued to expand, with informal housing now covering 390,085 m². The density of informal dwellings has increased significantly, leading to highly compact settlement patterns. The reduction in exposed land, as shown in Table 5.1, highlights the continued urbanisation of the area, driven by infrastructure expansion, including tar roads and the construction of a school. Informal settlements now comprise 18% of the area, reflecting the ongoing housing demand and challenges of accommodating a growing population in an economically constrained environment.

The spatial distribution of settlements suggests a lack of coordinated urban planning in the region. Informal settlements have expanded into previously less developed areas, particularly in the northern and central regions, resulting in increased population density and greater pressure on the local infrastructure. Becker (2019) argues that unregulated urban growth presents governance challenges, often leading to inadequate infrastructure and essential services provision. The clustering of informal dwellings has exacerbated the strain on local resources, including roads, sanitation, and access to clean water, contributing to poor drainage systems and limited service delivery.

Furthermore, the environmental consequences of these changes are significant. The reduction of exposed and vegetated land, which plays a crucial role in maintaining ecological balance and supporting biodiversity, raises concerns regarding environmental sustainability. The continued loss of vegetation, regardless of its extent, may increase the risk of flooding and other environmental hazards, further complicating settlement management and reducing urban resilience.

Conclusion

This study utilised spatio-temporal analysis and GIS techniques to monitor the expansion of the Middelpos settlement between 2004 and 2024. The analysis revealed a significant increase in settlement density, predominantly due to informal housing expansion. This trend raises critical concerns regarding safety and security, as competition over limited resources within settlements may escalate. Despite government efforts to curb informal housing growth by introducing formal housing, informal dwellings continue to proliferate, contributing to overcrowding and urban sprawl.

This study highlights the effectiveness of spatio-temporal analysis as a valuable tool for monitoring urban expansion. Prior studies suggest that areas with dense informal settlements are often linked to heightened crime rates, and Middelpos may face similar challenges as its population density increases. Further studies should explore the potential correlation between settlement expansion and crime rates to determine whether they influence each other.

While GIS and remote sensing techniques proved valuable in this study, limitations such as the resolution constraints of Google Earth Pro may affect the accuracy of the detected changes. Future research should focus on incorporating higher-resolution satellite imagery and additional geospatial datasets to refine settlement monitoring and comprehensively assess the socio-economic implications of urban expansion.

References

Adger, W.N., Arnell, N.W., Black, R., Dercon, S., Geddes, A. & Thomas, D.S.G. 2015. Focus on environmental risks and migration: Causes and consequences. *Environmental Research Letters*, 10(6). https://doi.org/10.1088/1748-9326/10/6/060201

Aymes, M. 2021. The fabric of Nizam: Uncertainty and the activation of economic norms in nineteenth-century provincial contexts. In: E. Boyar & K. Fleet (eds.). *Making a Living in Ottoman Anatolia*. Leiden: Brill. 206–227. https://doi.org/10.1163/9789004466982_010

Becker, H. 2019. Living politics in South Africa's urban shacklands by Kerry Ryan Chance (review). *Anthropological Quarterly*, 92(4): 1283–1285. https://doi.org/10.1353/anq.2019.0070

Carling, J., Czaika, M. & Erdal, M.B. 2020. Translating migration theory into empirical propositions. *Comparative Migration Studies*, 8(1): 1–28.

Cummings, C., Pacitto, J., Lauro, D. & Foresti, M. 2015. *Why people move: Understanding the drivers and trends of migration to Europe*. Overseas Development Institute. https://odi.org/en/publications/why-people-move-understanding-the-drivers-and-trends-of-migration-to-europe/ [31 May 2022].

Department of Trade and Industry. 2013. *Notice 1081 of 2013: Designation of the Saldanha Bay Industrial Development Zone (IDZ)*. Government Gazette No. 36988, 3. https://www.gov.za/sites/default/files/gcis_document/201409/36988gen1081.pdf [2 December 2025].

Duda, T. & Canty, M. 2002. Unsupervised classification of satellite imagery: Choosing a good algorithm. *International Journal of Remote Sensing*, 23(11): 2193–2212. https://doi.org/10.1080/01431160110078467

Google Earth Pro. 2022. Middelpos. https://earth.google.com/web/search/middelpos [6 June 2022].

Horn, A. 2019. The history of urban growth management in South Africa: Tracking the origin and current status of urban edge policies in three metropolitan municipalities. *Planning Perspectives*, 34(6): 959–977. https://doi.org/10.1080/02665433.2018.1503089

Horn, A. & Van Eeden, A. 2018. The application of an urban sprawl index: Comparing towns and cities in the Western Cape Province, South Africa. *South African Journal of Geomatics*, 7(3): 257–267. https://doi.org/10.4314/sajg.v7i3.7

Hurt, D.A. 2003. Defining American homelands: A creek nation example, 1828–1907. *Journal of Cultural Geography*, 21(1): 19–43. https://doi.org/10.1080/08873630309478273

Jaeger, J.A., Bertiller, R., Schwick, C. & Kienast, F. 2010. Suitability criteria for measures of urban sprawl. *Ecological Indicators*, 10(2): 397–406. https://doi.org/10.1016/j.ecolind.2009.07.007

Jog, S. & Dixit, M. 2016, June. Supervised classification of satellite images. *2016 Conference on Advances in Signal Processing (CASP)*. IEEE. 93–98. https://doi.org/10.1109/CASP.2016.7746169

Kantakumar, L.N., Kumar, S. & Schneider, K. 2016. Spatiotemporal urban expansion in Pune metropolis, India, using remote sensing. *Habitat International*, 51: 11–22. https://doi.org/10.1016/j.habitatint.2015.10.001

Lu, Q. & Weng, D. 2005. Urban classification using full spectral information of Landsat ETM+ imagery in Marion County, Indiana. *Photogrammetric Engineering & Remote Sensing*, 71(11): 1275–1284. https://doi.org/10.14358/PERS.71.11.1275

Moses, D. 2017. Saldanha Bay Industrial Development Zone: Coordination as a determining factor of opportunity lost or gamechanger. Master's dissertation. Cape Town: University of Cape Town.

Nkeki, F.N. 2016. Spatio-temporal analysis of land use transition and urban growth characterization in Benin metropolitan region, Nigeria. *Remote Sensing Applications: Society and Environment*, 4: 126–137. https://doi.org/10.1016/j.rsase.2016.08.002

Ramachandra, T.V. & Kumar, U. 2009. Geoinformatics for urbanisation and urban sprawl pattern analysis. In: P.K. Joshi, P. Pani, S.N. Mohapatra & T.P. Singh (eds.). *Geoinformatics for Natural Resource Management. New York, NY: Nova Science Publishers.* 235–272.

Saldanha Bay Local Municipality. 2023. Annual Performance Report, 2023. Saldanha Bay: Saldanha Bay Local Municipality.

Saldanha Bay Municipality. 2014. Saldanha Local Plan Final Report. SSCWQ60/2013DRDLR. Saldanha Bay: Saldanha Bay Local Municipality.

Saldanha Bay Municipality. 2018. Saldanha Bay IDZ Annual Report 2017/18. Cape Town: South Africa Auditor General.

Saldanha Bay Municipality. 2025. *Final Third (3rd) Review and Amendment of the 5th Generation Integrated Development Plan 2022–2027*, May 2025, Saldanha Bay Municipality, viewed 4 December 2025. https://sbm.gov.za/wp-content/uploads/Pages/IDP/Final-3rd-Review-Amendment-25-26-IDP-.pdf [December 2025].

Saldanha Bay Municipality. 2025. *Industrial Development Zone (IDZ)*. https://sbm.gov.za/industrial-development-zone-idz/ [02 December 2025].

SAHO [South African History Online]. 2019. Land, labour, and apartheid. https://www.sahistory.org.za/article/land-labour-and-apartheid [7 June 2022].

Statistics South Africa. 2001. Census 2001: Saldanha Bay Municipality. https://www.statssa.gov.za/?page_id=993&id=saldanha-bay-municipality [27 March 2025].

Welman, L. & Ferreira, S.L.A. 2014. Regional development of Saldanha Bay region, South Africa: The role of Saldanha Steel. *Bulletin of Geography. Socio-economic Series*, 26: 219–231. https://doi.org/10.2478/bog-2014-0055

CHAPTER 6

Using hotspot analysis to geo-visualise criminal activity in the Saldanha Military Area: 2017–2023

T.M. Tlhekwe and H.A.P. Smit

Abstract

Over the past two years, South Africa's criminality score has worsened to third in Africa and seventh globally for mafia-style criminal networks and organised crime syndicates. The Saldanha Military Area (SMA) has not been immune to this trend, with numerous crimes reported over the past decade. In this study, hotspot analysis was used to geo-visualise criminal activity to provide a comprehensive spatial understanding of criminal patterns in the SMA from January 2017 to December 2023. The primary objective is to visualise crime areas and provide an overview of the current crime situation in the SMA. Prominent criminal activities in the area are mostly minor, such as burglary, general theft, theft from motor vehicles, trespassing, housebreaking, and malicious property damage. The Getis-Ord Gi* statistic was used to identify statistically significant hot and cold spots. The crime data were processed and analysed using Geographic Information System (GIS) software to create visual maps which identify crime hotspots. The study found relatively low levels of significant criminal activity, largely due to the limited availability of crime data for the SMA. Consequently, it faced challenges in fully demonstrating the potential of the Getis-Ord Gi* statistical tool in ArcGIS for mapping crime hotspots. This study sought to make a practical contribution by highlighting the benefits of GIS techniques for crime analysis and visualisation within a military context. Additionally, the study sought to illustrate the relevance and potential impact of these approaches.

Keywords: ArcMap, crime, Getis-Ord Gi* statistic, hotspot analysis, Saldanha military area

Introduction

Crime is an everyday problem in South Africa (NPR Africa, 2024). According to the 2023 Global Organized Crime Index (GOCI), South Africa's criminality score has worsened over the last two years. The country is currently ranked third in Africa and seventh globally (DefenceWeb, 2023). This is not a new trend. Schonteich and Louw (2001) reported that crime has increased markedly in South Africa and that the police have failed to address this issue adequately. Compared to most other countries, South Africa has an extremely high rate of homicides, assaults, rapes, and other crimes (Statistics South Africa, 2024).

Henry and Lanier (1998) define crime as activities prohibited, prosecuted, and punished by criminal law. Thotakura (2011) adds that crime refers to unlawful acts or omissions that are punishable by imprisonment or a fine. Crimes include actions such as murder, robbery, burglary, rape, drunk driving, child neglect, and unpaid taxes (Thotakura, 2011). Kramer (1985) argues that crime is not just a legal construct but also a socio-legal one. According to Kramer (1985), crime is any behaviour that violates criminal law. Crime can be viewed as any action, omission, or behaviour that is deemed unlawful and punishable by law.

Crime has a detrimental impact on people of all ages in society. Most South African emigrants claim that crime played a significant role in their choice to emigrate (Ballard, 2004). As a problem in South Africa, crime has varying effects on individuals. Consequently, all social and ethnic groups in South Africa are affected by crime (Silber & Geffen, 2009). The conventional approach to fighting crime has proven ineffective over time, as evidenced by a significant increase in crime rates (Newham et al., 2023). According to Kedia (2016), it appears as if the police cannot keep up with highly sophisticated criminals and criminal organisations, mainly because of the outdated methods they rely on. As most police agencies in underdeveloped countries still rely on old-fashioned, traditional filing systems, most police operations are conducted using trial-and-error methods, tip-off information, or intuition (Kedia, 2016).

Crime management on military bases involves collaboration between the Military Police (MPs) and the South African Police Service (SAPS). MPs are responsible for internal security, including investigating crimes involving military personnel and base resources, whereas the SAPS may provide external support for crimes that extend beyond military unit perimeters (South African Government, 2025; GCIS, 2025).

The Military Police Division (MPD) plays a vital role in maintaining law and order within the South African National Defence Force (SANDF) and among the public. Their functions include patrolling, making arrests, testifying in court, investigating crimes, and responding to emergencies (Xobiso, 2018; SAQA, 2021). Military Police Agency Instruction 12/00 serves as a guideline for crime prevention within the SANDF. The MPD operates under the Defence Act and Military Disciplinary Code (MDC), which enforces adherence to military laws and strict discipline (South African Government, 2025; Military Police Agency, 2001). Despite their crucial role, the MPD faces challenges such as understaffing and limited resources, which can impede their crime prevention efforts. To address these limitations, MPs conduct regular inspections and cooperate with law enforcement agencies to ensure effective law enforcement within the SANDF (Xobiso, 2018).

Technology and crime

Advancements in technology have made it possible to identify, deter, and combat crime. Technologies such as GIS can be used to reduce crime by identifying trends in criminal activity, pinpointing crime hotspots, and directing resources towards high-risk areas (SATPALDA, 2024). GIS can be used as a crime management tool. It can be employed by agencies such as law enforcement agencies as a data management system to store, process, analyse, and reveal spatial data related to crimes (Wang, 2012). The crime data includes various types of crimes committed, the locations of the crimes, the perpetrators, and the victims of the crimes (Reid et al., 2011).

Furthermore, these data are crucial for gathering, analysing, and mapping crime incidents, as they assist in decision-making and the efficient use of resources, including the police, to improve community safety and protection (Hunt, 2019). GIS enables organisations to enhance planning, minimise expenses, and promote transparency (Kedia, 2016). Analysts map criminal scenes using GIS technology. Moreover, crime analysts can use GIS technologies to advise the police on criminal investigations, resource optimisation and deployment, planning, assessment, and crime prevention (Henrico et al., 2022).

Kedia (2016) argues that most developing countries' law enforcement agencies (such as those in India) are not technologically advanced in terms of efficient record keeping, case analysis, basic reference and retrieval of data, and information storage. Henrico et al. (2022) point out that a few studies have examined the spatial distribution of crime in South Africa using geospatial approaches, but none have specifically addressed crime in the Saldanha Bay Municipal area. This is a clear gap in the literature that needs to be addressed. Henrico et al. (2022) emphasise that little is known about geospatial crime patterns in the Saldanha Bay Municipality region, although a great deal is known about the causes, effects, and application of GIS to map the spatial distribution of crime. Previous research has concentrated on mapping and analysing the spatial distributions of crime for certain nations, areas, and cities using GIS tools (Henrico et al., 2022). A similar problem was experienced in the crime dataset from the Langebaanweg Military Police Station. For instance, data regarding property crimes are not made public, even to the military community.

This study aimed to geo-visualise crime-related activities that occurred in the past seven years (January 2017 to December 2023) in the Saldanha Military Area (SMA). This will be done using hotspot analysis to map the crime hotspots of common crimes within the SMA. This study provides an overview of the current crime situation to showcase the relevance and potential of GIS tools and techniques in helping to manage crime.

Study area

The town of Saldanha is located approximately 110 km northwest of Cape Town on the West Coast of South Africa (Saldanha Bay Municipality, 2023; Welman & Ferreira, 2016). Saldanha is home to two military units: the South African Ship Saldanha (SAS Saldanha) Naval Base and the South African Military Academy. The two military units are located approximately 2.5 km south of Saldanha, at the SMA (Bezuidenhout, 2012). The SMA is shown in Figure 6.1.

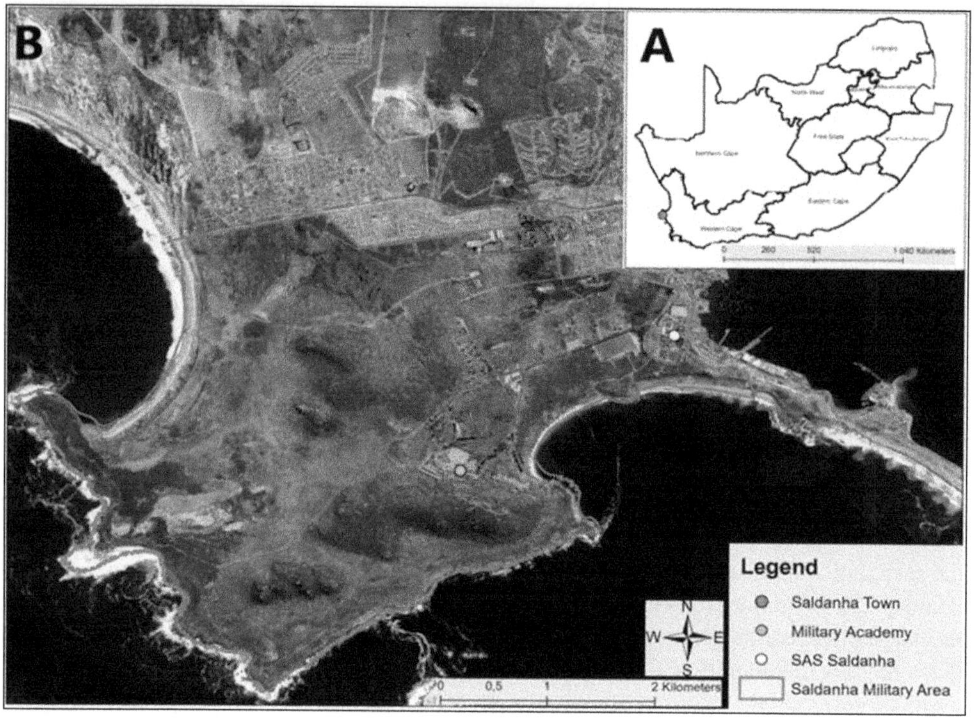

Figure 6.1: Map showing the Saldanha Military Area (B) and its location in South Africa (A). (*Source:* Authors' own work using ArcGIS 10.8.1)

The SMA was the designated area of interest for this study. Figure 6.1 (A) represents the location of the town of Saldanha in the provincial map of the Republic of South Africa, and Figure 6.1 (B) represents the demarcation of the SMA and the location of the SAS Saldanha Naval Base and the South African Military Academy.

Geographically, the study area is bordered to the south and west by the Atlantic Ocean, to the east by Saldanha Bay Harbour, and to the north by White City and the town of Saldanha. The northern section of the SMA features several low-income residential areas and informal settlements, including Diazville, Middelpos, and White City, as illustrated in Figure 6.2.

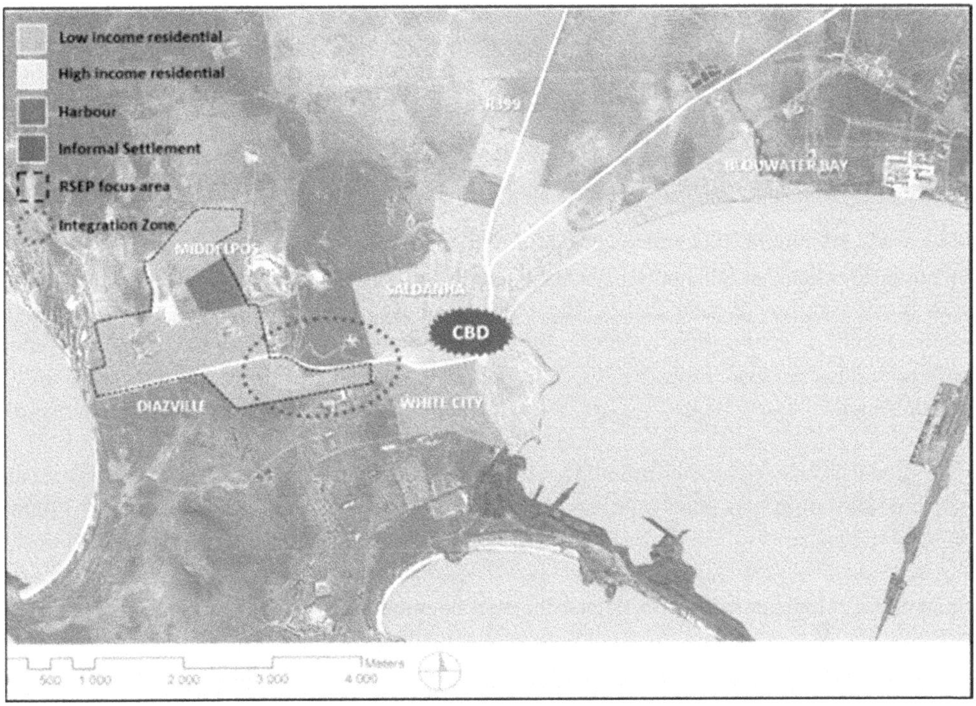

Figure 6.2: Areas bordering the study area.
(*Source:* RSEP Programme, Western Cape Government, 2024)

The population of Saldanha has been increasing steadily since 2019, with most of the incoming residents moving to low-income neighbourhoods such as Diazville, Middelpos, and White City. Saldanha Bay is the second fastest-growing municipal area in the West Coast District with 123,070 residents as of 2021. By 2025, this total is expected to increase to 129,209, with an average annual growth rate of 1.6% (Saldanha Bay Municipality, 2021). The rapid increase in population has caused low-income residences with high density and few basic services to expand quickly, posing a potential challenge to law enforcement (Western Cape Government, 2025).

According to the Western Cape Government (2017), alcohol misuse and drug abuse were identified by the SAPS as the two main social and economic risk factors, as well as the two primary sources of crime in the Saldanha Bay area. The SAPS further stated that unemployment is another contributing factor to crime. The high unemployment rate of approximately 56% worsens the community's drug and alcohol abuse issues, leading to a further increase in criminal activities in the study area (Western Cape Government, 2017). Given the high level of criminal activity in the Saldanha Bay area, the Department of Public Works proposed constructing a concrete wall at the SMA. Work started in 2013, and the construction of the concrete wall was completed in 2015. The primary goal of the wall was to safeguard and ensure the safety of the people living inside the SMA (SAHRIS, 2013). Despite these efforts, the SMA continues to encounter a range of criminal activities.

This study aims to bolster the deteriorated safety and security infrastructure by employing geo-visualisation techniques to pinpoint prevalent crime hotspots in the SMA. This study analysed the criminal incidents reported in the SMA over the last seven years, focusing on identifying the most frequent types of criminal behaviour. Utilising hotspot analysis, this study mapped the distribution of these crimes from 2017 to 2023 and provided a visual representation of high-risk areas. The culmination of this research will be the formulation of actionable recommendations, offering insights into potential, effective, and economically viable security enhancements.

Methodology

This study adopted a quantitative research approach, which entails using certain statistical techniques for analysis and numerical data to address questions such as who, what, where, when, how many, and how (Apuke, 2017). The study made use of spatial statistics tools such as the Getis-Ord Gi* to locate crime hot and cold spots within the SMA. The application of computer software and geographic tools in ArcGIS formed the foundation of this study. The locations of criminal incidents were plotted and are shown on maps and graphs using GIS computer software.

The spatial distribution of criminal activity and assessment of whether each crime dataset reflected clustering were conducted using the ArcGIS spatial statistics tool. Chainey et al. (2008) state that hotspot analysis tools use vector data to pinpoint the locations of statistically significant hot and cold areas. The methodology involves a case study approach to analyse a unique dataset from SMA. This dataset was collected through document analysis from the Langebaanweg Military Police Station and the SAS Saldanha Naval Base's protection unit. The data cover crime data from January 2017 to December 2023, categorised by crime type, location, and date. Spatial patterns were visualised through point mapping, and crime characteristics were analysed.

The locations of all crime incidents were plotted using the ArcGIS software. A spatial statistical tool was employed to examine the spatial distribution of criminal activity and assess whether there was any clustering within each crime dataset (Argun & Dağlar, 2016). Computer programmes such as Microsoft Excel, Google Earth, and Google Maps were used to aid the analysis. The crime data were analysed and presented using ArcGIS 10.8.1 tools. These tools include spatial autocorrelation and hotspot analyses.

Spatial autocorrelation

According to Haining (2001), spatial autocorrelation assesses the degree to which spatial attributes and their corresponding data values are clustered. In this study, Moran's I statistic was used to measure the spatial autocorrelation within the dataset. Moran's I is calculated using the following formula (Esri, 2020):

$$I = \frac{n}{w} \frac{\sum_{b=1}^{n} \sum_{c=1}^{n} w_{bc}\left(x_b - \bar{x}\right)\left(x_c - \bar{x}\right)}{\sum_{b=1}^{n}\left(x_b - \bar{x}\right)^2}$$

where:

I represent Moran's I statistic.

n is the number of spatial units.

x_b and x_c are the values of the variable at locations b and c.

$\bar{x}$ is the variable's mean.

w_{bc} is the spatial weight between locations b and c.

w is the sum of all spatial weights.

The values of Moran's I indicate whether the spatial arrangement of the features is clustered, dispersed, or random (Manap et al., 2019). A Moran's I value close to +1 suggests clustering of similar values, whereas a value closer to -1 indicates dispersion. The spatial autocorrelation tool was used to analyse spatial patterns within the crime data and calculate Moran's I statistics (Esri, 2020). The significance of Moran's I was determined using z-scores and p-values. High positive z-scores and low p-values indicate statistically significant spatial clustering, whereas high negative z-scores and positive p-values indicate significant dispersion. These results provide valuable insights into the geographic dynamics of the phenomenon being studied (Mitchell, 2005).

Hot spot analysis

The Getis-Ord Gi* statistic was employed to identify and analyse spatial crime patterns within the study area (Esri, 2020). This spatial analysis tool detects clusters of high or low values, referred to as hot and cold spots, respectively, within the given data. Getis-Ord Gi* is calculated using the following formula (Esri, 2020):

$$G_i^* = \frac{\sum_{c=1}^{n} w_{bc}x_c - \overline{A}\sum_{c=1}^{n} w_{bc}}{S\sqrt{\frac{n\sum_{c=1}^{n} w_{bc}^2 - \left(\sum_{c=1}^{n} w_{bc}\right)^2}{n-1}}}$$

where:

G_{i*} is the Getis-Ord Gi* statistic for feature b.
x_c is the attribute value for feature c.
w_{bc} is the spatial weight between features b and c.
$\overline{A}$ is the mean of the attribute values.
S is the standard deviation.
n is the total number of features.

A fixed distance band was used to create the spatial weights matrix, ensuring that each feature had an equal number of neighbours within the defined distance (Allen, 2009). The results were visually represented through graduated colours, where high Gi* values indicated hot spots and low Gi* values signified cold spots (Esri, 2020). Statistically significant hot spots can be identified by high positive z-scores and low p-values, while significant cold spots can be identified by high negative z-scores and low p-values (Mitchell, 2005). z-scores and p-values were used to assess the statistical significance of Gi* values. The spatial weight matrix output (Figure 6.3) was generated using the spatial matrix tool, and the hotspot analysis (Getis-Ord Gi*) tool was used to compute the Gi* statistic.

Figure 6.3: Spatial weights matrix output of crime incidents in the Saldanha Military Area. (*Source:* Authors' own work using ArcGIS 10.8.1)

Figure 6.3 depicts how spatial entities are related to each other based on geographic proximity. The spatial weights matrix tool was used to quantify the relationship between crime incidents, as shown in the output.

The confidence levels produced by the hotspot analysis tool indicate whether clustering exists in the data. According to Choate (2015), confidence levels are based on the z-score and p-value. These measures of statistical significance determine whether the null hypothesis is rejected for each feature. For validity and reliability, this technique assesses each location according to its surrounding points and identifies any statistically significant regions of high or low values (Esri, 2020; Chaote, 2015). The hotspot analysis tool enables the visualisation of hotspot clusters (high data values) and cold spots (low data values).

Results

This study employed a quantitative research approach, utilising statistical techniques and numerical data to analyse crime patterns in the SMA from 2017 to 2023. ArcGIS software was used to assess the spatial distribution of crime. The spatial weights matrix and hot spot analysis (Getis-Ord Gi*) tools were employed to conduct the hotspot analysis. The results were visualised using colour-scaled maps, highlighting prominent hot and cold spots. Cross-referencing the locations of the hotspots with credible crime reports and the SMA's past crime tendencies served as validation. This guarantees the accuracy and reliability of the findings. The data were refined and analysed using ArcGIS's data management tools and spatial statistics capabilities, such as

Integrate, Collect Events, and Incremental Spatial Autocorrelation, to enhance the reliability of the hotspot analysis (Mitchell, 2005). Table 6.1 shows the crime statistics of the SMA from 2017 to 2023.

Table 6.1: Types of criminal occurrences over the study period

Saldanha crime statistics for the SMA	**2017**	**2018**	**2019**	**2020**	**2021**	**2022**	**2023**	**Total**
Housebreaking and theft	3	5	7	2	0	0	0	17
Trespassing	0	0	1	4	6	7	2	20
Burglaries at business or residential premises	0	0	0	1	10	8	0	19
General theft	1	4	4	1	5	7	11	33
Theft from a motor vehicle	0	2	1	0	0	0	2	5
Malicious damage to property	0	0	2	2	1	3	0	8
Total	**4**	**11**	**15**	**10**	**22**	**25**	**15**	**102**

There were 102 criminal incidents during the period. These crime statistics are SMA-specific, excluding any other crime-related activities outside of this study area. Four incidents occurred in 2017, 11 in 2018, 15 in 2019, 10 in 2020, 22 in 2021, 25 in 2022, and 15 in 2023. The relatively low number of recorded crime incidents during the period under investigation is expected, given that SMA is a military area and a national key point where crime should be at a minimum. However, the occurrence of some minor or petty crimes is concerning, especially in view of this expectation.

Figure 6.4 depicts the annual records of the most common criminal activities that occurred in the SMA over seven years. Housebreaking and theft had the highest incidences in 2017, 2018, and 2019. Trespassing had the highest incidence in 2020. In 2021 and 2022, burglaries at business and private premises had the highest incidence. General theft had the highest incidence in 2023. Burglary at business or residential premises and malicious damage to property are the least common crimes in the study area, whereas general theft is the most prominent criminal activity in the SMA over the seven-year period. It is important to note that the limited number of recorded incidents may have impacted the statistical analysis of the results. Nevertheless, this is acceptable, as the primary objective of the study was to demonstrate the application of GIS techniques and tools, particularly to geo-visualise crime locations.

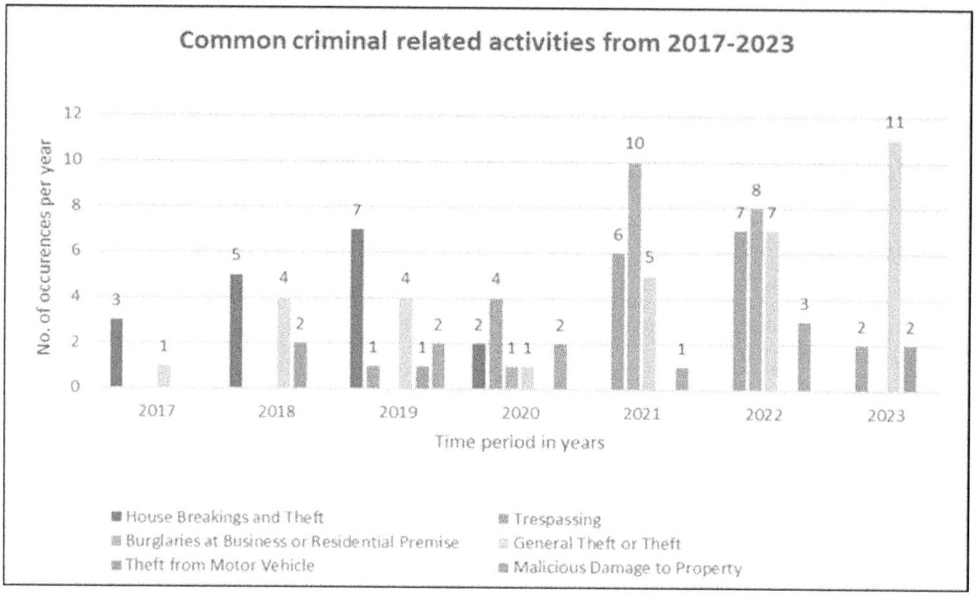

Figure 6.4: Annual records of crimes that commonly occurred in the Saldanha Military Area from 2017–2023.

Discussion

This study aimed to address a research gap by graphically presenting crime statistics for the SMA to make crime areas easily identifiable. The primary focus was on crime trends in the SMA over seven years, from January 2017 to December 2023.

This study identified relatively low levels of significant criminal activity, primarily due to the limited availability of crime data for the SMA. Consequently, it faced challenges in effectively illustrating the innovative application of the Getis-Ord Gi* statistical tool in ArcGIS for mapping crime hotspots. Additionally, the study did not adequately identify statistically significant hot and cold spots where crime clusters were concentrated. The Moran's Index spatial autocorrelation report indicated that the crime data is dispersed, with the value of 0.012693 indicating dispersion, and not clustering. These results can be seen in Figure 6.5.

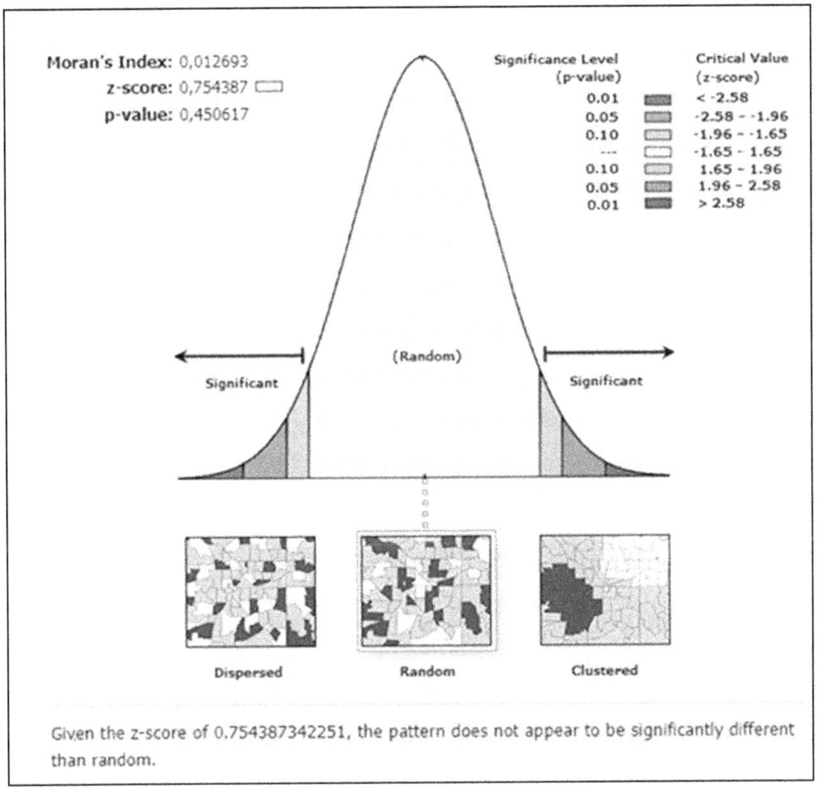

Figure 6.5: Spatial autocorrelation results for criminal activity in the Saldanha Military Area.

Figure 6.5 illustrates a spatial autocorrelation report with a significantly low z-score of 0.754387. The p-value represents the probability score for the pattern analysis tools. The spatial autocorrelation report illustrated a p-value of 0.450617, which was greater than 0.05. This indicates that no spatial pattern was identified in the crime data. The Moran's Index of 0.012693 indicates that the crime incidents are located randomly across the study area. Nevertheless, there has been a notable increase in crime incidents within the military area over the past five years, particularly during the Covid-19 pandemic.

The spatial autocorrelation results showed a p-value of 0.450617, which was greater than 0.05. The p-value represents a probability score for the pattern analysis tools, ranging from 0 to 1. According to Field (2013), a p-value less than 0.05 indicates statistical significance, whereas a value greater than 0.05 is not significant. The observed p-value suggests that the spatial pattern was likely created by random processes. This indicates that the spatial pattern was dispersed and not clustered. The z-score of 0.754387 indicates that the pattern did not significantly differ from random. This means that it is not possible to predict where criminal incidents would be prevalent, as they occurred randomly in space.

However, the rise in crime within the SMA over the past five years, particularly during the Covid-19 pandemic, raises questions regarding the underlying causes. Existing research has shown that Covid-19 restrictions significantly impacted crime rates globally (Hoeboer et al., 2023). Overall, studies suggest that the global decline in crime rates was largely temporary, with evidence pointing to stay-at-home orders during the Covid-19 pandemic as the primary driver of the reduction. The decrease in crime has been partially explained by reduced mobility (Holze, 2021; Díaz et al., 2022).

Government lockdown restrictions had some impact but did not prevent criminal activities from occurring (ISS, 2020). The persistence of crime in the SMA despite movement restrictions suggests the potential involvement of individuals with authorised access, such as military personnel or employees, as entry to this area was restricted during the pandemic (South African Soldier Magazine, 2023). Furthermore, the failure to substantially reduce crime rates during this period suggests that the restrictions may not have been as effective as anticipated, with offenders finding ways to circumvent security measures (ISS, 2020; Ashby, 2020)

Economic strain and increased stress during the pandemic could also have contributed to opportunistic crimes (Holze, 2021), even in the controlled environments of military bases. The observed increase in crime may also reflect undetected or underreported activities, which are more difficult to monitor in such a unique environment (ISS, 2020). Finally, the ability of criminals to adapt to the constraints imposed by the pandemic may enable them to continue their activities despite the challenges posed by the restrictions (Payne & Morgan, 2020).

As alluded to above, MPs in SANDF military units are responsible for crime prevention, investigation, and maintaining law and order, as well as ensuring appropriate conduct among personnel regardless of rank. MPs are also responsible for the rehabilitation of military offenders (Xobiso, 2018; Military Police Agency, 2007).

Security measures at the SMA include concrete walls and steel palisade fencing as external perimeter fences, patrols on foot and in military vehicles, guards, illumination at main gate entrances and roads, and the placement of roving guards and CCTV systems at key military installations (Esn M.J. Sikosana, personal communication, 27 March 2024). In accordance with the Defence Act 42 of 2002, security measures at military bases involve regular and sporadic inspections, internal audits, stocktaking, and the use of minimum force or self-defence when necessary (South African Government, 2025).

The factors contributing to crime within military areas include insufficient personnel, inadequate perimeter protection, and outdated security technologies (Esn M.J. Sikosana, personal communication, 27 March 2024). Weak surveillance and poor access control further increase the risk of undetected criminal activities, particularly when the staff are overstretched or lack proper training (Fennelly, 2012; Welsh & Farrington, 2009). To address these vulnerabilities, plans have been implemented to strengthen perimeter defences, upgrade security technologies, increase the number of trained personnel, and conduct regular security audits. These measures aim to reduce crime by addressing current security risks and ensuring more effective prevention strategies (Esn M.J. Sikosana, personal communication, 27 March 2024).

A review of the Defence Act has been initiated to introduce stricter penalties and corrective measures as deterrents to strengthen security within the Department of Defence and SANDF (Parliamentary Monitoring Group, South African Government, 2025). However, budget constraints, particularly at the tactical level, limit the ability to acquire and implement advanced security technologies such as biometrics and intruder detection systems (DefenceWeb, 2018).

Challenges were encountered in retrieving crime data, which fell under the jurisdiction of the Langebaanweg Military Police Station located at the Langebaanweg Air Force Base. This station is the sole provider of crime statistics for the SMA, which limits data verification from other sources. Additionally, crime data were provided based on the researcher's security clearance. To the best of our knowledge, this dataset is the only one accessible for this study. Over the years, crime data were manually captured in incident reports and occurrence books before being uploaded onto computer systems by various Langebaanweg Military Police officials (Sgt S.N. Zwane, personal communication, 18 April 2024). This process may introduce errors, such as spelling mistakes and omissions of physical addresses, which negatively affect the reliability and validity of the data. The crime data were provided in hard copy, which further complicated the analysis process.

Generally, the findings indicate that there are no primary hot and cold spots for criminal activities in the SMA. This result was statistically significant because the p-value was greater than 0.05 (Figure 6.6). The study found insignificant concentrations of criminal activity in the specified study area. This result seems possible, as this is a military facility and a strategically important location where criminal activity is not anticipated to occur, at least not regularly.

Figure 6.6: Hotspot analysis map of the Saldanha Military Area.
(*Source*: Authors' own work using ArcGIS 10.8.1)

Figure 6.6 displays a hotspot analysis map that presents the findings through a colour-scaled representation, highlighting the virtual absence of prominent hot and cold spots within the military area. The map illustrates the geographic distribution of criminal incidents. Areas marked with a dark blue dot represent cold spots with a 95% confidence level, while those marked with a lighter blue denote cold spots with a 90% confidence level. The yellow dots indicate hotspots with a 90% confidence level. Most dots are shaded green, indicating that most crime incidents are statistically insignificant in terms of clustering.

However, the hotspot analysis map indicated three areas where most of the crime occurred. These are areas in the south, far east, and northeast of the map. Although this indicates areas with notable incidences of criminal activity, it does not constitute statistically significant clustering as identified by the software.

Conclusion

The primary goal of this research was to geo-visualise crime incident hotspots and cold spots within the SMA. This study aimed to provide protection and law enforcement agencies with detailed maps of crime patterns. These maps were intended to assist in the strategic placement and utilisation of resources, as well as in predicting possible criminal activities. The study revealed that the SMA showed predominantly insignificant clusters of high and low crime concentrations over seven years when analysed using the Getis-Ord Gi* statistical tool for hotspot analysis. The analysis indicated that trespassing incidents were more prevalent along the periphery of the SMA, whereas general theft was more common within the unit lines, although it was not notably clustered. As previously indicated, the crime increase in SMA suggests that crime during the Covid-19 period may have been driven by internal actors, ineffective restrictions, and the evolving tactics of offenders.

The geographic distribution of criminal activity in the SMA from January 2017 to December 2023 appeared random, suggesting that criminal incidents could occur throughout the area. The lack of statistically significant clusters implied that the crime levels in the individual locations were insufficient to generate effective crime hotspot maps. Despite the absence of statistically significant clustering, three areas were responsible for most of the criminal activities reported. This result can be used by law enforcement agencies to focus on crime prevention strategies. This study represents the first effort to map crime patterns and trends in the SMA and highlights issues such as inadequate crime data recording, storage, and retrieval. This is an important result that should be noted by law enforcement agencies operating in the study area.

This study emphasises the need for law enforcement agencies within military communities to improve their record-keeping methods. It advocates for the computerisation of data storage, reference, and retrieval systems to provide more reliable and comprehensive data, which can support decision-making and assist in predicting criminal trends. Further research is required to develop viable solutions to the identified crime problems. Expanding the study area and obtaining additional crime data would allow for a more successful application of the Getis-Ord Gi* tool in future studies. Additionally, researchers should conduct more comparable research that would benefit military communities in their efforts to prevent and manage crime.

References

Allen, D.W. 2009. *GIS Tutorial 2: Spatial Analysis Workbook*. Redlands, CA: Esri Press.

Apuke, O.D. 2017. Quantitative research methods: A synopsis approach. *Arabian Journal of Business and Management Review (Kuwait Chapter)*, 6: 40–47. https://doi.org/10.12816/0040336

Argun, U. & Dağlar, M. 2016. Crime mapping and geographical information systems in crime analysis. *Journal of Human Sciences*, 13(1): 2208–2221. https://doi.org/10.14687/ijhs.v13i1.3736

Ashby, M.P. 2020. Initial evidence on the relationship between the coronavirus pandemic and crime in the United States. *Crime Science*, 9(1). https://doi.org/10.1186/s40163-020-00117-6

Ballard, R. 2004. Assimilation, emigration, semigration and integration. In: N. Distiller & M. Steyn (eds.). *Under Construction: "Race" and "Identity" in South Africa Today*. Johannesburg: Heinemann Publishers. 1–24.

Bezuidenhout, J. 2012. Mapping of historical human activities in the Saldanha Bay Military Area. *Scientia Militaria: South African Journal of Military Studies*, 40(2): 89–101. https://doi.org/10.5787/40-2-998

Chainey, S., Tompson, L. & Uhlig, S. 2008. The utility of hotspot mapping for predicting spatial patterns of crime. *Security Journal*, 21: 4–28. https://doi.org/10.1057/palgrave.sj.8350076

Choate, C. 2015. A spatial statistical analysis of the effect of Gentlemen's Clubs on crime in the City of Los Angeles. Unpublished master's dissertation. Northridge, CA: California State University.

DefenceWeb. 2018. Lack of budget hindering security of SANDF bases, property. https://www.defenceweb.co.za/sa-defence/sa-defence-sa-defence/lack-of-budget-hindering-security-of-sandf-bases-property/ [20 August 2024].

DefenceWeb. 2023. The crime index shows deterioration in South Africa, which is now the third worst-affected African nation. https://www.defenceweb.co.za/featured/crime-index-shows-deterioration-in-south-africa-which-is-now-third-worst-affected-african-nation/ [12 March 2024].

Díaz, C., Fossati, S. & Trajtenberg, N. 2022. Stay at home if you can: COVID-19 stay-at-home guidelines and local crime. *Journal of Empirical Legal Studies*, 19(4): 1067–1113. https://doi.org/10.1111/jels.12339

Esri [Environmental Systems Research Institute]. 2020. How hot spot analysis (Getis-Ord Gi*) works. https://pro.arcgis.com/en/pro-app/latest/tool-reference/spatial-statistics/h-how-hot-spot-analysis-getis-ord-gi-spatial-stati.htm [18 July 2024].

Fennelly, L.J. 2012. *Handbook of Loss Prevention and Crime Prevention*. Fifth edition. Waltham, MA: Butterworth-Heinemann.

Field, A. 2013. *Discovering Statistics Using IBM SPSS Statistics*. Fourth edition. Los Angeles, CA: SAGE.

GCIS [Government Communication and Information System]. 2025. South Africa Yearbook (SAYB). *South Africa Yearbook 2023/24 Police, Defence and Military Veteran.* https://www.gcis.gov.za/sites/default/files/docs/gcis/pdf/yearbook2023-24-15%20Police%2C%20Defence%20and%20Military%20Veterans%202023-24.pdf [20 August 2025].

Haining, R.P. 2001. Spatial autocorrelation. In: *International Encyclopedia of the Social & Behavioral Sciences*. 14763–14770. https://doi.org/10.1016/B0-08-043076-7/00950-5

Henrico, I., Mayoyo, N. & Mtshawu, B. 2022. Understanding crime in the context of COVID-19: The case of Saldanha Bay Municipality. *SA Crime Quarterly*, 71(1). https://doi.org/10.17159/2413-3108/2022/vn71a9539

Henry, S. & Lanier, M.M. 1998. The prism of crime: Arguments for an integrated definition of crime. *Justice Quarterly*, 15(4): 609–627. https://doi.org/10.1080/07418829800093921

Hoeboer, C., Kitselaar, W., Henrich, J.F., Międzobrodzka, E., Wohlstetter, B., Giebels, E. & de Kogel, K. 2023. The impact of COVID-19 on crime: A systematic review. *American Journal of Criminal Justice*, 49: 1–19. https://doi.org/10.1007/s12103-023-09746-4

Holze, J. 2021. The effect of the Covid-19 lockdown on crime rates in South Africa. CIES Research Brief 8. https://www.graduateinstitute.ch/sites/internet/files/2021-10/CIES_Research%20brief%208.pdf [25 August 2024].

Hunt, J. 2019. From crime mapping to crime forecasting: The evolution of place-based policing. *National Institute of Justice Journal*, 9: 1–6.

ISS [Institute for Security Studies]. 2020. SA crime reductions during COVID-19 lockdown may be short lived. https://issafrica.org/about-us/press-releases/sa-crime-reductions-during-covid-19-lockdown-may-be-short-lived [4 September 2024].

Kedia, P. 2016. Crime mapping and analysis using GIS. *International Institute of Information Technology*, 1(1): 1–15.

Kramer, R.C. 1985. Defining the concept of crime: A humanistic perspective. *Journal of Sociology & Social Welfare*, 12: 469–486. https://doi.org/10.15453/0191-5096.1715

Manap, N., Borhan, M.N., Yazid, M.R.M., Hambali, M.K.A. & Rohan, A. 2019. Determining spatial patterns of road accidents at expressway by applying Getis-Ord Gi* spatial statistic. *International Journal of Recent Technology and Engineering*, 8(3S3): 345–350. https://doi.org/10.35940/ijrte.C1004.1183S319

Military Police Agency. 2001. *The Military Police Agency Instruction 12/00.* First edition. Pretoria: Military Police Agency.

Military Police Agency. 2007. *The Military Police Agency Instruction 12/00: Crime Prevention Guidelines and Procedures*. Second edition. Pretoria: Military Police Agency.

Mitchell, A. 2005. *The Esri Guide to GIS Analysis, Volume 2: Spatial Measurements and Statistics*. Redlands, CA: Esri Press.

Newham, G., Van der Heyde, J., Lancaster, L. & Machabaphala, T. 2023. South Africa does have the means to reduce violence. *Institute for Security Studies (ISS Today)*. https://issafrica.org/iss-today/south-africa-does-have-the-means-to-reduce-violence [26 May 2024].

NPR Africa. 2024. As police lose the war on crime in South Africa, private security companies step in. *The Associated Press*. https://www.npr.org/2024/01/07/1223358578/as-police-lose-the-war-on-crime-in-south-africa-private-security-companies-step- [24 February 2024].

Parliamentary Monitoring Group, South Africa. 2015. South African Defence Review 2015. https://static.pmg.org.za/170512review.pdf [5 September 2024].

Payne, J. & Morgan, A. 2020. Property crime during the COVID-19 pandemic: A comparison of recorded offence rates and dynamic forecasts (ARIMA) for March 2020 in Queensland, Australia. https://doi.org/10.31235/osf.io/de9nc

Reid, S., Tita, G. & Valasik, M. 2011. The mapping and spatial analysis of crime. Oxford Bibliographies. https://doi.org/10.1093/obo/9780195396607-0123 [26 August 2024].

SAHRIS [South African Heritage Resources Information System]. 2013. Draft basic assessment report proposed Saldanha Naval Base – replacement of an existing security fence. https://sahris.sahra.org.za/sites/default/files/heritagereports/SALDANHA%20BAR.compressed.pdf [14 March 2024].

Saldanha Bay Municipality. 2021. The 2021 Socio-Economic Profile of Saldanha Bay Municipality. https://sbm.gov.za/wp-content/uploads/Pages/Statistics/2021-SEP-LG-Saldanha-Bay-Municipality-.pdf [14 March 2024].

Saldanha Bay Municipality. 2023. Saldanha Bay Municipal Area. https://sbm.gov.za/towns-of-sbm/ [12 March 2024].

SAQA [South African Qualifications Authority]. 2021. *Occupational Certificate: Military Police Official*. https://allqs.saqa.org.za/showQualification.php?id=110351 [5 September 2024].

SATPALDA. 2024. GIS for crime mapping. https://satpalda.com/blogs/gis-for-crime-mapping/ [13 March 2024].

Schonteich, M. & Louw, A. 2001. Crime in South Africa: A country and cities profile. *Institute for Security Studies Papers*, 2001(49).

Silber, G. & Geffen, N. 2009. Race, class and violent crime in South Africa: Dispelling the 'Huntley thesis'. *SA Crime Quarterly*, 2009(30): 35–43.

South African Government. 2025. Defence Act 42 of 2002. https://www.gov.za/documents/defence-act [6 December 2025].

South African Soldier Magazine. 2023. Anti-criminality conference aims to protect the SANDF's integrity. *South African Soldier Magazine*, 30(4): 18–19.

Statistics South Africa. 2024. Lifting the veil on violence against children in South Africa. https://www.statssa.gov.za/?p=17023 [12 March 2024].

Thotakura, S. 2011. Crime: A conceptual understanding. *Indian Journal of Applied Research*, 4(3): 196–198. https://doi.org/10.15373/2249555X/MAR2014/58

Wang, F. 2012. Why police and policing need GIS: An overview. *Annals of GIS*, 18(3): 159–171. https://doi.org/10.1080/19475683.2012.691900

Welman, L. & Ferreira, S.L. 2016. The co-evolution of Saldanha Bay (town and hinterland) and its port. *Local Economy*, 31(1–2): 219–233. https://doi.org/10.1177/0269094215623849

Welsh, B.C. & Farrington, D.P. 2009. *Making Public Places Safer: Surveillance and Crime Prevention.* New York, NY: Oxford University Press. https://doi.org/10.1093/acprof:oso/9780195326215.001.0001

Western Cape Government. (2017). Policing needs and priorities (PNP) report 2016–17. https://www.wcpp.gov.za/sites/default/files/policing_needs_and_priorities_pnps_report_16-17.pdf [6 December 2025].

Western Cape Government. (2024). Socio-economic profiles 2024. https://www.westerncape.gov.za/treasury/socio-economic-profiles-2024 [4 December 2025].

Western Cape Government. (2025). Provincial strategic plan (PSP) 2025–2030. https://www.westerncape.gov.za/files/wcg-blob-files?file=2025-03/psp-2025-30.pdf&type=file [4 December 2025].

Xobiso, L. 2018. *The critical analysis of the role of the military police in crime prevention.* Doctoral dissertation, University of South Africa [UNISA]. https://ir.unisa.ac.za/handle/10500/24520

CHAPTER 7

Evaluating crime patterns in South Africa: The imperative for enhanced police presence in Mamelodi East

L.D. Pila and P. Schmitz

Abstract

South Africa constantly struggles with high rates of violent crime, including murder, assault with the intention to cause grievous bodily harm, and sexual assault. These crimes require law enforcement operations, alongside a wide variety of theft, carjacking, harassment, and non-criminal challenges, such as car accidents. This study analysed actual crime statistics over an eleven-year period, 2013/2014 to 2024/2025, in which the South African Police Service (SAPS) Mamelodi East was among the top 30 police stations for certain types of crime. The second part of the study focused on how to analyse mapped crime data to determine patterns that can be used to plan appropriate responses to reduce crime in the area. The limitation of this study was the absence of actual mapped crime incidents. Simulated mapped crime incidents were used to demonstrate the concept of analysing the crime problem to determine suitable responses to reduce crime. The study looked at various policing models, such as community-oriented policing (COP) and problem-oriented policing (POP) and focused on routine activity theory to determine possible solutions to reduce the number of burglaries in SAPS Mamelodi East. The findings underscore the value of geospatial crime analysis for planning effective law enforcement strategies and highlight the need for collaborative efforts between the police, local government, and communities to enhance public safety in high-risk urban areas.

Keywords: community, crime rate, law enforcement officers, law enforcement operations, safety and security

Introduction

Law enforcement operations play an important role within communities as they constitute a public good responsible for regulating society, maintaining order, preventing criminal activity, responding to crime, and restoring order in communities. Law enforcement operations ensure accountability and public protection within communities (Nkwenyane, 2011). Furthermore, the South African Constitution requires the South African Police Service (SAPS) to combat and prevent crime for the wellbeing of the South African population (Machethe & Mofokeng, 2022; Manamela et al., 2024). Violent crimes such as murder, rape, assault, and gender-based violence are considered serious social problems, causing direct and indirect impacts on people (Breetzke, 2020; Ross & Rasool, 2019).

The Global North and Global South, including South Africa, experience crime. Crime involves actions that cause harm to others, damage to or loss of finances and properties, and are against the law. Threats and harassment, domestic abuse, theft, unlawful possession of firearms, drugs, cybercrime, and violent acts such as terrorism, kidnapping, and human trafficking are some of the offences committed (Machethe & Mofokeng, 2022; Statista Research Department, 2025).

Crime rates differ widely around the world, with some countries having significantly higher crime rates than others. Countries such as the Democratic Republic of the Congo (DRC), Venezuela, Honduras, and Papua New Guinea have some of the highest crime rates in the world, with violent crime and homicide (criminal act of murder) rates far above the global average. Regional trends can also be seen in crime rates, with Latin America and the Caribbean having some of the highest rates of crime in the world, while Western Europe and East Asia have some of the lowest. These trends are mostly associated with factors such as poverty, political instability, and the presence of organised crime. Crime rates are typically expressed in the form of an index that considers the number of reported crimes per capita or per 100,000 of the population in each country (Global Organized Crime Index, 2023).

With a crime rate of 7.75, the DRC has the highest crime rate globally. The DRC faces several issues related to crime and violence, including armed warfare, sexual violence, robbery, theft, murder, and rape. Decades of conflict and instability in the DRC have led to an increase in criminal activities. The DRC is also home to many armed groups, some of which engage in illegal activities, such as poaching and trafficking of natural resources, which worsen the issues of crime. (Global Organized Crime Index, 2023).

According to the 2023 Global Organized Crime Index, South Africa has the third-highest crime rate in the world, indicating that the country continues to struggle with a high level of crime. Burglary, street robbery, carjacking, murder, rape, and sexual assault are common crimes in South Africa (Statistics South Africa, 2020). According to Statistics South Africa (2020), the number of persons who are victims of these crimes has increased over the years, from 2.1 million in 2015/2016 to 2.3 million in 2019/2020. This study was influenced by the increasing crime rates in South Africa.

Numerous causes of crime, such as high levels of poverty, inequality, unemployment, social exclusion, and the normalisation of violence, are to blame for South Africa's high crime rates (Bhorat et al., 2017; Breetzke, 2018). According to crime figures compiled by SAPS, South Africa has one of the highest rates of rape in the world (SAPS, 2023). According to Randa et al. (2023), half of the women in South Africa experience violence or sexual abuse by their partners during their lifetime.

Law enforcement officers are responsible for responding to, detecting, and preventing crime (Machethe & Mofokeng, 2022; Manamela et al., 2024). From this perspective, law enforcement officers play a crucial role in adapting to and responding to events that are known and unknowable, such as theft or domestic disputes, as well as scenarios that are known to occur (Dervin 1983, 1992, 1996).

Hinkle et al. (2020) contend that simply increasing the number of officers on duty or responding promptly to calls from the public does not always result in a decrease in crime or an increase in public confidence. Crime can be decreased by dispatching law enforcement officers on foot patrol to crime hotspots. Additionally, it gives the public confidence that the police are acting on their behalf. Police visibility through patrol can reduce crime, but only if it is targeted in small geographic locations or hotspots where crime is concentrated (Sherman & Eck, 2002; Weisburd & Eck, 2004). The SAPS's policing objectives, in accordance with the provisions of Chapter 11 of the Constitution (Act 108 of 1996), are as follows:

- prevent, combat, and investigate crime;
- maintain public order;
- protect and secure the inhabitants of South Africa and their property; and
- uphold and enforce the law.

This study analysed the importance of dispatching law enforcement officers in SAPS Mamelodi East, motivated by two factors. First, according to Mnguni (2021), SAPS Mamelodi East has high crime rates, a trend reflected in the latest crime statistics, which rank SAPS Mamelodi East among the top 30 police stations for certain crimes (SAPS, 2024). Second, young unemployed men form groups that terrorise local businesses and the community of Mamelodi. For example, there is a group in Mamelodi called the Boko Haram gang, which has been involved in a spate of robberies, as well as extortion, intimidation, and other serious and violent crimes in Mamelodi and the surrounding areas over the past few years (Makhetha, 2024). In the context of these factors, this study sought to analyse the importance of dispatching/allocating law enforcement officers in Mamelodi to reduce or prevent crime.

Literature review

The importance of law enforcement operations

Law enforcement operations are of public interest because they are responsible for regulating society, maintaining order, preventing criminal activities, responding to crime, restoring order in communities, and protecting the public (Nkwenyane, 2011). Violence and crime are considered serious social problems in South Africa, having direct and indirect impacts on communities (Ross & Rasool, 2019). Households in different regions of South Africa often experience common crimes such as theft (Statistics South Africa, 2020).

According to Kappeler et al. (2020), law enforcement operations are more philosophical than academically theoretical. He emphasised that socio-economic issues in the community lead to criminal behaviour and that these issues may be remedied with the help and support of the general population. Law enforcement operations as a philosophy call for law enforcement operation approaches and alternative techniques. These approaches ensure police accountability and transparency by increasing the public's knowledge of law enforcement strategies. They also foster positive working relationships and public trust between law enforcement officials and the public.

To examine the efficacy of various law enforcement operation strategies, 83 street-violent hot locations in the US were randomly assigned to receive high-intensity police patrols, problem-solving tactics, or normal police response (Taylor et al., 2011). Compared to other locations, intense patrol hotspots had a much lower initial level of violence during the 90-day intervention period.

The presence of law enforcement officers in high-crime areas enhances public perceptions of crime, feelings of safety, and public faith in law enforcement officers. (Tuffin et al., 2006). According to Braga et al. (2014), the presence of law enforcement officers in high-crime areas can generally be helpful in lowering crime.

Concerns with law enforcement officers

According to Goldstein (1979), law enforcement officers seem to be more preoccupied with means than with aims. Goldstein (1979) also contends that police personnel should avoid using the conventional, standard model of policing and should use a proactive rather than a reactive strategy when dealing with crime and disturbances, which is known as problem-oriented policing (POP). POP is based on the scanning, analysis, response, and assessment (SARA) model. Scanning (S) involves looking for and identifying problems that could lead to crime and disorder. Analysis (A) was performed to understand the problem using various tools to develop a response. Response (R) is based on the results of the analysis and is implemented by the police to solve a particular problem. Assessment (A) is used to assess how successful the intervention was and how it can be improved in the future (Hinkle et al., 2020). This approach was further developed and researched by Weisburd et al. (2010) and Hinkle et al. (2020).

Community-oriented policing (COP) is a comprehensive policing method that mainly relies on community involvement and cooperation and the presence of law enforcement officials in the neighbourhood to combat local crime and disorder. To prevent and reduce crime and disorder, law enforcement professionals can use POP to establish analytical tactics, including problem identification, analysis, reaction, and assessment (National Academies of Sciences, Engineering, and Medicine, 2018).

There are several differences between COP and POP, including attention intensity and method diversity (National Research Council, 2004); however, there are also many parallels. For instance, COP and POP are proactive forms of policing that emphasise preventing crime rather than merely responding to it once it has occurred. Collaboration across various organisations and partners, including community members, is also necessary for COP and POP (National Academies of Sciences, Engineering, and Medicine, 2018). POP and COP involve communities in identifying issues and choosing solutions, which is another similarity between them (Green, 2000).

Crime prevention

Kruger et al. (2016:5) state that "[g]overnment policy and legislation urge local government to take the lead in implementing local level crime prevention programmes". Law enforcement operations, social crime prevention, and situational awareness are some of the interventions used in crime prevention that attempt to address some of the fundamental causes of crime in South Africa. Closed-circuit television (CCTV), private security companies, community safety forums, and the presence of law enforcement officers in local communities are all measures used to reduce and prevent criminal activities. While situational prevention includes installing video surveillance systems in city centres and lowering the danger of street crimes, including assault, theft, and rape, social prevention includes enhancing street lighting in cities and urban regions (Kruger et al., 2016).

According to Kruger et al. (2016), protecting vulnerable populations from crime and enhancing the operation of the criminal justice system are two recommendations made in Chapter 12 (Building Safer Communities) of the 2012 National Development Plan, which prioritises public safety in South Africa. The manual for community-based crime prevention developed by Kruger et al. (2016) is one approach that can be used to reduce and prevent crime in a community.

Theoretical framework of law enforcement operations

Several theoretical frameworks for law enforcement operations exist, such as predictive policing by Perry et al. (2013) and situational crime prevention by Clarke (1983). In this section, the frameworks listed by Todd et al. (2011) are discussed. Todd et al. (2011) discuss in more detail the models and theories that support law enforcement operations, such as the disorder models related to broken windows theory, disorganisation models related to systems theory, and informal social control models related to collective efficacy.

According to Todd et al., (2011) the 'broken windows' theory is one of the most well-liked theories regarding how crime occurs in urban settings. The broken windows hypothesis connects crime to urban problems or anarchy. Littering, noise, and unruly groups of men and boys are examples of urban disorder, which is how the term 'broken windows' was coined. It assumes that unmaintained areas are prime targets for criminal activities. Law-abiding citizens leave these regions because they are uncomfortable. Therefore, the broken windows theory urges the resolution of urban disorder (Todd et al., 2011).

The shattered windows theory is based on two guiding principles: the fear of crime or the fear of being harassed by disorderly people and the fear of being in public areas. Disorganised behaviour is a type of social disorder, and the psychological notions of disorder and crime are connected. According to Zimbardo (1969), an unmaintained property, whether in a good or bad neighbourhood, is an easy target for theft and vandalism. Wilson and Kelling (1982) added that undesirable behaviour is not under control.

The idea of social disorganisation created by Clifford Shaw and Henry McKay in the 1940s, according to Todd et al. (2011), is the original theory of crime and place. This idea emphasises three social factors: poverty, ethnic diversity, and movement. The authors, as indicated by Todd et al. (2011), further contend that these factors cause society to become disorganised when they occur. Crime is always considered to be caused by poverty. The goal of social disorganisation theory is to develop strategies to create social groups and political elements within neighbourhoods to improve their circumstances, which will counter the disorganisation in the community and have a knock-on effect in reducing crime in the identified neighbourhood (Todd et al., 2011).

According to Todd et al. (2011), the collective efficacy thesis adopts a distinct approach to socio-economic issues affecting neighbourhoods. This thesis contends that effective formal and informal social controls can be implemented in neighbourhoods to reduce crime. Family, relatives, and friends exert informal social control over young people's lives. For instance, some local families have close adult-adolescent relationships, which may deter young people from participating in criminal activities. Other families, on the other hand, choose to sever their relationships, which can lead to adolescents' out-of-control behaviour and criminal activity. Therefore, collective efficacy promotes crime reduction by fostering social connections in troubled neighbourhoods, including the formation of organisations and centres for residents to participate in and build strong social networks and connections (Todd et al., 2011).

Methodology

This study focuses on SAPS Mamelodi East, as indicated in the introduction. The study was conducted in two phases. The first phase was analysing SAPS Mamelodi East's crime over a ten-year period using official statistics from SAPS. The second phase was to demonstrate the use of mapped crime incidents to determine proactive policing activities. Unfortunately, actual mapped crime data were not available to the authors, and it was decided to create a simulated crime incident map with data such as the type of incident, month, day, and time of the incident to demonstrate the use of mapped crime data for policing activities.

Study area

Mamelodi was established in June 1953, when 16 houses were built. It is located northeast of Pretoria, as shown in Figure 7.1. Mamelodi is a residential suburb approximately 45.2 km^2 in size. It was originally established as a township for Black workers. Over time, Mamelodi began to expand to the east. To support effective policing, Mamelodi is served by two policing precincts: SAPS Mamelodi and SAPS Mamelodi East (SAPS, 2025b). The township is composed of formal (free-standing detached and semi-detached housing) and informal (shacks and backyard shacks) settlements. Its population was an estimated 330,000 in 2011, and the township continues to grow.

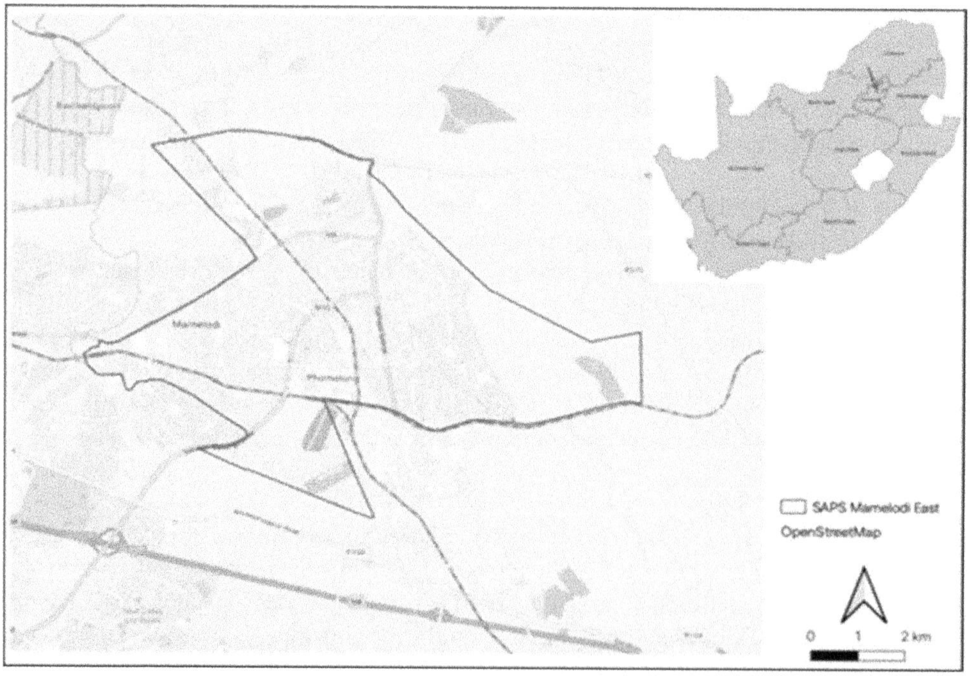

Figure 7.1: Map showing the area of Mamelodi and its location in South Africa. (*Source:* Map supplied by SAPS Mamelodi East)

The informal settlements, which were mostly built via land invasion, have greatly contributed to Mamelodi's expansion, especially in Mamelodi East (as indicated in Figure 7.2). Due to this growth, Mamelodi is now experiencing high rates of unemployment, crime, and drug use, as well as high living expenses and a lack of essential utilities, including housing, electricity, water, and sanitation. Most residents either work as independent contractors or are unemployed.

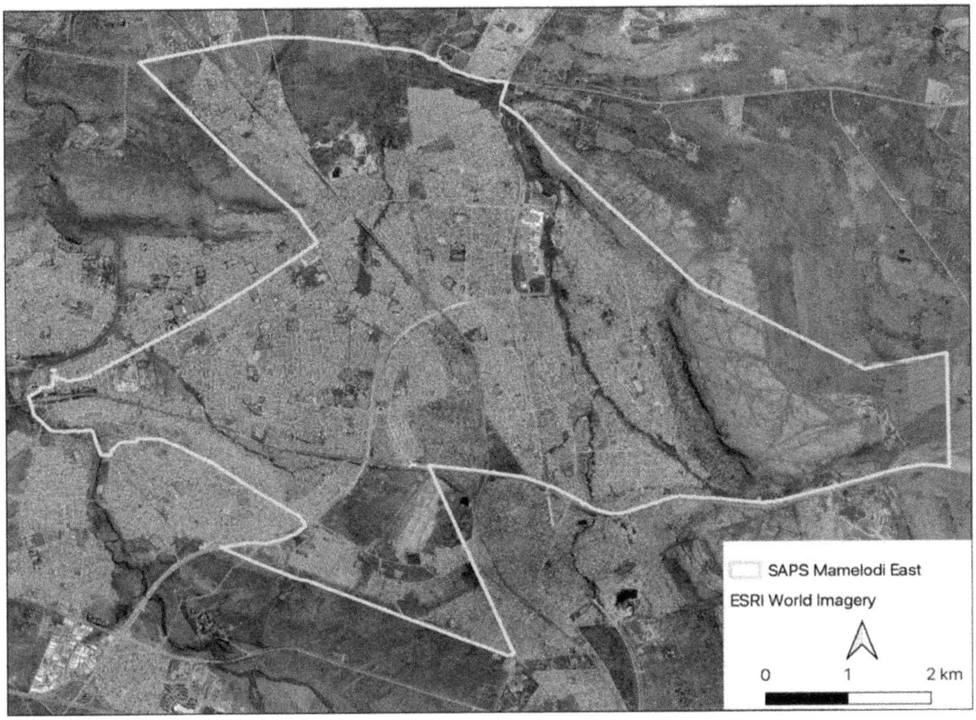

Figure 7.2: A map showing the population distribution within Mamelodi East.
(*Source:* Map supplied by SAPS Mamelodi East)

Data collection procedure and method analysis

SAPS publishes quarterly crime statistics for South Africa, including data on the areas with high crime rates. The highest-crime areas are highlighted in the 'top 30 police stations' lists. Using SAPS datasets, SAPS Mamelodi East was identified as one of the stations with the highest crime rates.

For this analysis, crime statistical datasets were collected from the SAPS website for SAPS Mamelodi East and included: the top 30 police stations for the year 2023/2024, the crime data for the top 30 police stations for Quarter 1, Quarter 2, and Quarter 3 for 2024/2025, and the annual statistics for SAPS Mamelodi East from 2013/2014 to 2022/2023 (SAPS, 2024, 2025a). The crime statistics follow SAPS's financial year, which is from 1 April to 31 March of the following year.

The second part of the methodology was to map the simulated crime incidents, as indicated in the introduction to this section. Mapping and analysis were performed using QGIS and MS Excel. The incidents were mapped using the road network to simulate incidents occurring at an address. QGIS was used to map the distribution of crime incidents in SAPS Mamelodi East, and hotspot maps were created in QGIS for all crime incidents and for burglary of residential premises (burglary), which was the highest number of incidents in this simulated example. The analysis in MS Excel included pivot tables to determine the number of burglaries by month, day of the week, and combinations of day of the week and time of day. These were necessary to determine seasonality, peak weeks in a month, peak weekdays, and peak hours within a weekday. These analyses were conducted to demonstrate how crime data can be analysed to prioritise policing activities. The results of these analyses are discussed in the next section.

Results

This section presents the findings. This study aimed to analyse the importance of dispatching law enforcement officers to reduce or combat the high crime rate in Mamelodi. First, the study aimed to analyse the crime rate in SAPS Mamelodi East, exploring how law enforcement operations by means of dispatching law enforcement officers in SAPS Mamelodi East can be used to fight crime in Mamelodi and analysing how this operation will be effective at reducing and fighting crime. Statistical crime data on different crimes committed in SAPS Mamelodi East between 2013/2014 and 2024/2025 were acquired from SAPS recorded crime statistics. These results were analysed and are presented below.

SAPS Mamelodi East Crime Statistics

The crimes recorded by the SAPS Mamelodi East are summarised in Table 7.1 and Figure 7.3. The total number of crimes committed in 2013/2014 was 8,876. The total number of crimes committed in 2014/2015 was 10,403, which is an increase of 17.2%. The total number of crimes committed in 2015/2016 was 9,613, a decrease of 7.6%. The total number of crimes committed in 2016/2017 was 10,726, an increase of 11.6%. The total number of crimes committed in 2017/2018 was 12,134, which is an increase of 13.1%. In 2018/2019, the total number of crimes committed was 12,648, an increase of 4.2%. In 2019/2020, the total number of crimes committed was 13,525, an increase of 6.9%. In 2020/2021, the total number of crimes committed was 11,106, a decrease of 17.9%. This decrease in crime incidents could be ascribed to the Covid-19 lockdown. The total number of crimes committed in 2021/2022 was 12,272, which is an increase of 10.5%. The total number of crimes committed in 2022/2023 was 12,413, a slight increase of 1.2%. In summary, there was a significant increase in criminal activity over the years leading up to the Covid-19 lockdown, resulting in a 17.9% decrease in crime, which increased and stabilised in 2021/2022 and 2022/2023, respectively.

Table 7.1: Ten-year crime statistics recorded by SAPS Mamelodi East

Crime category	2013–2014	2014–2015	2015–2016	2016–2017	2017–2018	2018–2019	2019–2020	2020–2021	2021–2022	2022–2023
Murder	54	80	55	53	75	75	79	69	76	77
Sexual offences	122	113	128	136	167	200	198	210	187	194
Attempted murder	81	76	73	72	66	53	65	43	108	92
Assault with the intent to inflict grievous bodily harm	664	766	732	691	758	835	852	755	733	657
Common assault	504	554	598	694	755	872	992	896	1 111	1 124
Common robbery	106	130	148	150	166	166	194	126	130	129
Robbery with aggravating circumstances	378	450	473	578	630	704	815	621	748	674
Contact crime (crime against the person)	1 909	2 169	2 207	2 374	2 617	2 905	3 195	2 720	3 093	2 947
Carjacking	62	79	112	122	123	137	230	174	248	164
Robbery at residential premises	44	31	32	60	65	56	62	52	66	67
Robbery at non-residential premises	89	87	73	60	84	81	86	76	105	65
Robbery of cash in transit	1	0	0	0	0	0	1	2	1	1
Bank robbery	0	0	0	0	0	0	0	0	0	0
Truck hijacking	4	1	2	6	4	7	4	1	8	4
Arson	28	31	20	14	14	26	21	14	16	6
Malicious damage to property	375	427	365	443	533	611	598	453	533	543
Contact-related crime	403	458	385	457	547	637	619	467	549	549
Burglary at non-residential premises	69	120	93	133	171	183	136	103	51	92
Burglary at residential premises	540	718	556	605	581	518	548	355	409	448

Crime category	2013–2014	2014–2015	2015–2016	2016–2017	2017–2018	2018–2019	2019–2020	2020–2021	2021–2022	2022–2023
Theft of motor vehicle and motorcycle	156	129	137	184	180	188	143	94	96	101
Theft out of or from motor vehicle	251	231	246	322	265	288	223	130	141	142
Stock-theft	0	0	0	0	0	0	0	1	3	1
All theft not mentioned elsewhere	694	848	673	659	845	887	812	610	643	785
Commercial crime	45	76	70	107	124	132	186	203	213	239
Shoplifting	41	59	51	71	47	130	155	201	135	183
Other serious crime	780	983	794	837	1 016	1 149	1 153	1 014	991	1 207
Illegal possession of firearms and ammunition	72	58	88	111	97	86	67	66	31	58
Drug-related crime	578	722	527	674	816	648	654	571	636	708
Driving under the influence of alcohol or drugs	77	103	153	144	212	141	336	207	243	165
Sexual offences detected as a result of police action	3	0	2	2	0	1	1	0	2	4
Crime detected as a result of police action	730	883	770	931	1 125	876	1 058	844	912	935
Kidnapping	16	21	50	36	51	56	42	28	54	52
Total	**8 876**	**10 403**	**9 613**	**10 726**	**12 134**	**12 648**	**13 525**	**11 106**	**12 272**	**12 413**

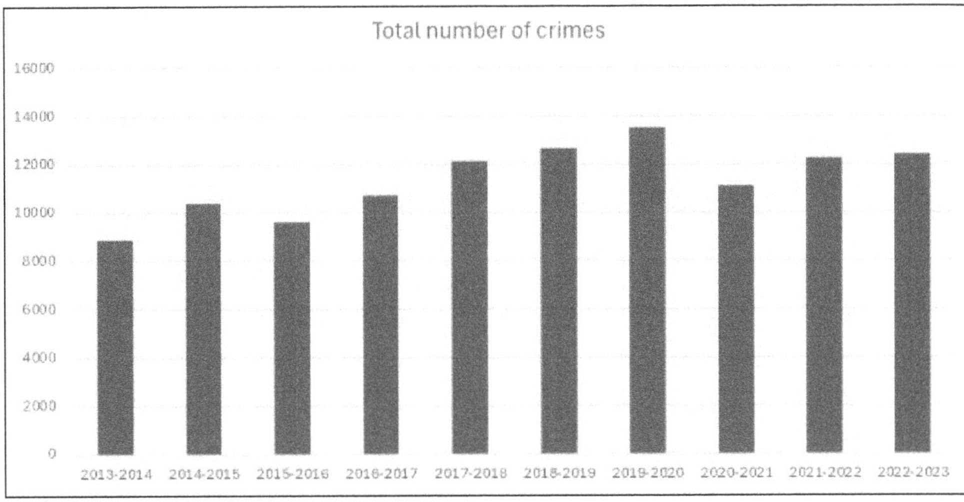

Figure 7.3: Number of crime incidents recorded by SAPS Mamelodi East.

Table 7.2 shows the crime types that caused SAPS Mamelodi East to be in the top 30 police stations in South Africa over a ten-year period (2014/2015 to 2023/2024), as extracted from SAPS (2024). The trend over the ten-year period is illustrated in Figure 7.4.

Table 7.2: Crime types over a ten-year period for SAPS Mamelodi East

Crime type	2014/ 2015	2015/ 2016	2016/ 2017	2017/ 2018	2018/ 2019	2019/ 2020	2020/ 2021	2021/ 2022	2022/ 2023	2023/ 2024
Contact crime	2 169	2 207	2 374	2 617	2 905	3 195	2 720	2 093	2 947	2 901
Sexual offences	113	128	136	167	200	198	210	187	194	206
Rape	103	115	115	148	175	164	183	158	167	176
Assault GBH	766	732	691	758	835	852	755	733	657	708
Common assault	554	598	694	755	872	992	896	1 111	1 124	1 048
TRIO crimes	197	217	242	272	274	378	302	419	296	338
Carjacking	79	112	122	123	137	230	174	248	164	195
Contact-related crime	458	385	457	547	637	619	467	549	549	500
Malicious damage to property	427	365	443	533	611	598	453	533	543	495

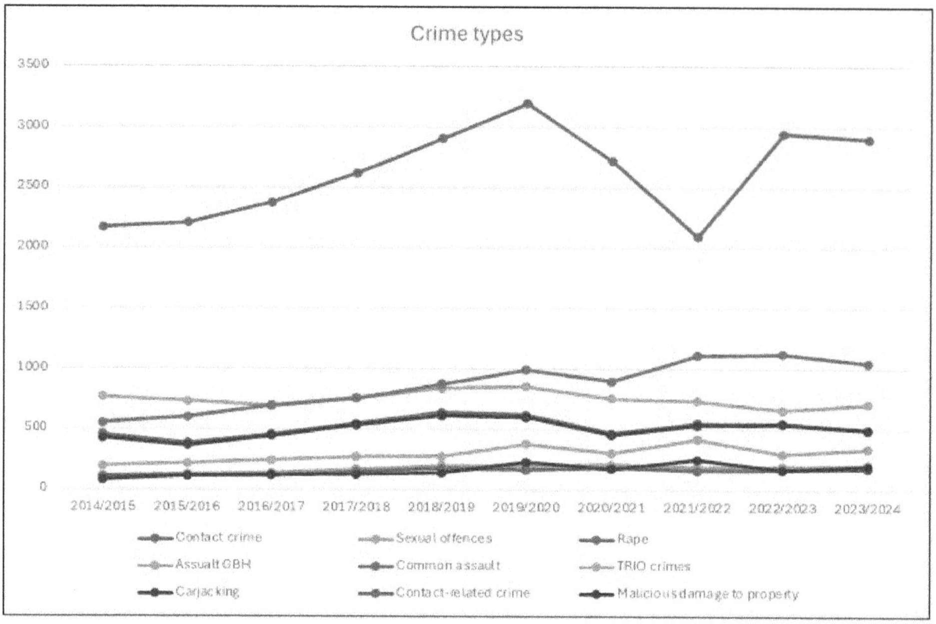

Figure 7.4: Trends of the identified crime types in SAPS Mamelodi East.

An interesting feature of this analysis is that there was a significant increase in contact crime up to the Covid-19 lockdowns (2020/2021 and 2021/2022), which showed a significant decrease and then a sharp increase in 2022/2023 and 2023/2024 after the lockdowns. Furthermore, the other crime types shown in Table 7.2 and Figure 7.5 do not follow the same drastic trend as contact crimes. According to SAPS (2024), contact crimes refer to "crimes in which the victims are the targets of violence or instances where the victims are in the vicinity of property that criminals target and are subjected to the use of/or threats of violence by perpetrators".

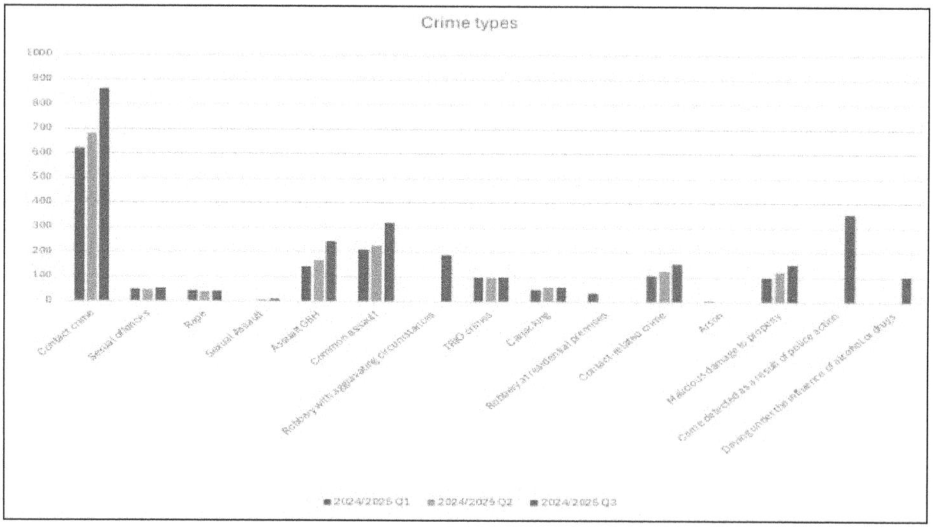

Figure 7.5: Quarterly data for SAPS Mamelodi East.

Table 7.3 and Figure 7.6 show the trends of the identified crime types which caused SAPS Mamelodi East to be in the top 30 police stations in South Africa for Quarter 1 (Q1), Quarter 2 (Q2), and Quarter 3 (Q3) for the year 2024/2025. Data were extracted from SAPS (2025a).

Table 7.3: Number of crime incidents for Q1, Q2, and Q3 in 2024/2025

Crime	**2024/2025 Q1**	**2024/2025 Q2**	**2024/2025 Q3**
Contact crime (CC)	621	683	863
Sexual offences (SO)	49	47	53
Rape (R)	44	39	42
Sexual assault (SA)	8	8	11
Assault – Grievous Bodily Harm (AGBH)	141	168	243
Common assault (A)	210	227	318
Robbery with aggravating circumstances (RAC)	0	0	189
TRIO crimes (carjackings, house robberies & business robberies)	101	97	101
Carjacking (CJ)	50	58	59
Robbery at residential premises (RRP)	34	0	0
Contact-related crime (CC)	106	126	154
Arson (A)	5	0	0
Malicious damage to property (MDP)	101	122	152
Crime detected as a result of police action (PA)	0	0	356
Driving under the influence of alcohol or drugs (DUI)	0	0	104

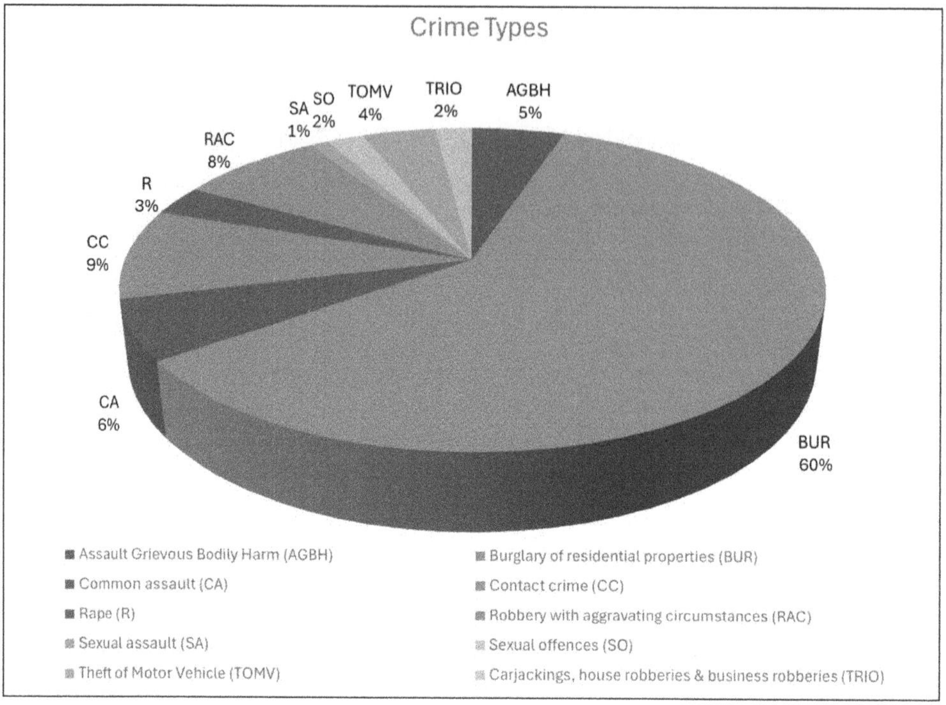

Figure 7.6: Frequencies of various simulated crime types in SAPS Mamelodi East.

With regard to contact crime, there is a sharp increase from 683 incidents in Q2 to 863 incidents in Q3. Sexual assault also increased in Q2 and Q3, although low in numbers, it was still sufficient to list SAPS Mamelodi East as one of the 30 top police stations with regard to the number of incidents, which should be of concern. This increase in sexual offences and common assault reflects the findings of Mazibuko and Umejesi (2019), who discussed domestic violence in Mamelodi. The authors indicate that this violence occurs not only behind closed doors but also openly in the community. Although still underreported, it is a real concern (Mazibuko & Umejesi, 2019). Robbery with aggravating circumstances was significant in Q3, also contributing to the station being listed in the top 30 police stations.

A positive indication is that SAPS Mamelodi East is conducting policing activities, as evidenced by the number of crimes detected as a result of police actions (356 incidents) and driving under the influence of alcohol or drugs (104 incidents). This section provided an overview of the actual crime situation in SAPS Mamelodi East. In summary, there was an increase in crime incidents in Q3, which SAPS Mamelodi East should consider and plan for crime reduction activities.

Crime mapping and analysis

POP includes the analysis of crime patterns. This is for two reasons: to reduce crime and to develop methods to assist and advise the community to reduce crime (Hinkle et al., 2020). In this section, the authors, in the absence of actual mapped crime incidents in SAPS Mamelodi East, simulated the mapping of crime incidents using the following crime types: assault with the intent to inflict grievous bodily harm (Assault GBH), burglary of residential premises (BUR), common assault (CA), contact crimes (CC), rape (R), robbery with aggravating circumstances (RAC), sexual assault (SA), sexual offences (SO), theft of/out of motor vehicle (TOMV), and TRIO crimes (carjacking, robbery at residential premises, robbery at non-residential premises) (TRIO). Acronyms are used in some of the figures in the discussion. Figure 7.7 shows the distribution of the listed crime types in SAPS Mamelodi East. This map can lead to information overload, and it is necessary to conduct exploratory analysis to ascertain the number of crimes that were mapped. This is the 'scan (S)' part of the SARA model for POP.

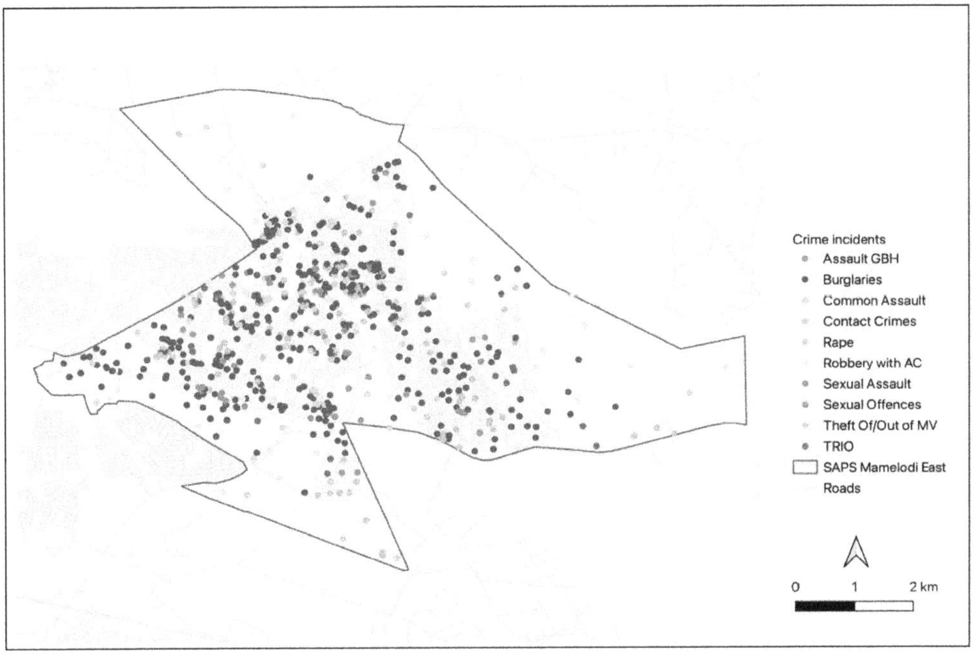

Figure 7.7: Mapped simulated crime incidents in the SAPS Mamelodi East.

As shown in Figure 7.6, burglary of residential premises (BUR) comprises 60% of the mapped crime incidents. Although the other mapped crime types are serious, the authors decided to further analyse burglary of residential properties. Mamelodi is a commuter village with a large number of residents using public transport, such as minibus taxis and buses, to commute to work and back. The next step is the 'analysis (A)' part of the SARA model to determine patterns such as time of day, day of the week, month-day pattern, and month. These patterns will enable the police to develop a suitable response(s) to decrease the number of burglaries in SAPS Mamelodi East. This analysis was performed using pivot tables in MS Excel. The first analysis was performed to establish the number of incidents per day per month. This indicates a seasonal variation, namely a high number of incidents over the summer months (December, January, and February) and the late spring month of November. It also coincides with the South African summer break (December to January), which means that a large number of families go on holiday or visit their relatives in the rural areas of South Africa. This pattern is illustrated in Table 7.4 and Figure 7.8, respectively.

Table 7.4: Number of burglaries per day, per month

Day	Jan	Feb	Mar	Apr	May	Jun	Jul	Aug	Sep	Oct	Nov	Dec	Total
Sunday (Su)	10	6	2	0	0	0	1	0	0	1	4	12	36
Monday (Mo)	19	17	7	2	3	5	0	0	1	7	11	23	95
Tuesday (Tu)	17	9	7	5	4	4	1	0	0	0	9	10	71
Wednesday (We)	12	13	10	7	5	1	1	1	3	3	13	20	89
Thursday (Th)	13	13	2	4	4	1	1	3	3	3	7	9	63
Friday (Fr)	17	9	1	3	2	2	1	0	1	2	7	21	66
Saturday (Sa)	7	6	2	3	1	3	1	1	0	0	9	10	43
Grand total	**95**	**73**	**31**	**24**	**19**	**16**	**6**	**5**	**8**	**21**	**60**	**105**	**463**

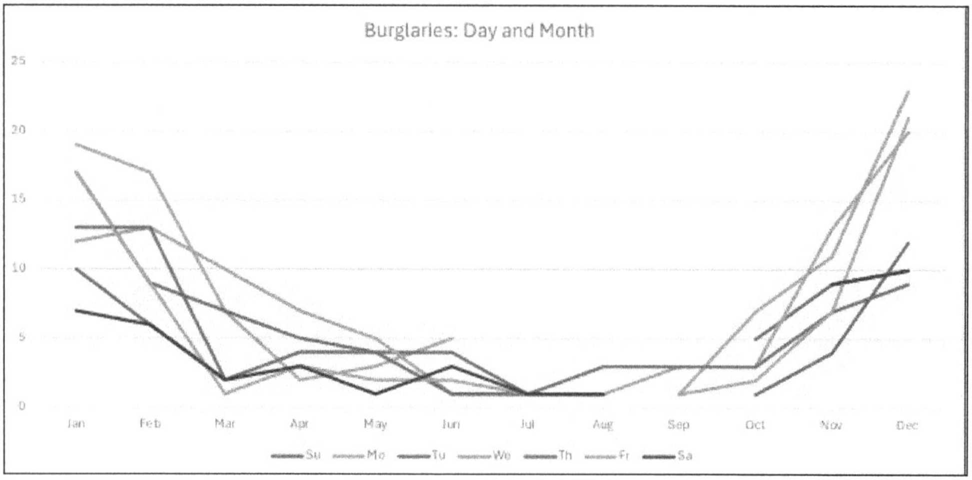

Figure 7.8: Daily burglary trends per month.

Further analysis of Table 7.4 shows that most burglaries occur during the working week from Monday to Friday. This relates to Mamelodi being a commuter village where the inhabitants travel to work early in the morning via public transport to their places of work. The absence of people during working hours at their homes provides opportunities for criminals to burgle residential premises.

The next step in the analysis was to determine the time of day when most incidents occurred. The highest number of incidents occurred between 9 and 10 o'clock (Timeslot 9), followed by 10 to 11 o'clock (Timeslot 10), and between 3 and 4 o'clock. The latter could indicate that the resident(s) are doing the night shift, which further creates opportunities to break into a house. This is illustrated in Table 7.5 and Figure 7.9. This pattern coincides with the movement of people to work and back to their homes. In Mamelodi, people have to leave their homes early to be on time at their place of work, thus providing ample time for criminals to commit burglaries in the mornings when people are away at work. This assumption is made on routine activity theory, where there is a motivated burglar to commit the crime, the residences where the people have to commute by bus or minibus taxi to work are suitable targets, and the possibility of the absence of a suitable guardian makes these residences a suitable target for the criminal to commit the crime (Kitteringham & Fennelly, 2020).

Table 7.5: Number of burglaries per timeslot per day

Timeslot	Su	Mo	Tu	We	Th	Fr	Sa	Total
1	7	6	6	4	4	4	0	31
2	4	6	7	8	8	3	2	38
3	16	10	6	6	9	5	8	60
5	3	0	0	2	1	0	2	8
9	7	19	21	21	15	21	12	116
10	4	14	11	14	9	10	3	65
11	0	11	8	10	5	6	4	44
15	1	4	3	3	5	4	2	22
16	0	2	1	2	2	0	1	8
17	0	3	2	1	1	1	1	9
22	0	2	0	2	2	1	0	7
23	0	5	1	0	5	5	3	19
24	1	12	5	9	3	4	2	36
Total	**36**	**95**	**71**	**89**	**63**	**66**	**43**	**463**

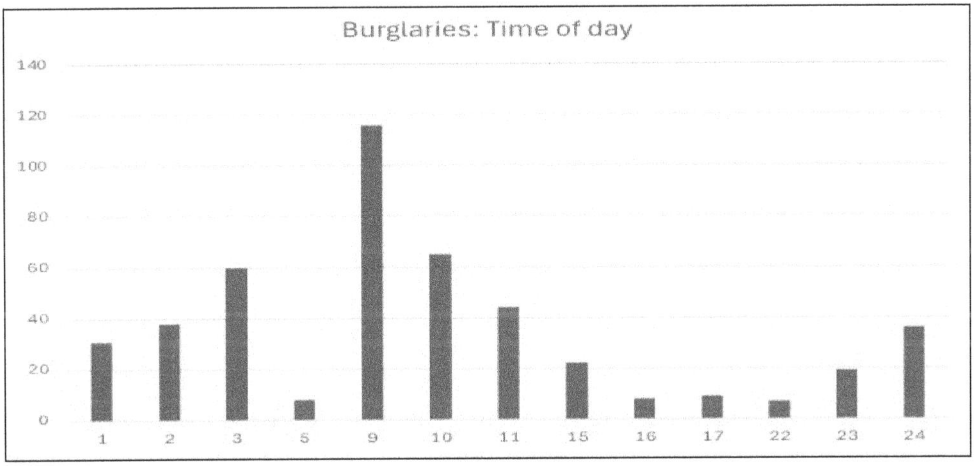

Figure 7.9: Number of burglaries per time slot.

A heat map (hotspot analysis) was created to determine the concentration of burglaries. These hot spots can be targeted first and then expanded into other areas where burglaries occur to reduce crime. The distribution of burglary incidents and the heat map are shown in Figure 7.10.

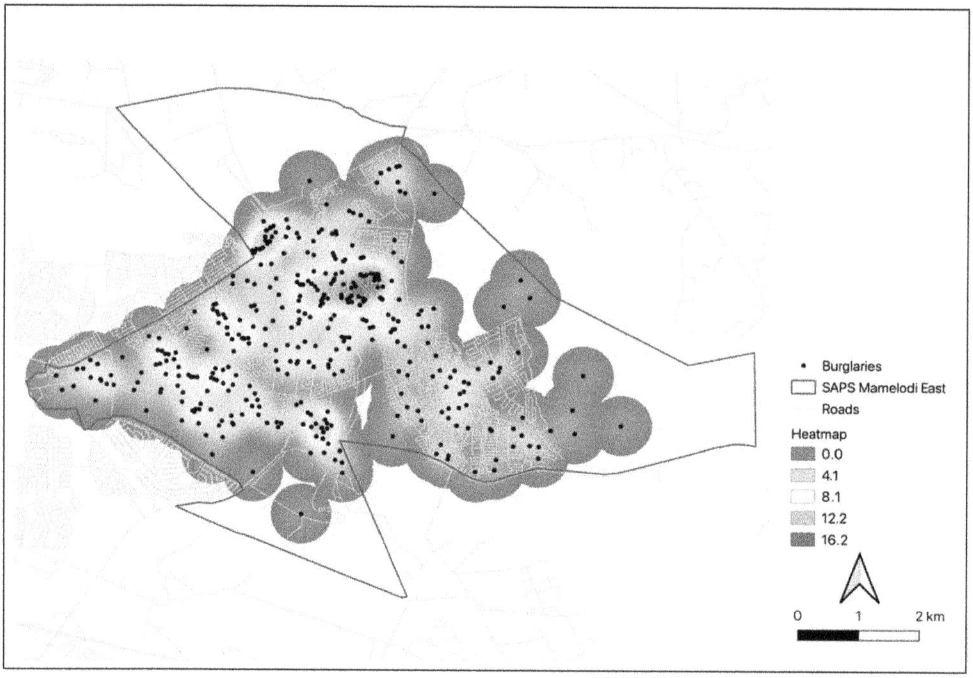

Figure 7.10: Heat map and distribution of burglary of residential premises incidents.

Based on the information and trends regarding burglaries, the 'response (R)' part of the SARA model can be determined. Figure 7.11 illustrates this concept. The authors acknowledge that the example may be naïve, as it does not account for the number of personnel and vehicles available to conduct such a response, which could help reduce the number of burglaries. However, the authors hope that it introduces the concept.

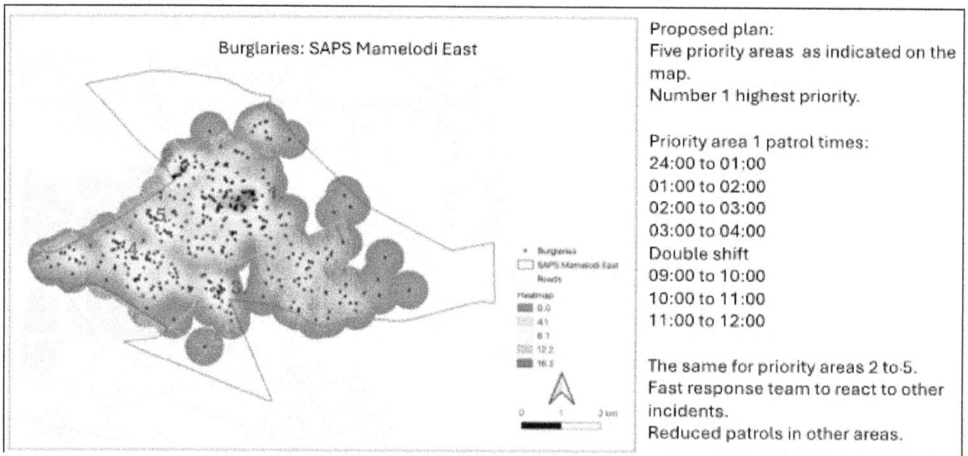

Figure 7.11: Illustrative concept of a response to reduce burglaries.

The example in Figure 7.11 is only a part of the response to reducing burglaries. Other planned responses include assisting and advising the community via community forums on how to improve home security to prevent burglaries. The enforcement of by-laws, such as those prohibiting loitering, and improved community involvement ('eyes on the ground') to report incidents and suspicious activity is crucial. These are based on the discussions by Hinkle et al. (2020). Communities can be empowered in partnership with the SAPS using the manual developed by Kruger et al. (2016). The manual was specifically developed for South African conditions.

The 'assessment (A)' part of the SARA model was used to determine the impact of the responses developed by SAPS Mamelodi East. It is proposed that these should be conducted quarterly, coinciding with the SAPS reporting of crime statistics. The assessment should answer the following question: Was there a drop in the number of incidents? What worked and did not work during the three-month period? Both are very important to understand, as well as the response to them. Another important aspect that police stations should analyse is the displacement of crime due to police action in the identified priority areas. The next round of responses can be part of the plan, focusing on those areas.

As the National Research Council (2004: 250) review of police effectiveness noted: "studies that focused police resources on areas with high crime rate provided the strongest collective evidence of police effectiveness that is now available". A Campbell systematic review by Braga et al. (2019) comes to a similar conclusion. Although not every study about high crime rate has shown statistically significant findings, the majority of such studies have from experimental or quasi-experimental evaluations reported noteworthy crime or disorder reductions, suggesting that when police focus in on areas with high crime rates, as indicated in the aforementioned example, they can have a significant beneficial impact on reducing crime in these areas.

Conclusion and limitations

This study analysed the actual and simulated crime statistical data of SAPS Mamelodi East. During the analysis process, it was discovered that SAPS Mamelodi East covers most of the Mamelodi area, with the highest criminal activities. This is reflected in the data analysis that SAPS Mamelodi East is among the top 30 police stations in South Africa for certain crime-type features.

With regard to the simulated crime data to determine areas, seasonality, day of the week, and time of day of criminal activities, it is possible to plan responses to reduce crime. However, the police in SAPS Mamelodi East should also get the communities involved via various forums to assist and advise on strengthening the security of residences to improve the presence of a guardian (whether a person or technical) at the residence to reduce the attractiveness for a criminal to commit a burglary. Another activity that the police can do is to cooperate with the City of Tshwane to enforce by-laws which include the prohibition of loitering and littering.

The limitation of this study was the unavailability of mapped actual crime incidents that could be used for further analysis, as demonstrated. It is recommended that further research be conducted to deepen the understanding of COP and POP, as well as the various theories about crime, such as routine activity theory and environmental criminology. The latter looks at the interface between the offender and victim's activity space, and if the situation is suitable, such as the absence of a guardian, the crime may occur (Kitteringham & Fennelly, 2020).

References

Bhorat, H., Lilenstein, A., Monnakgotla, J., Thornton, A. & van der Zee, K. 2017. *The Socio-Economic Determinants of Crime in South Africa: An Empirical Assessment*. Development Policy Research Unit Working Paper 201704. DPRU, University of Cape Town, Cape Town, South Africa.

Braga, A.A., Papachristos, A.V. & Hureau, D.M. 2014. The effects of hot spots policing on crime: An updated systematic review and meta-analysis. *Justice Quarterly*, 31(4): 633–663. https://doi.org/10.1080/07418825.2012.673632

Braga, A.A., Weisburd, D. & Turchan, B. 2019. Hot spots policing of small geographic areas effects on crime. *Campbell Systematic Reviews*, 15(3): 289–311. https://doi.org/10.1002/cl2.1050

Breetzke, G.D. 2018. The concentration of urban crime in space by race: Evidence from South Africa. *Urban Geography*, 39(8): 1195–1220. https://doi.org/10.1080/02723638.2018.1440127

Breetzke, G.D. 2020. Crime and spatiality in South African Cities. In: R. Massey & A. Gunter (eds.). *Urban Geography in South Africa: Perspectives and Theory.* Cham: Springer. 155–172. https://doi.org/10.1007/978-3-030-25369-1_10

Clarke, R.V. 1983. Situational crime prevention: Its theoretical basis and practical scope. *Crime and Justice*, 4: 225–256. https://doi.org/10.1086/449090

Dervin, B. 1983. An overview of sense-making research: Concepts, methods, and results to date. International Communications Association Annual Meeting Proceedings. 1– 68.

Dervin, B. 1992. From the mind's eye of the user: The sense-making qualitative-quantitative methodology. In: J.D. Glazier & R.R. Powell (eds.). *Qualitative Research in Information Management*. Englewood, CO: Libraries Unlimited. 61–84.

Dervin, B. 1996. Sense-making theory and practice: An overview of user interests in knowledge seeking and use. *Journal of Knowledge Management*, 2(2): 36–46. https://doi.org/10.1108/13673279810249369

Global Organized Crime Index. 2023. Crime rates by country. https://globalinitiative.net/analysis/ocindex-2023/ [15 January 2025].

Goldstein, H. 1979. Improving policing: A problem-oriented approach. *Crime & Delinquency*, 25(2): 236–258. https://doi.org/10.1177/001112877902500207

Green, D.P. 2000. Community policing in the United States: An evaluation of community safety programmes. *Journal of Policy Analysis and Management*, 19(3): 453–474.

Hinkle, J.C., Weisburd, D., Telep, C.W. & Petersen, K. 2020. Problem-oriented policing for reducing crime and disorder: A systematic review. *Campbell Systematic Reviews*, 16(1): 1–86. https://doi.org/10.1002/cl2.1089

Kappeler, V.E., Gaines, L. K., & Schaefer, B.P. 2020. *Community Policing: A Contemporary Perspective*. Eighth edition. London: Routledge.

Kitteringham, G. & Fennelly, L.J. 2020. *Handbook of Loss Prevention and Crime Prevention*. Sixth edition. Waltham, MA: Butterworth-Heinemann. https://doi.org/10.1016/C2019-0-00006-4

Kruger, T., Lancaster, L., Landman, K., Liebermann, S., Louw, A. & Robertshaw, R. 2016. *Making South Africa Safe: A Manual for Community-Based Crime Prevention*. Pretoria: The Council for Scientific and Industrial Research.

Machethe, P. & Mofokeng, J.T. 2022. The impact of illicit drug networks on the effectiveness of law enforcement in South Africa. *Technium Social Sciences Journal*, 27: 338–360. https://doi.org/10.47577/tssj.v27i1.5355

Makhetha, T. 2024. Boko Haram extortion gang is fleecing us and we have the receipts, say Mamelodi residents, *News 24,* 22 August. https://www.news24.com/news24/southafrica/news/boko-haram-extortion-gang-is-fleecing-us-and-we-have-the-receipts-say-mamelodi-residents-20240822 [31 January 2025].

Manamela, M.S., Mokwena, R.J. & Matlala, M.M. 2024. Police intervention strategies in the enforcement of liquor policies in South Africa. *Pakistan Journal of Criminology*, 16(3), July-September: 1225–1240. https://doi.org/10.62271/pjc.16.3.1225.1240

Mazibuko, C.N. & Umejesi, I. 2019. The public accounts of a 'private' act: Domestic violence in the eyes of Mamelodi, a South African Township. *Multidisciplinary Journal of Gender Studies*, 8(1): 48–76. https://doi.org/10.17583/generos.2019.3307

Mnguni, T. 2021. The role of community policing forums in addressing crime in Mamelodi. Unpublished master's dissertation. Pretoria: University of Pretoria.

National Academies of Sciences, Engineering and Medicine. 2018. *Proactive Policing: Effects on Crime and Communities.* Washington, DC: The National Academies Press. https://doi.org/10.17226/24928

National Research Council. 2004. *Fairness and Effectiveness in Policing: The Evidence.* Washington, DC: The National Academies Press. https://doi.org/10.17226/10419

Nkwenyane, E.B. 2011. Community participation in the establishment of community policing forum: A case study of Nelspruit police station, Mpumalanga province. Master's dissertation. Polokwane: University of Limpopo.

Perry, W.L., McInnis, B., Price, C.C., Smith, S.C. & Hollywood, J.S. 2013. *Predictive Policing: The Role of Crime Forecasting in Law Enforcement Operations*. RAND Corporation. https://doi.org/10.7249/RR233 [18 January 2025].

Randa, M.B., McGarry, J., Griffiths, S. & Hinsliff-Smith, K. 2023. Accessing care services after sexual violence: A systematic review exploring experiences of women in South Africa. *Curationis*, 46(1). https://doi.org/10.4102/curationis.v46i1.2405

Ross, E. & Rasool, S. 2019. 'You go to campus with fear and come back with fear': University students' experiences of crime. *South African Crime Quarterly*, 68(June): 7–20. https://doi.org/10.17159/2413-3108/2019/v0n68a4895

SAPS [South African Police Service]. 2023. *Annual Crime Statistics 2022/2023 Document*. https://www.saps.gov.za/services/downloads/2022-2023-Annual-Crime-Statistics-Report.pdf [18 January 2025].

SAPS [South African Police Service]. 2024. *Police recorded crime statistics, Republic of South Africa: 2023-2024 Financial year - Annual figures (April 2023 to March 2024).* https://www.saps.gov.za/services/downloads/2024/2023-2024%20_Annual_Financial%20year_WEB.pdf

SAPS [South African Police Service]. 2025a. *Crime statistics: integrity.* https://www.saps.gov.za/services/crimestats.php.

SAPS [South African Police Service]. 2025b. Contact us. Province: Gauteng. https://www.saps.gov.za/contacts/provdetails.php?pid=1

Sherman, L.W. & Eck, J.E. 2002. Policing for crime prevention. In: L.W. Sherman, D.P. Farrington, B.C. Welsh & D.L. MacKenzie (eds.). *Evidence-Based Crime Prevention*. London: Routledge. 295–329. https://doi.org/10.4324/9780203166697_chapter_8

Statista Research Department. 2025. Crime worldwide - statistics & facts. https://www.statista.com/topics/780/crime/#topicOverview [21 January 2025].

Statistics South Africa. 2020. *Governance, Public Safety, and Justice Survey GPSJS 2019/20.* Pretoria: Government Printers.

Taylor, R.B., Koper, C.S. & Woods, D.J. 2011. Randomized experiments for police decision-making: The challenges of implementing the evidence-based policing agenda. *Criminal Justice and Behavior*, 38(5): 493–509. https://doi.org/10.1177/0093854811399389

Todd, C.R., Hamilton, J.R. & Cadora, E. 2011. *Community Justice.* New York, NY: Routledge.

Tuffin, R., Morris, J. & Poole, A. 2006. *An Evaluation of the Impact of the National Reassurance Policing Programme.* London: Home Office Research.

Weisburd, D. & Eck, J.E. 2004. What can police do to reduce crime, disorder, and fear? *The Annals of the American Academy of Political and Social Science*, 593(1): 42–65. https://doi.org/10.1177/0002716203262548

Weisburd, D., Telep, C.W., Hinkle, J.C. & Eck, J.E. 2010. Is problem-oriented policing effective in reducing crime and disorder? *Criminology & Public Policy*, 9(1): 139–172. https://doi.org/10.1111/j.1745-9133.2010.00617.x

Wilson, J.Q. & Kelling, G.L. 1982, March. The police and neighbourhood safety: Broken windows. *Atlantic Monthly*, 29–38.

Zimbardo, P.G. 1969. The human choice: Individuation, reason, and order versus deindividuation, impulse, and chaos. *Nebraska Symposium on Motivation*, 17: 237–307.

CHAPTER 8

Evolving approaches to human security: Geographic insights into contemporary challenges

A.R. Govender and I. Henrico

Abstract

Human security has evolved to encompass a range of complex and interconnected challenges, extending beyond traditional military threats to include climate change, migration, environmental degradation, and armed conflict. This study examines the pivotal role of geography in addressing these challenges by integrating geospatial technologies, tools, and data into security frameworks. Through a literature review supplemented by visual examples from global platforms, this study explores how geospatial technologies enhance disaster preparedness, conflict monitoring, environmental security, and migration management. The findings highlight the capacity of geographic information systems (GIS) and remote sensing techniques to provide spatial intelligence, improve early warning systems, and support data-driven decision-making, thereby facilitating more effective responses to transnational security threats in the future. Furthermore, this study underscores the importance of multilateral collaboration, illustrating how international organisations operationalise geospatial tools, such as the Urban Adaptation Map Viewer and the UN Geospatial Hub, to strengthen global security efforts. As human security continues to evolve, this study affirms that geospatial technologies are not merely supplementary tools but essential components in shaping adaptive and resilient security strategies in the 21st century.

Keywords: conflict monitoring, geographic information systems (GIS), geospatial technologies, human security, multilateral security cooperation, remote sensing

Introduction

Human security is an evolving, multidimensional concept that extends beyond the traditional focus on military threats (Soomro, 2011). Unlike conventional state-centric security, which prioritises national defence and sovereignty, human security shifts the focus to individuals and communities, addressing their wellbeing, dignity, and survival. This comprehensive approach integrates economic, environmental, social, and political dimensions to address diverse and interconnected global challenges. Contemporary threats such as climate change, transnational migration, pandemics, and resource conflicts necessitate innovative security strategies that go beyond military solutions.

Geography plays a pivotal role in advancing human security by providing a range of analytical approaches and methodologies that help in understanding and mitigating security threats. Traditional methods for addressing human security challenges include political analysis, economic modelling, social vulnerability assessments, and risk management frameworks. These approaches have been instrumental in shaping policies and interventions aimed at safeguarding the health of individuals and communities in the region. However, as global security threats become increasingly complex and interconnected, geographic techniques, particularly geospatial studies, have emerged as critical tools for enhancing human security. Geospatial studies, encompassing geographic information systems (GIS) and remote sensing, provide powerful capabilities for analysing and visualising security-related data through spatial modelling (Soomro, 2011; Longley et al., 2015). GIS facilitates the collection, storage, processing, analysis, and representation of geographic information (Chang, 2008), enabling the identification of vulnerabilities, monitoring of crisis zones, and detection of spatial patterns of security threats (Longley et al., 2015). Remote sensing, through satellite imagery and aerial reconnaissance, offers real-time and historical data that enhance situational awareness and inform strategic decision-making. These geospatial tools allow for more effective security-related interventions by integrating spatial intelligence into security planning and response.

Military geography, a specialised subfield of geography, further strengthens human security by applying geospatial techniques within the defence and security contexts. It plays a crucial role in military operations, disaster response, infrastructure protection, and strategic planning by assessing terrain, mapping conflict zones, and evaluating the spatial distribution of threats in an area. Military geographers support security efforts by analysing troop movements, modelling logistical routes, and identifying high-risk areas for conflict or humanitarian crises. By bridging the gap between geospatial sciences and security strategies, military geography contributes to the development of adaptive and data-driven solutions to contemporary human security challenges.

This study examines the significance of geography in addressing evolving human security challenges, with particular emphasis on the roles of GIS and spatial analysis. In addition, it explores how these tools contribute to mapping threats, identifying vulnerable populations, and supporting sustainable development. Furthermore, it highlights how multilateral organisations and alliances leverage geographic data to address transnational security threats, including climate change, resource scarcity, and migration crises. By integrating geographic insights into security strategies, this study underscores the necessity of flexible approaches to safeguard human security in an increasingly globalised and interconnected world. In this context, this study seeks to answer the following question: How can geospatial technologies enhance human security in response to evolving global security challenges?

Methodology

To ensure a structured and rigorous approach, the methodology adopted for this research is outlined in the following sections, detailing the research design, data collection, thematic analysis, and synthesis of the findings (Figure 8.1).

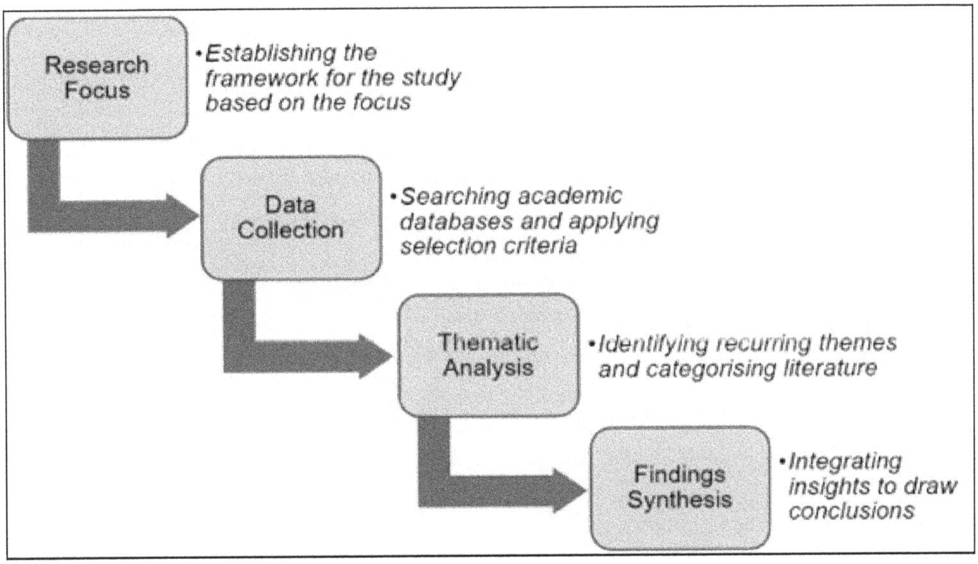

Figure 8.1: Methodology flow diagram.

Research approach and design

This study employs a qualitative research approach, utilising a systematic literature review to examine the evolving role of geography in human security. Qualitative methodology is particularly suited to this research as it enables the integration of diverse academic perspectives, policy frameworks, and empirical case studies to provide a comprehensive understanding of how geographic tools contribute to security.

This study follows an integrative review framework that allows for a structured synthesis of the literature from multiple disciplines, including geography, remote sensing, security studies, political science, and environmental science. This approach ensures a rigorous evaluation of the theoretical underpinnings of human security, the practical applications of geospatial technologies, and the role of multilateral institutions in implementing GIS-based security measures.

Data collection process

This study relied exclusively on secondary data sources, including peer-reviewed journal articles, books, policy documents, and institutional reports. A structured, systematic search strategy was applied across multiple academic databases and institutional repositories to ensure a comprehensive and methodologically rigorous review of the literature.

Search strategy and databases

The literature search was conducted using the databases and repositories listed in Table 8.1.

Table 8.1: Databases and repositories used in the literature search

Database/source	Purpose
JSTOR	Theoretical perspectives on human security
Google Scholar	Broad academic search across disciplines
ScienceDirect	Recent GIS and remote sensing applications in security
SpringerLink	Geography and environmental security research
Web of Science	High-impact studies on spatial security analysis
IEEE Xplore	Engineering and GIS applications in disaster management
United Nations (UN) Reports	Policy integration of GIS in human security
World Bank, IFRC Reports	Case studies on GIS applications in security and humanitarian contexts
Political Geography Journal	Political and geopolitical dimensions of human security

The search strategy incorporated key Boolean operators to refine the search results. These queries ensured a broad yet focused collection of relevant studies across multiple themes of interest. The following search terms were applied in different combinations: "human security" AND "geography" AND "GIS", "spatial analysis" OR "environmental security", "military geography" AND "conflict mapping", "disaster risk reduction" AND "remote sensing", and "geospatial intelligence" AND "security policy".

Inclusion and exclusion criteria

To ensure that only high-quality and relevant literature was included, the following inclusion and exclusion criteria were applied (Table 8.2).

Table 8.2: Inclusion and exclusion criteria applied to ensure the selection of high-quality, relevant literature for the study

Criteria	Inclusion	Exclusion
Language	English publications	Non-English articles
Time frame	1994–2023 (to include foundational theories and recent applications)	Studies older than 1994, unless foundational
Content focus	GIS, remote sensing, geography, human security applications, disaster management, migration studies	Solely military-focused studies without a human security perspective
Publication type	Peer-reviewed journal articles, books, institutional reports from the UN, World Bank, IFRC	Non-peer-reviewed blogs, Wikipedia, opinion pieces

The expanded timeframe (1994–2024) was included to accommodate foundational theoretical work on human security (e.g. Alkire, 2003; UNDP, 1994) and recent applications of GIS in security (e.g. Eid et al., 2024; Tomaszewski, 2020; Ngetar & Rwandarugali, 2022). Reports from international organisations, such as the UN, World Bank, and IFRC, were included because of their significance in shaping real-world security strategies.

Data analysis methods

A thematic content analysis approach was applied to synthesise the findings from the literature. This method allowed us to identify recurring themes, conceptual patterns, and relationships within the selected studies. The literature review and analysis were structured around six key thematic areas to ensure a comprehensive examination of geography's contribution to human security. These themes encompass theoretical perspectives and practical examples to highlight the application of geospatial tools and techniques across various security threats such as climate change, migration, and armed conflict.

The six themes identified in this study are as follows:

(1) The first theme explores the evolution of human security, tracing its development from traditional state-centric models to more inclusive, people-centred approaches. This thematic area highlights how human security has expanded to encompass economic stability, health, environmental sustainability, and political inclusion, reflecting broader security challenges beyond military ones alone.

(2) The second thematic area focuses on GIS in disaster management, examining how geospatial tools are employed to map vulnerabilities, enhance disaster preparedness, and support response efforts. By leveraging remote sensing and spatial analysis, GIS enables real-time tracking of natural disasters, such as hurricanes, earthquakes, and floods, thereby improving decision-making in crisis situations.

(3) Closely related is the theme of environmental security and climate change, which investigates the role of GIS in assessing environmental risks, monitoring climate-related hazards, and facilitating resource management. This section highlights the crucial role of geographic tools in identifying regions at risk of climate-induced displacement, desertification, and water scarcity, which increasingly contribute to human insecurity in the region.

(4) Another critical area of analysis is geospatial technologies in armed conflict monitoring, which explores how geospatial technologies are used to track conflict zones, analyse patterns of violence and assess post-conflict damage. The application of geospatial technologies in this context has proven essential for humanitarian organisations, military strategists, and peacekeeping missions in conflict-affected regions, helping document human rights violations and support reconstruction efforts.

(5) The theme of geospatial applications in migration addresses the use of spatial technologies to track displacement patterns, refugee movements, and border security management. GIS provides essential data for international agencies and humanitarian organisations, enabling more effective planning for displaced populations and resource allocation in crisis-affected areas.

(6) Finally, this study examines multilateral security efforts and analyses how international organisations incorporate geospatial frameworks into security planning and decision-making. This thematic area discusses the role of geographic intelligence in shaping global security frameworks, supporting diplomatic cooperation, and enhancing the effectiveness of multilateral responses to transnational threats.

Findings and thematic analysis

The findings of this study are structured around six key themes that illustrate the evolving role of geography in human security. Each theme explores a specific aspect of how GIS and geospatial technologies contribute to the understanding and addressing of human security challenges.

Theme 1 – Human security: A changing paradigm

The concept of human security represents a fundamental shift from traditional state-centric security models, focusing on the protection of individuals and communities (UNDP, 1994). Unlike conventional national security approaches, human security addresses economic stability, environmental sustainability, public health, and political inclusion, reflecting a broader range of challenges beyond military threats (Fukuda-Parr & Messineo, 2012).

Historically, security policies have prioritised military defence and sovereignty, particularly during the Cold War period (Owen, 2008). However, the post-Cold War period, coupled with the rise of globalisation, transnational threats, and complex intrastate conflicts, necessitated a more holistic perspective (Alkire, 2003). Internal conflicts, ethnic violence, terrorism, and environmental crises demonstrate that human security risks often emerge within states, linked to economic inequality, resource scarcity, and human rights violations (Bocchi et al., 2006; Fukuda-Parr & Messineo, 2012; Owen, 2008).

The UNDP (1994) identified "freedom from fear" and "freedom from want" as the core tenets of human security, encompassing protection from violence and access to education, healthcare, natural disasters, and sustainable resources. Given the increasing interconnectivity of global challenges, human security now requires adaptive and collaborative approaches that integrate governmental and multilateral organisational efforts (Fukuda-Parr & Messineo, 2012).

Theme 2 – Geospatial technologies in disaster management and contributions to human security

Discussions of human security often focus on political, economic, and social aspects, frequently overlooking the vital role of geography. However, geography is fundamental to understanding and addressing human security threats, as spatial relationships influence resource distribution, disaster vulnerability, conflict patterns, and access to essential services. One of the most direct contributions of geography to human security is environmental security, where geospatial technologies and techniques play a key role in assessing risks such as climate change, deforestation, desertification, and water scarcity (Convergne & Snyder, 2015; Coskun, 2008).

Geospatial technologies enable role players to anticipate vulnerabilities and implement proactive mitigation strategies, which are fundamental in disaster management, conflict monitoring, health care, and sustainable development planning, highlighting their growing significance in human security frameworks (Convergne & Snyder, 2015; McLafferty, 2003). These technologies, tools, and data play critical roles in risk assessment, early warning systems, disaster response coordination, and post-disaster recovery efforts (Tomaszewski, 2020). According to Tomaszewski (2020), geospatial software such as GIS enhances the real-time monitoring of natural disasters, including hurricanes, earthquakes, and floods (Kundu & Kundu, 2011), by integrating spatial data from remote sensing, satellite imagery, and ground-based sensors. This integration significantly improves the decision-making capabilities of governments, humanitarian agencies, and emergency responders. The ability to map high-risk areas and infrastructure vulnerabilities ensures efficient disaster preparedness planning and more effective disaster relief operations.

In flood vulnerability mapping, Isma'il and Saanyol (2013) demonstrated how GIS and remote sensing were applied to assess flood-prone regions along the Kaduna River, Nigeria. Their study emphasised how GIS enhances the ability to predict flood events, model water level changes, and develop emergency response strategies. By using geospatial analysis, hydrological data, and population distribution maps, GIS allows emergency planners to prioritise high-risk communities and allocate resources accordingly (Kundu & Kundu, 2011). Examples where GIS has been extensively deployed in disaster-prone regions for various applications are listed in Table 8.3.

Table 8.3: Examples of GIS deployment in disaster-prone regions for various applications

Description
Following the 2010 Haiti earthquake, GIS was used in the relief efforts as location was central to assessing danger, resources, safety, and communications, identifying affected areas, and optimising relief distribution (Clark et al., 2010: Ahmed & Mitchell, 2023). The use of satellite imagery and GIS overlays allowed humanitarian organisations to quickly identify the hardest-hit areas and coordinate aid efforts effectively.
During the 2011 Tōhoku earthquake and tsunami, GIS played a crucial role in damage assessment, resource allocation, and evacuation planning (Marx & Goward, 2013). By integrating seismic data, real-time satellite imagery, and population density maps, GIS supported disaster response teams in prioritising rescue operations and rebuilding efforts.
GIS is routinely used for hurricane tracking and evacuation planning, improving the management of hurricanes such as Hurricane Katrina in 2005 and Hurricane Harvey in 2017 (Shad, 2006). GIS applications enable storm path prediction, flood risk analysis (Kundu & Kundu, 2011), and emergency shelter planning, ensuring that communities receive real-time updates and accurate evacuation routes.

The United Nations Office for the Coordination of Humanitarian Affairs (OCHA) (2014) underscores the importance of integrating GIS with socio-economic data to enhance disaster early warning systems. By mapping high-risk populations, critical infrastructure, and available resources, GIS provides emergency responders with actionable intelligence that enables a more coordinated and timely response to crises such as earthquakes.

The following visual examples (Figures 8.2 to 8.4), sourced from open-access publications, showcase the diverse applications of GIS and geospatial data in enhancing human security, ranging from disaster response and refugee movement tracking to climate risk assessments.

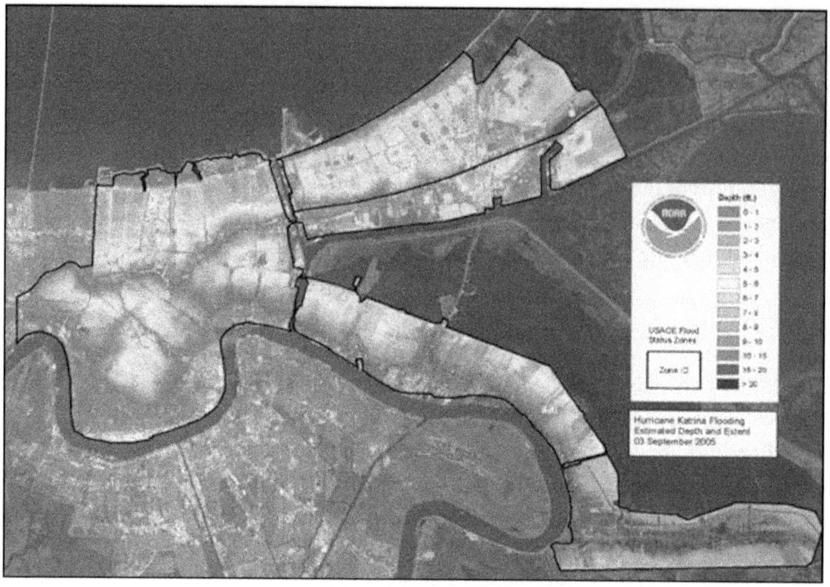

Figure 8.2: LIDAR-derived map showing the estimated flood depth and extent in New Orleans following Hurricane Katrina on 3 September 2005.
(*Source:* ©ESA, data courtesy of NOAA and USACE; https://www.esa.int [accessed 23 March 2025])

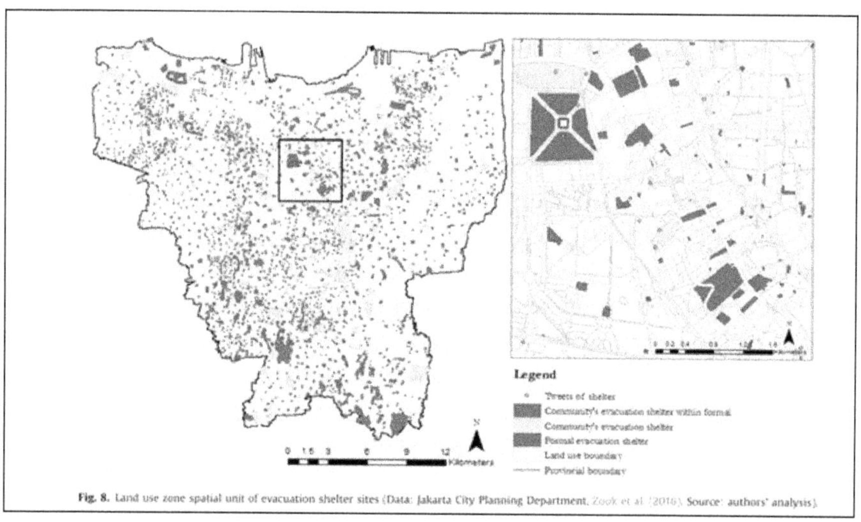

Figure 8.3: Spatial distribution of evacuation shelter sites in Jakarta based on land use zoning and social media data. The map combines formal and community evacuation shelters with tweets of shelter locations overlaid on land use data. The inset map shows a zoomed-in urban detail.
(*Source:* Kusumo et al. (2017). Reproduced under CC BY-NC-ND 4.0)

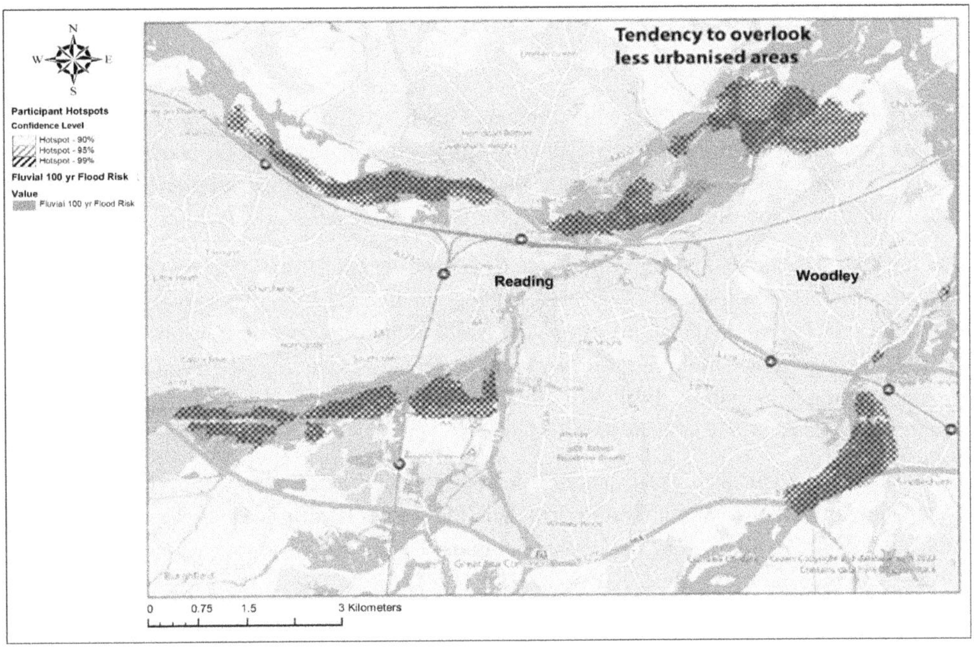

Figure 8.4: Statistically significant flood risk hotspots in Reading, UK, derived from the participants' responses using a participatory GIS approach. Hotspots were calculated based on areas identified by over 20% of the participants, overlaid with the 100-year fluvial flood risk zones.

(*Source:* Modified from Bullen and Miles (2024). Reproduced under CC BY-NC-ND 4.0)

Theme 3 – Environmental security and climate change

Environmental threats, such as climate change, natural disasters, starvation, and resource scarcity, pose significant risks to human security, especially in coastal and arid regions, and require sufficient environmental management programmes to address them (Droj, 2012). These challenges can exacerbate existing vulnerabilities, leading to conflicts over resources, displacement, and socio-economic instability (Coskun, 2008). Scheffran and Battaglini (2011) highlighted that climate-induced environmental changes can act as "threat multipliers", intensifying underlying sociopolitical tensions and impacting human livelihoods.

GIS has been instrumental in assessing environmental vulnerabilities and food security (Kumar et al., 2024), and enhancing global disaster preparedness (Table 8.4). By integrating spatial data with environmental and socio-economic parameters, GIS facilitates the identification of high-risk areas, supports early warning systems, and informs mitigation strategies (Longley et al. 2015).

Table 8.4: Examples of GIS applications in environmental vulnerability assessments

Description
Nobre et al. (2007) conducted a study utilising GIS-based risk index mapping to assess groundwater vulnerability in Brazil. By analysing factors such as soil properties, land use, and hydrogeological characteristics, the study identified regions susceptible to contamination, providing valuable insights for groundwater management and protection.
Guragain (2015) developed an Earthquake Risk Assessment System to enhance disaster preparedness and response in Nepal. The study compared HAZUS, OpenQuake, and CAPRA GIS for national and city-level risk assessments, focusing on fragility functions for Nepalese masonry buildings. GIS and remote sensing technologies played a key role in damage mapping, vulnerability assessments, and post-earthquake reconstruction efforts, ensuring more data-driven disaster management.
Isma'il and Saanyol (2013) utilised GIS for flood vulnerability assessment in Nigeria. By analysing topographical and hydrological data, the study identified flood-prone areas along the River Kaduna, supporting the implementation of effective flood management and mitigation measures (Kundu & Kundu, 2011).
A study in Turkey (Şahin & Kurum, 2002) employed GIS to analyse erosion risk in the Seyhan-Köprü Dam construction area. By assessing factors like slope, soil type, and land use, the research identified areas with higher erosion risk, contributing to environmental impact assessments and informing conservation strategies.
A GIS-based groundwater vulnerability assessment was conducted in the Midlands and South East of England (Hiscock et al., 1995). The study integrated data on soil properties, geology, and land use to create vulnerability maps, aiding in the protection of groundwater resources from potential contamination.
In the US, GIS techniques, software, and data are regularly used to support human and environmental security efforts. One such study highlights how geospatial analysis aids environmental protection and conflict mitigation, demonstrating the critical role of GIS in enhancing community resilience and managing security risks (Morrow et al., 2022).

By mapping high-risk areas and integrating topographical and climatic data, GIS enhances early warning systems and improves disaster preparedness efforts, ultimately contributing to the resilience and security of vulnerable communities in the Philippines. For instance, Rahman et al. (2015) demonstrated that combining spatial analysis with hazard and vulnerability assessments enables the identification of priority areas for intervention, thereby reducing potential disaster impacts.

The following visual examples (Figures 8.5 and 8.6) illustrate how GIS-based spatial analysis, when combined with hazard mapping and vulnerability data, supports evidence-based decision-making for disaster risk reduction. These cases highlight the practical value of integrating geospatial datasets to prioritise interventions, strengthen early warning systems, and safeguard at-risk populations.

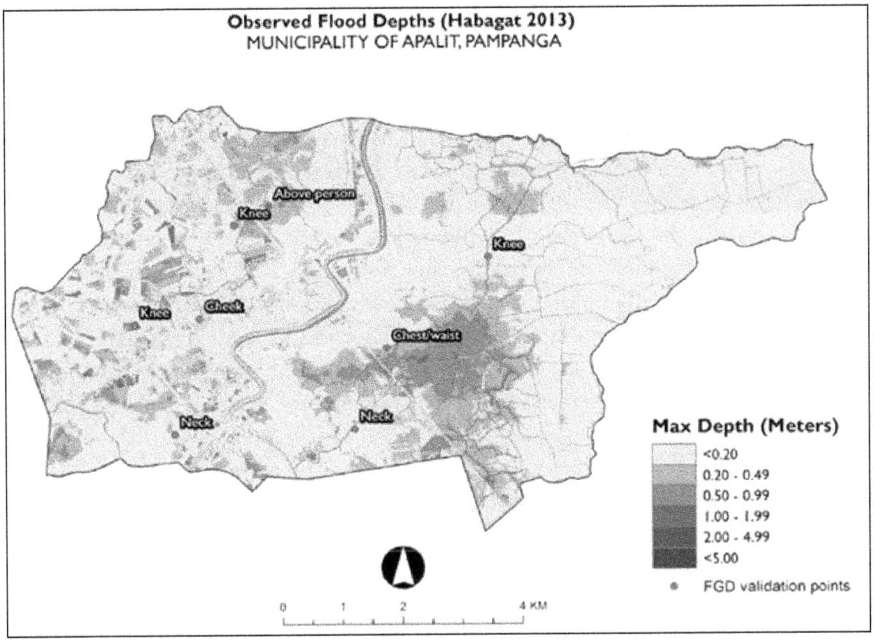

Figure 8.5: Flood depths observed during the Habagat 2013 monsoon in Apalit, Pampanga, Philippines, as reported by local participants through focus group discussions. The map shows spatial variations in the maximum flood depth and includes ground validation points used in the flood modelling process.
(*Source:* Toda et al., 2017. Reproduced under CC BY-NC-ND 4.0)

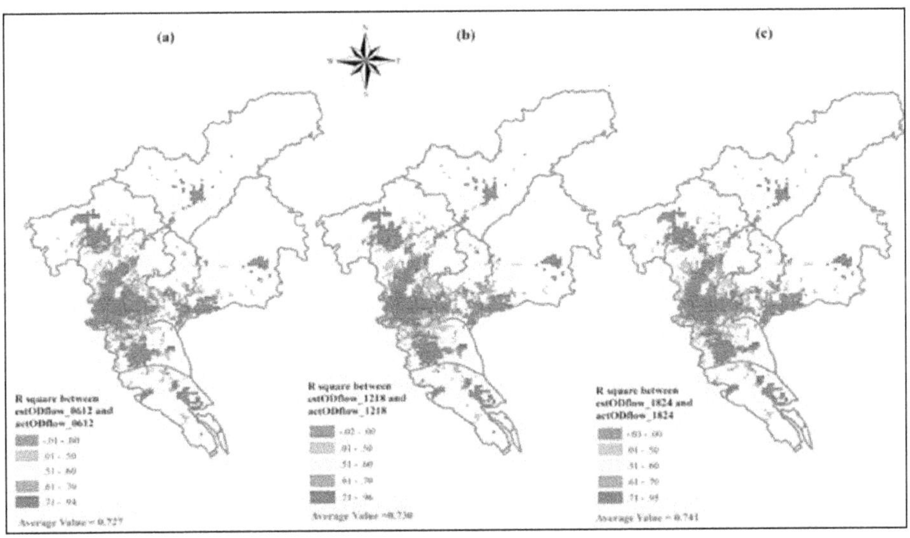

Figure 8.6: Spatial distribution of R^2 values comparing estimated and actual origin-destination (OD) flows using a gravity model across three time intervals in Guangzhou. The maps depict the model performance for intra-city travel from 06:00–12:00 (a), 12:00–18:00 (b), and 18:00–24:00 (c), highlighting the spatial variations in prediction accuracy.
(*Source:* Zhou et al., 2022. Reproduced under CC BY-NC-ND 4.0)

Theme 4 – Geospatial technologies for armed conflict monitoring

Geospatial technologies for spatial analysis and modelling have become indispensable tools for creating geospatial intelligence (GEOINT) in support of armed conflict monitoring (Ngetar & Rwandarugali, 2022), offering tactical and strategic insights for military operations, humanitarian aid, and post-conflict reconstruction (Convergne & Snyder, 2015). These technologies provide real-time spatial data that enable decision-makers, security analysts, and humanitarian organisations to track the evolution of conflicts, assess their impact on civilian populations, and develop early warning systems for conflict prevention. The application of geospatial technologies in armed conflict monitoring extends across multiple conflict zones worldwide, providing critical insights into military activities, civilian displacement, infrastructure damage, and humanitarian crises (Convergne & Snyder, 2015). Table 8.5 presents case studies in which geospatial technologies have been effectively used to monitor, analyse, and respond to armed conflicts.

Table 8.5: Applications of geospatial technologies in armed conflict monitoring

Description
Geospatial technologies have been pivotal in monitoring the conflict in Ukraine, providing critical insights into military activities, damage assessments, and humanitarian impacts (Yailymova et al., 2023: Butenko & Petrychenko, 2023).
In the Rohingya conflict, remote sensing has been utilised to detect and quantify conflict-related damages, aiding in human rights monitoring (Ren et al., 2020).
The Armed Conflict Location & Event Data Project (ACLED) (2022) employs geospatial data to track political violence and protests worldwide, offering valuable insights into conflict patterns. Some of these examples are: • Mapping military confrontations and civilian displacement in Syria (Ngetar & Rwandarugali, 2022). • Tracking escalating violence and refugee movements in Sudan and the Sahel region (Eid et al., 2024). • Assessing frequency and intensity of armed conflicts in East Africa (Seck, 2023).
Remote sensing and GIS have been used as counterterrorism tools in Afghanistan, providing valuable geospatial intelligence for military operations.
Satellite data has been used to assess the resilience of Tigrayan farmers in crop cultivation during the civil war, providing insights into agricultural impacts amidst conflict.
The FAO used Sentinel-1 SAR imagery and a baseline agricultural infrastructure layer to estimate agricultural infrastructure damage in Gaza by identifying sites within 15 metres of mapped damage on the 26 September 2025 proxy map. (Food and Agriculture Organisation of the United Nations (FAO), 2025).

Building on the real-world applications of geospatial technologies in conflict monitoring, it is essential to highlight the key capabilities that make these technologies indispensable for modern security and humanitarian operations. From real-time conflict mapping and damage assessment to predictive modelling and human rights monitoring, geospatial tools provide critical insights that enhance situational awareness, guide decision-making, and support intervention efforts in volatile regions. Examples of the core functionalities of GIS and remote sensing in tracking, analysing, and mitigating the impacts of armed conflicts include:

- **Real-time conflict mapping and event analysis:** GIS enables the collection, analysis, and visualisation of spatio-temporal data on conflicts, allowing for the interactive mapping of conflict zones. This includes tracking bombings, military engagements, civilian casualties, refugee movements, and humanitarian crises such as disease control (Christaki, 2015; Henrico et al., 2018). GIS databases, such as ACLED and UCDP, provide real-time spatial datasets on global armed conflicts, political violence, and insurgencies.

- **Remote sensing for infrastructure damage and humanitarian impact assessment:** High-resolution satellite imagery and remote sensing technologies allow for damage assessment in conflict zones. By analysing changes in land use, destroyed infrastructure, and displaced populations, remote sensing enables a quantitative assessment of the impact of conflict. For example, freely accessible satellite imagery from remote sensing platforms such as Sentinel-2, Landsat 8/9, and Synthetic Aperture Radar (SAR) has been employed in Ukraine, Syria, the Sahel, Sudan, and Myanmar. Imagery is used to identify urban destruction, monitor battlefront movements, support refugee relief operations, and track genocide in the context of armed conflict (Huang et al., 2023; Aimaiti et al., 2022; Avtar et al., 2021; Witmer, 2015).

- **GEOINT for military and security operations:** GEOINT combines satellite imagery, GPS tracking, and intelligence data to provide strategic insights into military planning, surveillance, and reconnaissance. Intelligence agencies and military forces use GEOINT for battlefield awareness, target detection, and logistical planning. This technology has played a crucial role in tracking insurgent groups, terrorist networks, and non-state actors in conflict zones, such as Afghanistan, the Sahel, and Colombia (Lutema, 2025).

- **Predictive modelling and early warning systems:** Spatial analysis techniques, such as hotspot mapping, machine learning, and Artificial Intelligence (AI)-based predictive models, are increasingly used to forecast the likelihood of future conflicts or humanitarian crises. By analysing variables such as resource competition, political instability, socio-economic disparities, and historical conflict data, GIS-based predictive models can help governments and NGOs allocate resources and implement preventive interventions before violence escalates (Novikarany, 2025; Panic, 2022).

- **Tracking refugee movements and displacement patterns:** Conflicts often result in the mass displacement of civilian populations. GIS and remote sensing assist in mapping migration corridors, identifying temporary refugee settlements and ensuring efficient aid distribution. Organisations such as the United Nations High Commissioner for Refugees (UNHCR), International Organization for Migration (IOM), and Red Cross use spatial data to optimise humanitarian assistance and improve logistics for displaced communities (Curry et al., 2019).

- **Monitoring human rights violations and war crimes:** Satellite imagery and GIS have been instrumental in documenting human rights abuses, war crimes, and violations of international humanitarian law. In regions such as Darfur, Syria, and Ukraine, geospatial evidence has been used to track mass executions, attacks on civilian infrastructure, and the destruction of historical sites, supporting legal proceedings at the International Criminal Court (ICC) and UNHRC (Witmer, 2015).

The integration of geospatial technologies into conflict monitoring has transformed the way military strategists, humanitarian organisations, and policymakers respond to crises. By providing real-time intelligence, predictive insights, and spatial analyses of affected regions, these tools contribute significantly to conflict resolution efforts, post-conflict reconstruction, and the protection of vulnerable populations.

The following visual examples (Figures 8.7 to 8.9) highlight how geospatial technologies have been applied in real-world conflict scenarios to map destruction, monitor troop movements, and assess the spatial dimensions of civilian displacement. These cases demonstrate the value of spatial intelligence in operational military planning and humanitarian responses.

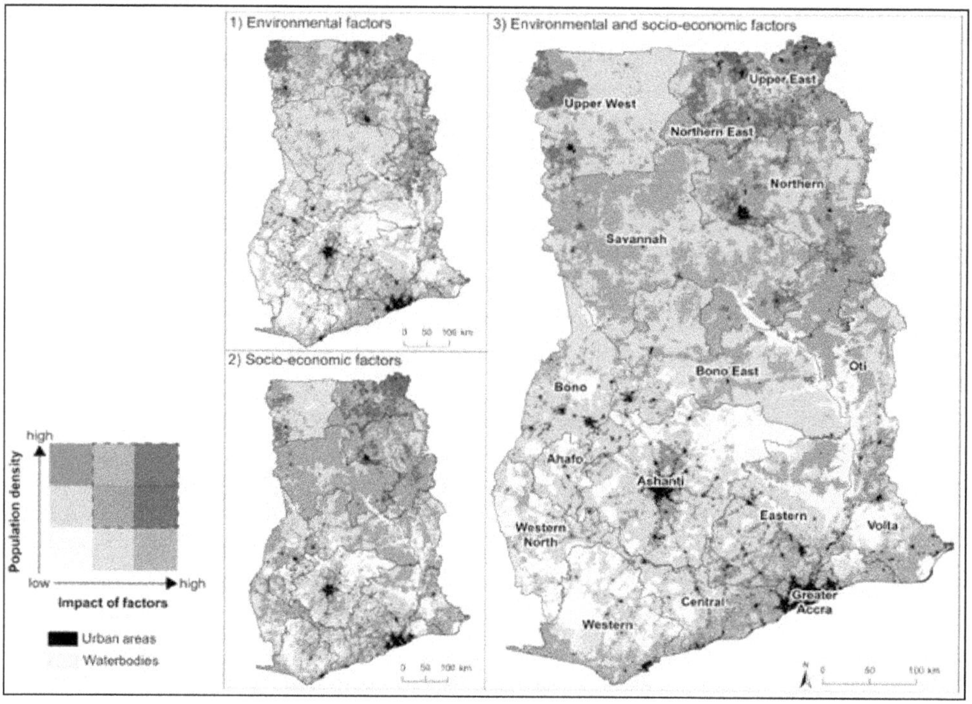

Figure 8.7: Impact maps showing the intersection of rural population densities with (1) environmental factors, (2) socio-economic factors, and (3) combined socio-environmental factors in Ghana. These maps identify areas of potential migration pressure by combining geospatial vulnerability indicators. The dashed box in the legend highlights the colour categories representing a higher likelihood of migration.
(*Source:* Schürmann et al., 2024. Reproduced under the CC BY 4.0 licence)

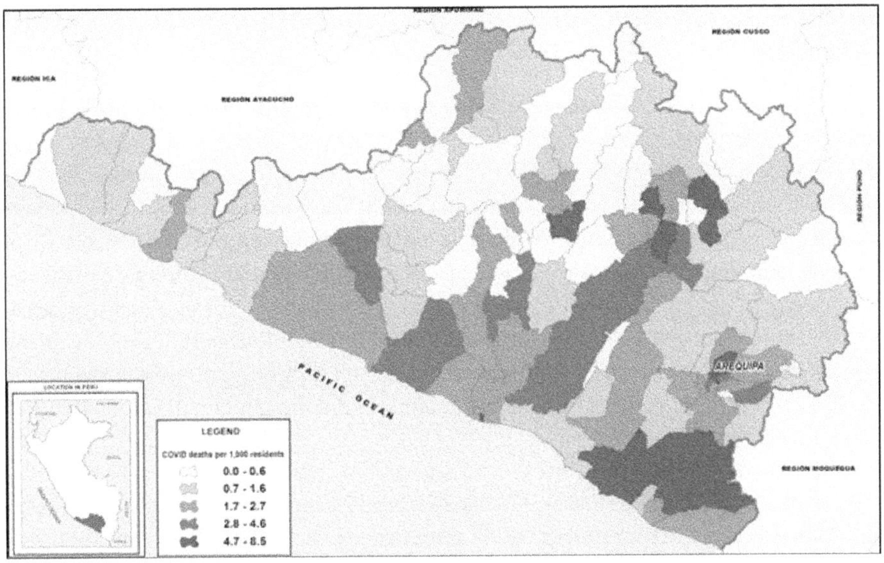

Figure 8.8. Spatial distribution of Covid-19 mortality in the Arequipa region, Peru, indicating cumulative deaths per 1,000 residents by district as of 9 July 2021. The map reveals pronounced geographic variation in mortality, with higher death rates concentrated in the southern and central districts.

(*Source:* Cabana et al., 2021. Reproduced under the CC BY-NC-ND 4.0 licence)

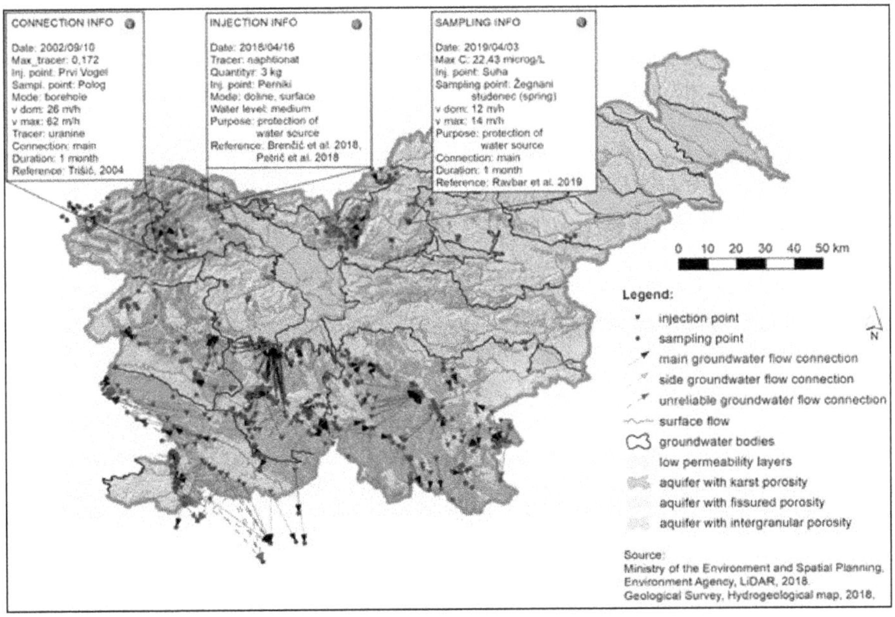

Figure 8.9: Visualisation of the GIS-based tracer test database over a hydrogeological map of Slovenia, showing the injection and sampling points, groundwater flow connections, and aquifer types. The map illustrates the key elements used to model groundwater movement through karst and porous aquifers, supporting source protection and hydrogeological analysis.

(*Source:* Petrič et al., 2020. Reproduced under CC BY-NC-ND 4.0)

Theme 5 – Geospatial applications in migration and displacement monitoring

Geospatial technologies are indispensable for monitoring and managing migration crises worldwide. These tools enable real-time tracking of displaced populations, optimisation of humanitarian responses, and facilitation of policy decisions regarding border security and resource distribution. The integration of geospatial data provides crucial insights into refugee movements, informal settlement growth, and environmental conditions affecting displaced populations, allowing for the visualisation of migration flows and offering detailed insights into displacement patterns and settlement locations (Curry et al., 2019). For instance, during the Syrian conflict, geospatial analysis was instrumental in monitoring refugee movements and camp locations. Quinn et al. (2018) highlighted the use of machine learning with remote sensing data to map and analyse refugee settlements, aiding humanitarian organisations in efficiently directing their efforts.

In sub-Saharan Africa, GIS has been used to assess border security challenges and displacement trends. The African Shifts report (2023) emphasises the role of geospatial analysis in understanding mobility decisions influenced by climate vulnerability, demonstrating how GIS is used to model migration flows across African regions (IOM, 2023). Similarly, in Europe, geospatial technologies have been applied to study social and geographic migration trails. A report by the Economic and Social Research Institute (Esri) explored data gaps in monitoring migrant integration by employing GIS to analyse movement patterns and policy impacts (Esri, 2019). Conversely, remote sensing technologies provide critical insights into the environmental conditions of refugee camps, aiding in resource planning and management. A study published in *Nature* introduced a systematic approach to measure climatic and environmental exposure in refugee camps using remote sensing and geospatial data, which is crucial for improving infrastructure and ensuring the wellbeing of displaced communities (Owen et al., 2023).

Beyond crisis response, many countries are increasingly turning to geospatial technologies as part of their broader efforts to monitor sustainable development and improve governance. Recent work by the World Bank highlights how Earth observation data, remote sensing, and GIS are being embedded in national SDG reporting systems to strengthen evidence-based decision-making (Rabiee, 2023). The United Arab Emirates, for instance, uses satellite imagery and night-time light data to track patterns of urban expansion linked to SDG 11.3.1, while the Hawaiian government has developed a detailed land-cover classification system to monitor carbon dynamics as part of its sustainability commitments. In East Africa, the city of Mombasa applies the Degree of Urbanisation method to identify and map public open-space access in support of SDG 11.7.1, integrating population density thresholds with geospatial datasets to understand local development pressures. These examples show how governments are adopting GIS not only to visualise change but also to produce more transparent, data-driven Voluntary National and Local Reviews. Through digital reporting tools, such as ArcGIS StoryMaps, countries are beginning to present their SDG progress with interactive geospatial layers, making complex spatial patterns more accessible to policymakers and the public alike (Rabiee, 2023).

In recent years, with advances in AI and machine learning applications, geospatial technologies combined with these applications have been instrumental in mapping informal settlements that emerge because of migration. During the Venezuelan migration crisis, researchers used Sentinel-2 satellite imagery and time-series analysis to locate new informal settlements, enabling timely humanitarian intervention (Tingzon et al., 2020). Together with the examples indicated previously, these studies showcase the vital role that these technologies play in migration management, border security, and optimising relief efforts (Table 8.6).

Table 8.6: Geospatial applications in humanitarian, migration, and urban monitoring

Description
During the Syrian conflict, geospatial analysis was instrumental in monitoring refugee movements and camp locations. Quinn et al. (2018) highlighted the use of machine learning with remote sensing data to map and analyse refugee settlements, aiding humanitarian organisations in efficiently directing their efforts.
A study by Tingzon et al. (2020) utilised machine learning and time-series satellite images to map new informal settlements resulting from the Venezuelan migration crisis. This approach effectively identified potential migrant settlements in Colombia that emerged between 2015 and 2020.
Geospatial data is increasingly being used by governments to monitor settlement growth, urban expansion, and environmental change, with satellite imagery, night-time lights, and land-cover analysis supporting national SDG reporting and offering insights relevant to migration dynamics and displacement pressures (Rabiee, 2023).

The following visual examples (Figures 8.10 and 8.11) illustrate how geospatial technologies have been applied to monitor forced migration, detect informal settlements, and identify displacement trends in crisis-affected regions. These maps and models demonstrate the analytical power of remote sensing, GIS, and AI in supporting evidence-based humanitarian responses and migration policy planning.

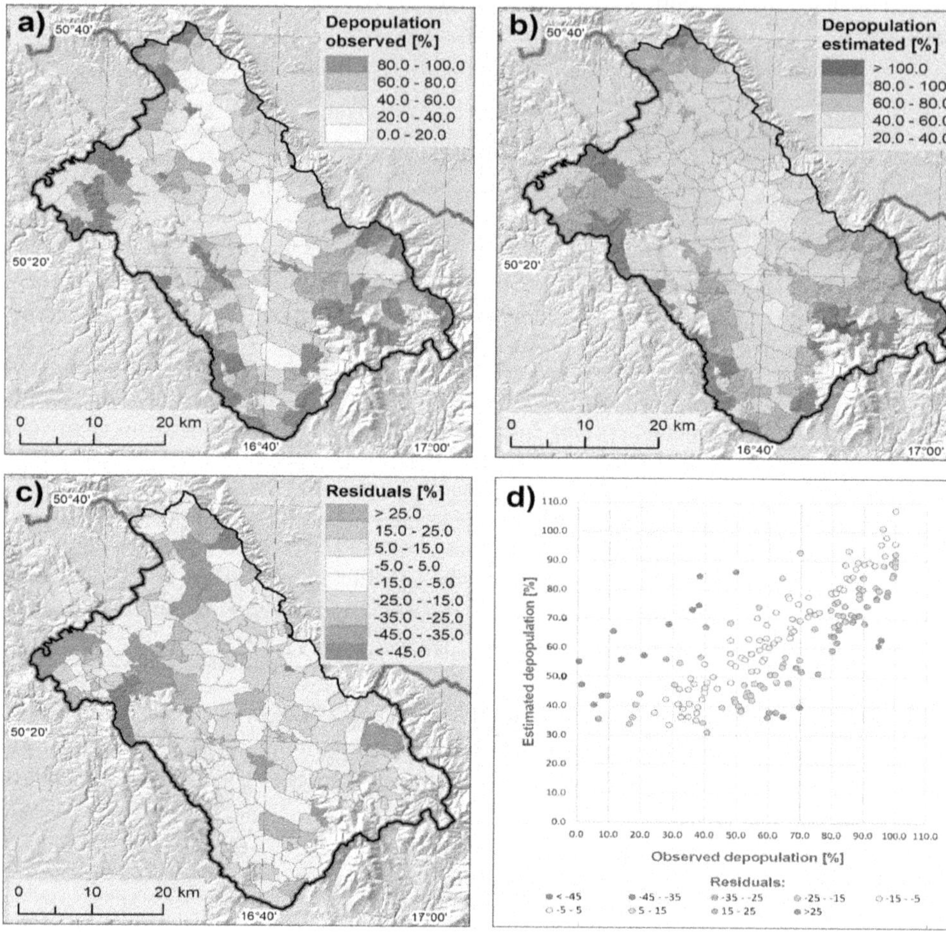

Figure 8.10: Observed (a) and MGWR-modelled (b) depopulation rates in the Kłodzko region, including residuals map (c) and scatter plot of estimated vs. observed values (d), illustrating model fit and spatial variation in depopulation patterns.
(*Source:* Szymanowski and Latocha, 2021. Reproduced under CC BY-NC-ND 4.0)

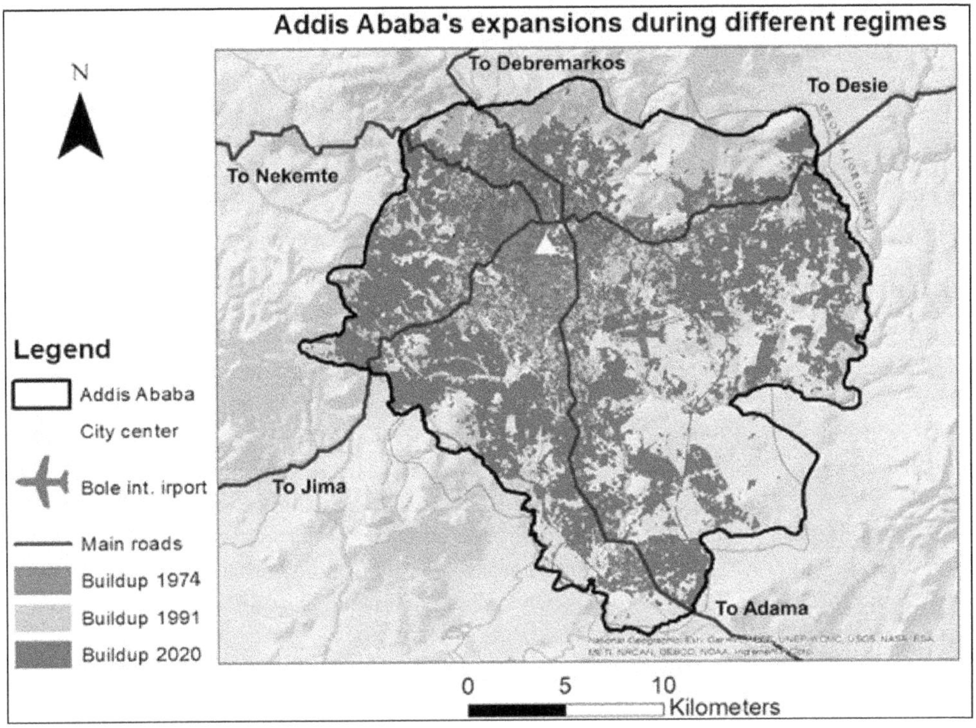

Figure 8.11: Urban expansion of Addis Ababa during the imperialist, socialist, and contemporary regimes (1974, 1991, 2020) mapped using Landsat satellite images. The figure highlights the major patterns of urban growth under shifting political and socio-economic contexts.

(*Source:* Weldegebriel et al., 2023. Reproduced under CC BY-NC-ND 4.0)

Theme 6 – Multilateral security and global collaboration

Geospatial technologies have become integral to multilateral security efforts, enhancing collaboration among governments, humanitarian organisations, and international agencies. These tools facilitate data-driven policy decisions, enabling more effective responses to transnational threats such as climate change, migration, natural disasters, and armed conflicts (Ngetar & Rwandarugali, 2022). The ability to integrate real-time remote sensing data and GIS-based analytics has significantly improved coordination among international stakeholders, ensuring better situational awareness and crisis management.

Enhancing intergovernmental collaboration in disaster response

OCHA (2014) aims to "bring the world together to tackle humanitarian emergencies and save the lives of people caught in crises". One key mechanism for achieving this is the Humanitarian Data Exchange (HDX) platform, which provides open-access data to support geospatial analysis in disaster response. By enabling humanitarian actors to access and share critical spatial datasets, the HDX facilitates real-time decision-making, improves situational awareness, and enhances coordination during crises. The platform hosts a wide range of geospatial data that can be used to map disaster-affected regions and guide global relief efforts, thereby strengthening international cooperation. The World Bank's Global Facility for Disaster Reduction and Recovery (GFDRR) employs geospatial intelligence to assist low- and middle-income countries in understanding and mitigating disaster risks, particularly in fragile and conflict-affected areas (The World Bank, 2020).

One of the most successful examples of GIS-enabled intergovernmental disaster collaboration is the Sendai Framework for Disaster Risk Reduction (2015–2030), which mandates the integration of geospatial technologies into disaster preparedness and response efforts worldwide (UNDRR, 2015). The European Union's (EU) Copernicus Emergency Management Service (EMS) provides satellite-derived maps and damage assessments to international disaster response agencies to aid in real-time crisis monitoring. During the 2023 Turkey-Syria earthquake, the Copernicus EMS delivered high-resolution geospatial imagery that allowed for the rapid assessment of the affected areas, informing relief operations led by the EU Civil Protection Mechanism (Joint Research Centre 2024). In addition to emergency response systems, such as the Copernicus EMS, the European Environment Agency's Urban Adaptation Map Viewer (Figure 8.12) offers a strategic planning tool for visualising climate vulnerabilities and urban risks, such as water scarcity and heat stress. It supports long-term adaptation planning by allowing users to explore geospatial data relevant to urban resilience in Europe.

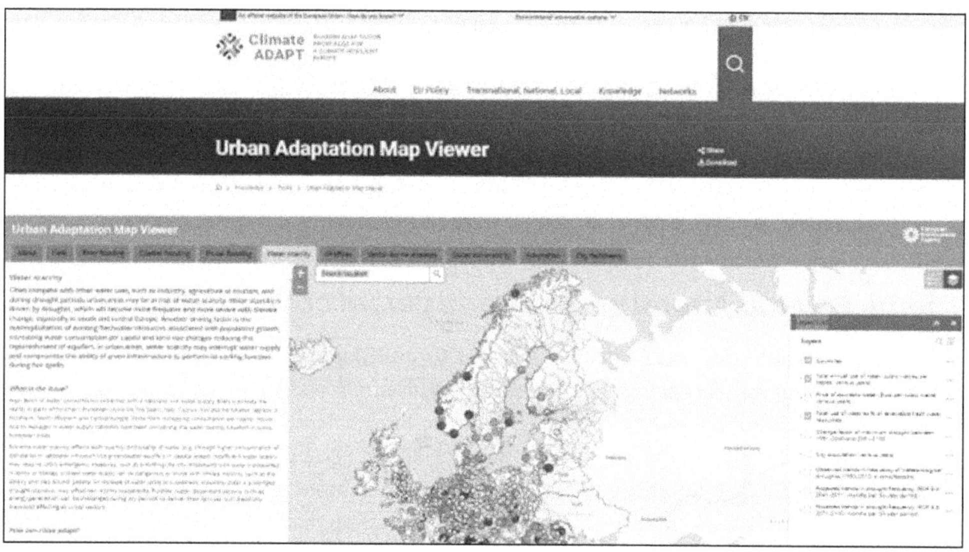

Figure 8.12: Urban Adaptation Map Viewer hosted by the European Environment Agency through the Climate-ADAPT platform. The tool enables users to visualise urban vulnerability and adaptation data across Europe, including indicators related to water scarcity, flooding, and heat stress in urban areas.
(*Source:* European Environment Agency, Urban Adaptation Map Viewer, https://climate-adapt.eea.europa.eu/en/knowledge/tools/urban-adaptation [accessed March 2025])

In support of the Sustainable Development Goals (SDGs), the UN established the UN Geospatial Hub (Figure 8.13), which serves as a global gateway to geospatial data, maps, and web services. This platform consolidates location-based information to guide international development, humanitarian actions, and environmental monitoring. It plays a vital role in enabling countries to visualise progress, conduct risk assessments, and strengthen evidence-based decision-making across sectors.

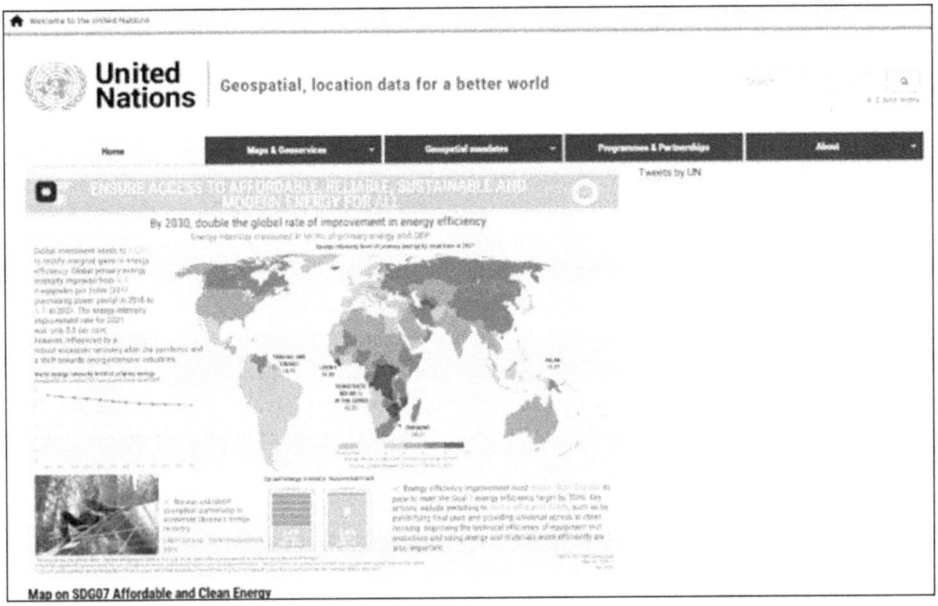

Figure 8.13: Screenshot of the United Nations Geospatial Hub homepage, including a visualisation of SDG 7 progress on energy efficiency, highlighting country-level energy intensity metrics for 2021. The UN Geospatial Hub provides access to geospatial data, maps, and services to support global development, environmental monitoring, and disaster response. This platform facilitates location-based monitoring, aligned with the 2030 Agenda for Sustainable Development.
(*Source:* United Nations Geospatial Hub, 2025, https://www.un.org/geospatial)

Geospatial technologies in migration management and border security

The increasing complexity of forced migration due to conflicts, climate change, and economic instability has made GIS and remote sensing indispensable tools for tracking displacement patterns, identifying border security risks, and supporting humanitarian responses. The International Organization for Migration (IOM) routinely applies GIS to monitor migration corridors and assist displaced populations, while recent research demonstrates how remote sensing and machine learning are being used to map and analyse refugee settlements, allowing humanitarian actors to assess conditions and allocate resources more effectively (Quinn et al., 2018). A notable initiative in EU migration management is the Frontex Risk Analysis Unit, which uses GIS and geospatial intelligence to monitor irregular migration trends and coordinate border security policies among EU member states (Frontex, 2024). GIS-assisted surveillance further helps identify migration hotspots, detect human trafficking routes, and guide policy interventions. In North Africa and the Mediterranean, geospatial intelligence has been instrumental in tracking migration flows from Libya and Tunisia to Southern Europe, providing early warnings for maritime rescue operations and informing border control strategies.

GIS in NATO and EU security policies

Geospatial technologies and data have become increasingly embedded in security policies and military operations in the North Atlantic Treaty Organization (NATO) and the EU member states. NATO has integrated geospatial analysis into its defence planning, conflict monitoring, and situational awareness frameworks. The NATO Communications and Information Agency (NCIA) leverages GIS-based intelligence tools to enhance battlefield awareness, track military assets, and support multinational operations (NATO, 2025b). During NATO's Enhanced Forward Presence (EFP) operations in Eastern Europe, geospatial technologies were used to monitor border security risks, assess terrain features for military deployment, and improve logistics planning. Additionally, in the summary of NATO's data exploitation framework strategic plan, geospatial analysis is used to achieve information superiority and data-driven decision-making (NATO, 2025b). Within the EU, the European Parliament regularly publishes GIS-based intelligence for crisis management, military deployment, and peacekeeping missions (European Parliament, 2025). The EU Satellite Centre (SatCen) provides geospatial intelligence for EU security operations and supports missions in Africa, the Middle East, and Eastern Europe (SatCen, 2025). In peacekeeping contexts, GIS has been instrumental in mapping ceasefire violations, analysing troop movements, and supporting diplomatic negotiations.

The role of geospatial technologies in climate security and international cooperation

Climate change has emerged as a global security threat, with its effects exacerbating resource conflicts, migration crises, and economic instability. Geospatial technologies, including GIS and remote sensing, are crucial for international cooperation in climate adaptation and mitigation strategies, as they enable the monitoring of environmental changes, support data-driven policy decisions, and foster coordinated global responses (Pirasteh and Varshosaz, 2019). The Paris Agreement (2015) and the European Green Deal have incorporated GIS-based monitoring systems to track carbon emissions, assess deforestation trends, and ensure compliance with climate policies (European Union, 2024). Geospatial analysis has also been used in conflict-prone regions, such as the Sahel and the Horn of Africa, where water scarcity and desertification are key drivers of instability. In collaboration with NASA, the UN Environment Programme (UNEP), and regional climate bodies, geospatial tools and techniques have been deployed to monitor drought conditions, predict agricultural losses, and assist communities in adapting to climate-related challenges in the region. Table 8.7 presents several examples of how GIS and remote sensing have been incorporated into global security and collaboration frameworks.

Table 8.7: Integration of GIS and remote sensing in multilateral security and global collaboration

Organisation / Initiative	Description
Copernicus Emergency Management Service (EMS)	The EU's Copernicus EMS provides satellite-derived maps and real-time disaster assessments, helping international agencies coordinate crisis monitoring and emergency response. It played a key role in the 2023 Turkey-Syria earthquake relief efforts (Joint Research Centre, 2024).
Remote Sensing & Machine Learning for Refugee Mapping	Uses satellite imagery and machine learning techniques to detect, map, and monitor refugee settlements, enabling humanitarian organisations to assess conditions, track population movements, and allocate resources more effectively (Quinn et al., 2018).
Frontex Risk Analysis Unit (EU Border Security Agency)	Applies geospatial intelligence to monitor migration trends, identify human trafficking routes, and support EU member states in border security coordination (Frontex, 2024).
NATO Deterrence and defence	Integrates geospatial analysis in defence planning and military operations, particularly in Eastern Europe and conflict-prone regions. GIS tools are used for battlefield awareness, troop movement analysis, and logistics planning (NATO, 2025a).
Paris Agreement & European Green Deal	Uses GIS-based monitoring systems to track carbon emissions, deforestation, and climate change impacts. Geospatial data informs global climate policies and adaptation strategies (European Union, 2024).

The geospatial technologies contribute to understanding and addressing human security challenges

From the six themes highlighted and described in this study, the following framework (Figure 8.14) highlights the interconnected nature of geospatial technologies, such as GIS and remote sensing, which serve as the central framework supporting various aspects of human security. Five interconnected domains were identified – disaster management, conflict monitoring and peacekeeping, environmental security, migration and displacement, and multilateral security and collaboration – to demonstrate the wide-ranging applications of spatial analysis in crisis response, security planning, and policy development. The arrows indicate the direct role of geospatial technologies in facilitating decision-making and enhancing resilience to global security challenges.

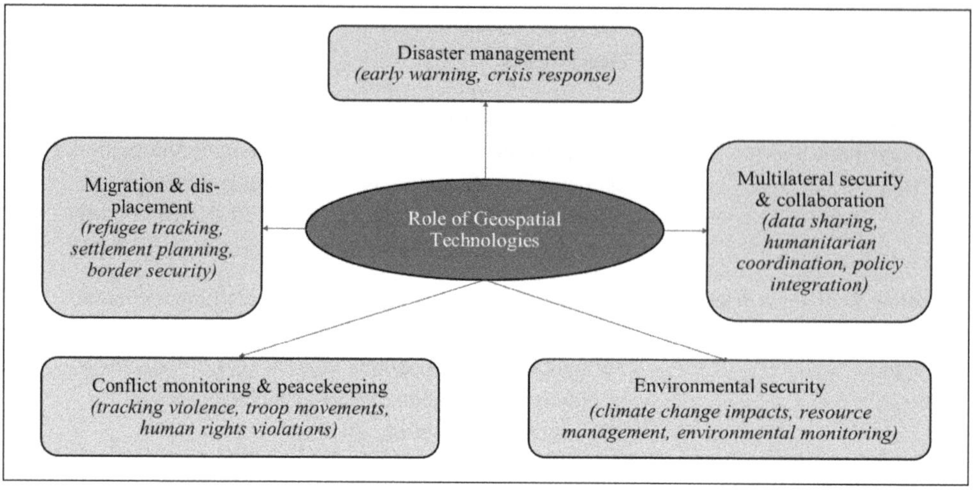

Figure 8.14: The role of geospatial technologies in human security.

Conclusion

This study demonstrates the pivotal role of geography and geospatial technologies in the evolving approaches to human security. By shifting from traditional state-centric security models to people-centred frameworks, geospatial technologies for spatial analysis have emerged as essential tools for understanding and addressing multidimensional security threats. This study illustrates how geography provides critical insights into conflict monitoring, disaster management, environmental security, migration patterns, and multilateral security efforts, enabling more proactive, data-driven decision-making.

A comparative thematic analysis highlights the diverse applications of GIS in the field of human security. In disaster response, GIS facilitates the mapping of high-risk zones, real-time hazard monitoring, and resource allocation, thereby enhancing preparedness and recovery efforts. In armed conflict monitoring, geospatial intelligence has been instrumental in tracking conflict zones, infrastructure damage, and displacement patterns and supporting peacekeeping and humanitarian interventions. Environmental security has also greatly benefited from GIS, with geospatial tools used to monitor the impacts of climate change, deforestation, and water scarcity and inform conservation and resilience strategies. Additionally, this study demonstrated the effectiveness of GIS in migration management, particularly in tracking refugee movements, settlement planning, and cross-border displacement trends. The integration of geospatial intelligence into multilateral security frameworks further underscores its importance in international collaboration, disaster coordination, and climate change adaptation. As transnational security threats continue to evolve, the need for spatially informed policies and cooperative global responses will become increasingly urgent.

By reinforcing the intersection between geography and security, this study affirms that geospatial technologies are not merely supplementary tools but are fundamental components of modern security strategies. GIS and remote sensing enhance spatial intelligence, predictive analytics, and situational awareness, enabling governments, international organisations, and humanitarian agencies to respond effectively to emerging security challenges. As global security concerns become increasingly complex and interconnected, the role of geospatial intelligence in shaping adaptive, sustainable, and human-centred security strategies will continue to expand. Future advancements in GIS, big data analytics, and AI-driven geospatial modelling will further enhance early warning systems, risk assessments, and crisis response capabilities, making geography an indispensable pillar of human security in the 21st century.

While this study provides valuable insights into the role of geography in human security, it has certain limitations that must be acknowledged. This study relied solely on secondary data, which means that no fieldwork or surveys were conducted. However, the inclusion of multiple case studies ensured empirical validation and strengthened these findings. Another limitation is the geographic scope variability, as case studies span different regions, which may affect the global generalisability of the conclusions. Some thematic areas, such as health security, are less represented in the literature reviewed, leading to potential data gaps in specific aspects of human security. Additionally, there is a technological bias in this study, as it primarily focuses on GIS and remote sensing applications, while not incorporating newer advancements such as AI and machine learning in security analysis. Despite these limitations, this research remains robust in its analysis, offering a comprehensive perspective on the evolving role of geography in human security.

References

ACLED [Armed Conflict Location & Event Data Project]. 2022. *ACLED 2022: The Year in Review.* https://acleddata.com/report/acled-2021-year-review [18 February 2025].

Ahmed, F. & Mitchell, E.H. 2023. Haiti's emergency preparation: Lessons learned from 2010 earthquake for future disasters. https://storymaps.arcgis.com/stories/a0d119fd23df49328f75eee584327554 [24 November 2024].

Aimaiti, Y., Sanon, C., Koch, M., Baise, L.G. & Moaveni, B. 2022. War related building damage assessment in Kyiv, Ukraine, using Sentinel-1 radar and Sentinel-2 optical images. *Remote Sensing*, 14(24). https://doi.org/10.3390/rs14246239

Alkire, S. 2003. A conceptual framework for human security. Working Paper. CRISE (Department of International Development, University of Oxford). https://ora.ox.ac.uk/objects/uuid:d2907237-2a9f-4ce5-a403-a6254020052d

Avtar, R., Kouser, A., Kumar, A., Singh, D., Misra, P., Gupta, A., Yunus, A.P., Kumar, P., Johnson, B.A., Dasgupta, R. & Sahu, N. 2021. Remote sensing for international peace and security: Its role and implications. *Remote Sensing*, 13(3). https://doi.org/10.3390/rs13030439

Bocchi, S., Disperati, S.P. & Rossi, S. 2006. Environmental security: A geographic information system analysis approach – The case of Kenya. *Environmental Management*, 37: 186–199. https://doi.org/10.1007/s00267-004-0331-3

Bullen, J. & Miles, A. 2024. Exploring local perspectives on flood risk: A participatory GIS approach for bridging the gap between modelled and perceived flood risk zones. *Applied Geography*, 163. https://doi.org/10.1016/j.apgeog.2023.103176

Butenko, Y. & Petrychenko, S. 2023. Monitoring and assessment of the scale of destruction by remote sensing methods during the war in Ukraine. International Conference of Young Professionals "GeoTerrace-2023". European Association of Geoscientists & Engineers, 1–5. https://doi.org/10.3997/2214-4609.2023510105

Cabana, Y.D.C., Malone, A., Zeballos, E.Z., Huaranca, N.O.H., Tinta, M.T., Beltrán, S.A.G., Arosquipa, A.A. & Pinedo, D. 2021. Pandemic response in rural Peru: Multi-scale institutional analysis of the COVID-19 crisis. *Applied Geography*, 134. https://doi.org/10.1016/j.apgeog.2021.102519

Chang, K.-T. 2008. *Introduction To Geographic Information Systems.* Boston, MA: McGraw-Hill.

Christaki, E. 2015. New technologies in predicting, preventing and controlling emerging infectious diseases. *Virulence*, 6: 558–565. https://doi.org/10.1080/21505594.2015.1040975

Clark, A., Holliday, P., Chau, R., Eisenberg, H. & Chau, M. 2010. Collaborative geospatial data as applied to disaster relief: Haiti 2010. In: T. Kim, W. Fang, M.K. Khan, K.P. Arnett, H. Kang & D. Ślęzak (eds.). *Security Technology, Disaster Recovery and Business Continuity*. International Conferences, SecTech and DRBC 2010, Held as Part of the Future Generation Information Technology Conference, FGIT 2010, Jeju Island, Korea, December 13–15, 2010. Proceedings. Heidelberg: Springer. 250–258.

Convergne, E. & Snyder, M. 2015. *Geospatial Technology as a Conflict Prevention and Management Tool in UN Peacekeeping*. International Peace Institute (IPI).

Coskun, H. 2008. An integrated approach for environmental security in NATO countries based on remote sensing and GIS technologies. In: H. Gonca Coskun, H. Kerem Cigizoglu & M. Derya Maktav (eds.). *Integration of Information for Environmental Security*. Dordrecht: Springer. 3–15. https://doi.org/10.1007/978-1-4020-6575-0_1

Curry, T., Croitoru, A., Crooks, A. & Stefanidis, A. 2019. Exodus 2.0: Crowdsourcing geographical and social trails of mass migration. *Journal of Geographical Systems*, 21: 161–187. https://doi.org/10.1007/s10109-018-0278-1

Droj, G. 2012. GIS and remote sensing in environmental management. *Journal of Environmental Protection and Ecology*, 13.

Eid, M.H., Kamel, M.S., Kamel Amer, A.S. & Szűcs, P. 2024. Spatial prediction of armed conflicts from the perspective of political geography using bivariate frequency ratio method (FR) in East African States. *Heliyon*, 10. https://doi.org/10.1016/j.heliyon.2024.e38684

Esri [Economic and Social Research Institute]. 2019. *Data for monitoring integration: Gaps, challenges and opportunities*. https://www.esri.ie/publications/data-for-monitoring-integration-gaps-challenges-and-opportunities [21 January 2025].

European Environment Agency. 2025. Urban Adaptation Map Viewer. https://climate-adapt.eea.europa.eu/en/knowledge/tools/urban-adaptation [23 March 2025].

European Parliament. 2025. Fact sheets on the European Union. https://www.europarl.europa.eu/factsheets/en/indexsearch?query=geographical+information&scope=ALL&ordering=RELEVANCE [3 January 2025].

European Space Agency. 2025. Data courtesy of NOAA and USACE. https://www.esa.int [23 March 2025].

European Union. 2024. *The European Green Deal: How open data supports green transition*, European Data Portal, 13 December. Available at: https://data.europa.eu/en/news-events/news/european-green-deal-how-open-data-supports-green-transition [Accessed: 5 December 2025].

FAO [Food and Agriculture Organisation of the United Nations]. 2025. Damage to agricultural infrastructure due to the conflict in the Gaza Strip as of 26 September 2025. https://openknowledge.fao.org/items/b4504d3b-53d8-47ea-a97c-3bb8d9f5722b. [23 October 2024].

Frontex. 2024. Who we are. https://www.frontex.europa.eu/about-frontex/who-we-are/tasks-mission/ [24 January 2025].

Fukuda-Parr, S. & Messineo, C. 2012. Human Security: A critical review of the literature. Centre for Research on Peace and Development (CRPD) Working Paper, 11: 1–19. https://doi.org/10.4337/9781781006313.00009

Guragain, R. 2015. *Development of earthquake risk assessment system for Nepal.* Doctoral dissertation. Tokyo: The University of Tokyo.

Henrico, S.J., Coetzee, S. & Cooper, A.K. 2018. Analysing the practical feasibility of FOSSGIS in military operations – A Cholera outbreak use case. *OSGeo Journal*, 16(1): 21–35. https://doi.org/10.7275/R59C6VMT

Hiscock, K., Lovett, A., Brainard, J. & Parfitt, J. 1995. Groundwater vulnerability assessment: Two case studies using GIS methodology. *Quarterly Journal of Engineering Geology and Hydrogeology*, 28: 179–194. https://doi.org/10.1144/GSL.QJEGH.1995.028.P2.09

Huang, Q., Jin, G., Xiong, X., Ye, H. & Xie, Y. 2023. Monitoring urban change in conflict from the perspective of optical and SAR satellites: The case of Mariupol, a city in the conflict between RUS and UKR. *Remote Sensing*, 15(12). https://doi.org/10.3390/rs15123096

IOM [International Organisation for Migration]. 2023. *African Shifts: The Africa Climate Mobility Report: Addressing Climate-Forced Migration & Displacement.* https://environmentalmigration.iom.int/sites/g/files/tmzbdl1411/files/documents/2023-03/African%20Shifts%20Report.pdf [15 January 2025].

Isma'il, M. & Saanyol, I.O. 2013. Application of remote sensing (RS) and geographic information systems (GIS) in flood vulnerability mapping: Case study of River Kaduna. *International Journal of Geomatics and Geosciences*, 3: 618–627.

Joint Research Centre (European Commission). 2024. *M7.8 and M7.5 Earthquakes in Türkiye and Syria: JRC scientific analysis, Update#2, 08 Feb 2023 19:00 UTC* (EUR 31882 EN). Publications Office of the European Union. https://op.europa.eu/en/publication-detail/-/publication/f391e021-e5bc-11ee-8b2b-01aa75ed71a1/language-en [3 December 2025].

Kumar, Y., Kumari, R. & Singh, N. 2024. Application of GIS to ensuring food security. *Food and Scientific Reports*, 5(12): 18–22.

Kundu, A. & Kundu, S. 2011. Flood vulnerability assessment using participatory GIS approach. *Journal of Remote Sensing & GIS*, 2: 8–22.

Kusumo, A.N.L., Reckien, D. & Verplanke, J. 2017. Utilising volunteered geographic information to assess resident's flood evacuation shelters. Case study: Jakarta. *Applied Geography*, 88: 174–185. https://doi.org/10.1016/j.apgeog.2017.07.002

Longley, P.A., Goodchild, M.F., Maguire, D.J. & Rhind, D.W. 2015. *Geographic Information Science and Systems*. Hoboken, NJ: John Wiley & Sons.

Lutema, E.K. 2025. *The Role of Geospatial Intelligence in Modern Military Operations*. https://d197for5662m48.cloudfront.net/documents/publicationstatus/247058/preprint_pdf/8dee55f7d9dc330c9d2367aa642620af.pdf [12 January 2025].

Marx, A. & Goward, S. 2013. Remote sensing in human rights and international humanitarian law monitoring: Concepts and methods. *Geographical Review*, 103: 100–111. https://doi.org/10.1111/j.1931-0846.2013.00188.x

McLafferty, S.L. 2003. GIS and health care. *Annual Review of Public Health*, 24: 25–42. https://doi.org/10.1146/annurev.publhealth.24.012902.141012

Morrow, N., Mock, N.B., Gatto, A., Lemense, J. & Hudson, M. 2022. Protective pathways: Connecting environmental and human security at local and landscape level with NLP and geospatial analysis of a novel database of 1500 project evaluations. *Land*, 11. https://doi.org/10.3390/land11010123

NATO. 2025a. *Joint Intelligence, Surveillance and Reconnaissance*. NATO (North Atlantic Treaty Organization) and OTAN (Organisation du traité de l'Atlantique Nord). https://www.nato.int/en/what-we-do/deterrence-and-defence/joint-intelligence-surveillance-and-reconnaissance 13 February 2025].

NATO. 2025b. *Summary of NATO's Data Exploitation Framework Strategic Plan*. https://www.nato.int/cps/en/natohq/official_texts_209999.htm?selectedLocale=en [13 February 2025].

Ngetar, S.N. & Rwandarugali, S. 2022. The application of Geographical Information Systems to armed violent conflicts: A literature review. *South African Journal of Geomatics*, 11. https://doi.org/10.4314/sajg.v11i2.5

Nobre, R., Rotunno Filho, O., Mansur, W., Nobre, M. & Cosenza, C. 2007. Groundwater vulnerability and risk mapping using GIS, modelling and a fuzzy logic tool. *Journal of Contaminant Hydrology*, 94: 277–292. https://doi.org/10.1016/j.jconhyd.2007.07.008

Novikarany, R. 2025. Geoint as a driver of national security. *International Journal of Scientific Multidisciplinary Research*, 3(2): 237–250. https://doi.org/10.55927/ijsmr.v3i2.66

OCHA [United Nations Office for the Coordination of Humanitarian Affairs]. 2014. Mapaction and OCHA build mapping skills for better disaster response. https://www.unocha.org/publications/report/world/mapaction-and-ocha-build-mapping-skills-better-disaster-response-enru [12 December 2024].

Owen, M., Kruczkiewicz, A. & Van Den Hoek, J. 2023. Indexing climatic and environmental exposure of refugee camps with a case study in East Africa. *Scientific Reports*, 13. https://doi.org/10.1038/s41598-023-31140-7

Owen, T. 2008. Measuring human security: Methodological challenges and the importance of geographically referenced determinants. In: P.H. Liotta, D.A. Mouat, W.G. Kepner & J.M. Lancaster (eds.). *Environmental Change and Human Security: Recognizing and Acting on Hazard Impacts*. Dordrecht: Springer. 35–64. https://doi.org/10.1007/978-1-4020-8551-2_3

Panic, **B. 2022.** *Can Emerging Technologies Lead a Revival of Conflict Early Warning/Early Action? Lessons from the Field*. New York, NY: Center on International Cooperation.

Petrič, M., Ravbar, N., Gostinčar, P., Krsnik, P. & Gacin, M. 2020. GIS database of groundwater flow characteristics in carbonate aquifers: Tracer test inventory from Slovenian karst. *Applied Geography*, 118. https://doi.org/10.1016/j.apgeog.2020.102191

Pirasteh, S. & Varshosaz, M. 2019. Geospatial information technologies in support of disaster risk reduction, mitigation and resilience: Challenges and recommendations. In: A. Rajabifard (ed.). *Sustainable Development Goals Connectivity Dilemma: Land and Geospatial Information for Urban and Rural Resilience*. Boca Raton: CRC Press. 93–108. https://doi.org/10.1201/9780429290626-6

Quinn, J.A., Nyhan, M.M., Navarro, C., Bromley, L.S. & Lall, R. 2018. Humanitarian applications of machine learning with remote sensing data: Review and case study in refugee settlement mapping. *Philos Trans A Math Phys Eng Sci*, 376(2128). http://dx.doi.org/10.1098/rsta.2017.0363

Rabiee, M. 2023. *How are countries and cities using geospatial data to monitor their SDGs?* World Bank Blogs. https://blogs.worldbank.org/en/opendata/how-are-countries-and-cities-using-geospatial-data-monitor-their-sdgs [5 December 2025].

Rahman, N., Ansary, M.A. & Islam, I. 2015. GIS based mapping of vulnerability to earthquake and fire hazard in Dhaka city, Bangladesh. *International Journal of Disaster Risk Reduction*, 13: 291–300. https://doi.org/10.1016/j.ijdrr.2015.07.003

Ren, C.X., Calef, M.T., Durieux, A.M., Ziemann, A. & Theiler, J. 2020. On the detectability of conflict: A remote sensing study of the Rohingya conflict. 2020 IEEE Southwest Symposium on Image Analysis and Interpretation (SSIAI). 86–89. https://doi.org/10.1109/SSIAI49293.2020.9094608

Şahin, Ş. & Kurum, E. 2002. Erosion risk analysis by GIS in environmental impact assessments: A case study – Seyhan Köprü Dam construction. *Journal of Environmental Management*, 66: 239–247. https://doi.org/10.1006/jema.2002.0574

SatCen. 2025. Working for EU's common foreign and security policy. https://www.satcen.europa.eu/ [15 February 2025].

Scheffran, J. & Battaglini, A. 2011. Climate and conflicts: The security risks of global warming. *Regional Environmental Change*, 11: 27–39. https://doi.org/10.1007/s10113-010-0175-8

Schürmann, A., Kleemann, J., Teucher, M. & Conrad, C. 2024. Mapping socio-environmental vulnerability to assess rural migration in Ghana. *Applied Geography*, 167. https://doi.org/10.1016/j.apgeog.2024.103283

Seck, N. 2023. The impact of digitalization on humanitarian aid: A systematic review about the use of GIS mapping in refugee camp management in Africa. Student essay. Enschede: University of Twente.

Shad, K.U.R. 2006. Disaster management and GIS. 2006 International Conference on Advances in Space Technologies. 6–10. https://doi.org/10.1109/ICAST.2006.313787

Soomro, T.R. 2011. Non-traditional security and role of GIS. *International Journal of Computer Science and Network Security,* 11(11).

Szymanowski, M. & Latocha, A. 2021. Does the environment matter? Depopulation in the Sudetes (case study of the Kłodzko region, SW Poland). *Applied Geography*, 135. https://doi.org/10.1016/j.apgeog.2021.102535

The World Bank. 2020. *Global Facility for Disaster Reduction and Recovery (GFDRR) – Bringing resilience to scale: Annual report 2020.* https://documents.worldbank.org/en/publication/documents-reports/documentdetail/099623504282325517 [7 December 2025].

Tingzon, I., Dejito, N., Flores, R.A., de Guzman, R., Carvajal, L., Erazo, K.Z., Cala, I.E.C., Villaveces, J., Rubio, D. & Ghani, R. 2020. Mapping new informal settlements using machine learning and time series satellite images: An application in the Venezuelan migration crisis. 2020 IEEE/ITU International Conference on Artificial Intelligence for Good (AI4G). 198–203. https://doi.org/10.1109/AI4G50087.2020.9311041

Toda, L.L., Yokingco, J.C.E., Paringit, E.C. & Lasco, R.D. 2017. A LiDAR-based flood modelling approach for mapping rice cultivation areas in Apalit, Pampanga. *Applied Geography*, 80: 34–47. https://doi.org/10.1016/j.apgeog.2016.12.020

Tomaszewski, B. 2020. *Geographic Information Systems (GIS) for Disaster Management*. London: Routledge. https://doi.org/10.4324/9781351034869

UNDP [United Nations Development Programme]. 1994. *Human Development Report, 1994: New Dimensions of Human Security*. https://hdr.undp.org/content/human-development-report-1994 [28 November 2024].

UNDRR [United Nations Office for Disaster Risk Reduction]. 2015. *Sendai Framework for Disaster Risk Reduction 2015–2030*. https://www.undrr.org/publication/sendai-framework-disaster-risk-reduction-2015-2030 [28 November 2024].

Weldegebriel, A.T., Tekalign, M. & Van Rompaey, A. 2023. Socio-spatial analysis of regime shifts in Addis Ababa's urbanisation. *Applied Geography*, 154. https://doi.org/10.1016/j.apgeog.2023.102918

Witmer, F.D. 2015. Remote sensing of violent conflict: Eyes from above. *International Journal of Remote Sensing*, 36(9): 2326–2352. https://doi.org/10.1080/01431161.2015.1035412

Yailymova, H., Yailymov, B., Kussul, N. & Shelestov, A. 2023. Geospatial analysis of life quality in Ukrainian rural areas. 2023 13th International Conference on Dependable Systems, Services and Technologies (DESSERT). IEEE. 1–5. https://doi.org/10.1109/DESSERT61349.2023.10416517

Zhou, S., Zhou, S., Zheng, Z., Lu, J. & Song, T. 2022. Risk assessment for precise intervention of COVID-19 epidemic based on available big data and spatio-temporal simulation method: Empirical evidence from different public places in Guangzhou, China. *Applied Geography*, 143. https://doi.org/10.1016/j.apgeog.2022.102702

SECTION 3

Military, Maritime, and Security Infrastructure (Planning and Decision-Making)

This section explores the application of geospatial intelligence in managing infrastructure and enhancing strategic decision-making in the military, nuclear, and maritime domains. The chapters showcase how spatial modelling and surveillance technologies can be operationalised to mitigate risks, improve service delivery, and secure borders.

Chapter 9 models the potential impact zones surrounding the Koeberg Nuclear Power Plant using Esri ArcGIS buffer and proximity tools. Drawing on international safety regulations (IAEA and NNR), this study identifies population clusters and critical facilities (e.g. hospitals) within the 5 km emergency zone, providing essential planning inputs for disaster response and evacuation modelling.

Chapter 10 presents a case study of utility infrastructure mapping at the South African Military Academy. By integrating GPS data collection with ArcGIS mapping, this chapter visualises key assets, including pipelines, fire hydrants, and sewerage systems, enhancing spatial awareness and streamlining utility management for campus operations.

Chapter 11 focuses on maritime surveillance in Saint Helena Bay using multisource integration of Sentinel-1 Synthetic Aperture Radar (SAR) and an Automatic Identification System (AIS). The chapter demonstrates how this fusion enables the detection of suspicious maritime activity even when transponders are disabled, offering a scalable model for monitoring illegal fishing and marine trafficking.

These chapters emphasise how geospatial technologies can enhance operational foresight, risk prevention, and tactical resource management in the security sector.

CHAPTER 9

Geospatial analysis of risk and consequences of an explosion incident at Koeberg Nuclear Power Station

T.L. Molefe and D. Putter

Abstract

Nuclear power plays a key role in the world's energy landscape, praised for its capacity to produce large amounts of electricity with small greenhouse gas emissions. Despite its advantages, the possible dangers linked to nuclear power in the event of a major accident raise serious concerns. Among these dangers, a nuclear power plant explosion stands out as a dangerous and complex threat that could affect people's health, the environment, and the stability of society and the economy. This study aims to map and visualise the places that are vulnerable to a nuclear radiation disaster around the Koeberg Nuclear Power Station in the event of a nuclear incident. The research utilised Esri ArcMestablish various buffer zones around the plant, based on the International Atomic Energy Agency (IAEA) and National Nuclear Regulatory (NNR). A statistical summary will be employed based on measures of central tendency to estimate the affected population percentages and the number of hospitals within the study area. The results revealed that two towns are situated within a 5 km radius of the emergency planning zones, impacting approximately 3,387 people of the Western Cape's population.

Keywords: buffer zones, International Atomic Energy Agency (IAEA), measures of central tendency, National Nuclear Regulatory (NNR), nuclear explosion, proximity model

Introduction

The Koeberg Nuclear Power Station (henceforth referred to as Koeberg), located in the Western Cape, South Africa, stands as a testament to the country's ingenuity in the field of nuclear energy technology. Built in 1976, Koeberg began operating with Unit 1 synchronised to the grid in April 1984, followed by Unit 2 in July. As Africa's only nuclear power station, it is more than just a facility; it is an important part of the region's identity. Featuring pressurised water reactors (PWR) and the largest turbine generators in the Southern Hemisphere, it is the southernmost nuclear power station in the world (Eskom, 2025). Koeberg is a structure that represents a crucial turning point in South Africa's energy journey.

As the nation's energy needs grew, so did the frequency of power outages and load shedding, particularly in the 1960s. Communities faced planned blackouts, which sparked the idea of harnessing nuclear power to meet the overwhelming demand. In response, the government set out to develop a nuclear power plant, and construction of Koeberg began in 1976. This engineering project became a vision for a more stable energy future for millions of South Africans (Schwikowski, 2017).

Koeberg is currently at a crossroads owing to the expiration of its operational licence. The utility company, Eskom, seeks to extend the plant's operating licence beyond its initial 40-year lifespan. The replacement of steam generators reflects a significant shift in the company's history and commitment to maintaining a steady energy supply. Eskom submitted this request to the National Nuclear Regulator (NNR) for an additional 20 years. This extension was approved on 15 July 2024 (KAA, 2024). Currently, Koeberg is set to operate for 60 years instead of the originally planned 40 years, which is consistent with the Integrated Resource Plan 2023. The NNR has approved the Unit 1 extension, and Unit 2 is expected to receive similar approval before its current license ends. Operating life extensions are a well-established practice internationally, with approximately 120 nuclear reactors successfully operating beyond their initial 40-year lifespan, and some even receiving an additional 20-year extension to their operating periods (Ramdass, 2024). This extension highlights Koeberg's vital role in providing reliable baseload power while meeting stringent international safety standards. Eskom has operated Koeberg safely for 40 years, investing significantly in safety improvements and extensive maintenance (Ramdass, 2024).

While Koeberg provides a reliable source of baseload electricity vital for Western Cape communities, extending the life of a nuclear facility is not without its concerns. As these plants age, there are growing concerns about the gradual damage and deterioration of critical components, such as reactor pressure vessels and containment structures. These issues increase the potential for safety violations. The accumulation of radioactive waste over the years is another concern (EIA, 2022). The management and disposal of radioactive waste are challenging endeavours that raise environmental and public health concerns. Additionally, whether in a normal or extended life cycle, the risk of human error and organisational failures is always present, underscoring the need for vigilant oversight and regulation to ensure that safety remains a top priority.

Moreover, there are serious concerns among the residents. The Koeberg Alert Alliance[1] has pointed out that Koeberg has not disclosed its emergency response plan for a catastrophic event, raising unsettling questions: Is there a feasible plan to ensure safety in emergencies?[2] These concerns make it more crucial for this study to utilise geospatial tools to analyse the risks and potential impacts of an explosion at Koeberg. The South African government has a responsibility to the communities surrounding Koeberg to better understand its implications and ensure their safety.

Geographic area of interest

Koeberg (33.6745° S, 18.4330° E) is a South African nuclear power station located in the Western Cape province (Figure 9.1), approximately 27 to 30 km north of Cape Town (33.9249° S, 18.4241° E) on the west coast of South Africa. The Atlantic Ocean forms its western border.

1 https://koebergalert.org/

2 A rudimentary plan is posted on the Koeberg webpage: https://www.eskom.co.za/wp-content/uploads/2024/01/2024_-Koeberg_EP_Calendar.pdf

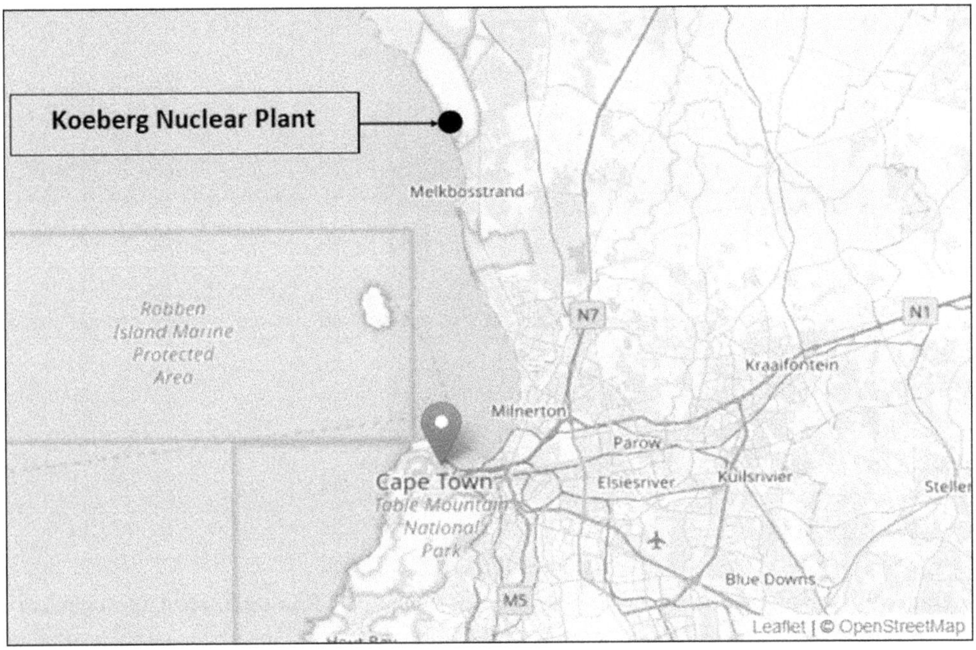

Figure 9.1: Cape Town city and Koeberg Nuclear Power Station showing the adjacent Atlantic seaboard.
(*Source:* Latlong.net, 2025)

Indicated in Figure 9.2 is a map of the study area and the location of Koeberg. Koeberg supplies electricity to the Western Cape. It is surrounded by a nature reserve of approximately 3,000 hectares, which is home to more than 150 fauna species (Eskom, 2025). Koeberg is situated near the residents of Duynefotein, approximately 4.6 km away via the R27 road.

Koeberg is exposed year-round to Mediterranean climatic conditions, with an average wind speed of 24 km/h (15 mph). Koeberg remains South Africa's only nuclear power station, with a total capacity of 1,860 MW, approximately 5% of the electricity generated by Eskom. The two reactors (Units 1 and 2) at Koeberg are the largest nuclear power generating units on the African continent and have been well maintained and safely operated for 40 years (Eskom, 2025). It is the only nuclear power station in Africa and the most southerly nuclear power station in the world.

Figure 9.2: Map of the study area delineated according to NNR criteria. The largest nuclear emergency zone extends 80 km. Consequently, the only Western Cape municipality located in this zone is the City of Cape Town.
(*Source:* Esri, Marxer, Earthstar Geographics, and the User GIS community)

Regulatory framework

Organisations such as the International Atomic Energy Agency (IAEA) provide recommendations and guidelines for emergency planning and response. The IAEA suggests general principles for emergency zones (IAEA, 2007); however, it leaves specific details to the NNR. Each country with a nuclear power plant has a regulatory body responsible for nuclear safety and emergency planning. Therefore, there is a lack of standardisation of nuclear power plant emergency planning zones. The international emergency planning zones around nuclear power plants are determined by a combination of international guidelines and national regulations. For example, in South Africa, the NNR sets regulations for emergency planning zones (NNR, 2022), but in the United States of America (USA), the Nuclear Regulatory Commission is responsible for setting regulations for emergency planning zones (USNRC, 2025).

According to the NNR, three nuclear emergency planning zones are defined. The first zone is the Precautionary Action Zone (PAZ), which consists of an area with a 5 km radius. Within this zone, urgent evacuation is required owing to high radiation exposure, and people within this zone must be capable of being evacuated within four hours. The second zone is the Urgent Protective Action Planning Zone (UPZ), which consists of an area that is 16 km in radius, and the people in this zone must be capable of being evacuated within 16 hours. Erasures, such as sheltering and drinking iodine tablets, are also recommended in this area. The third zone is the long-term protective action planning zone (LPZ), which consists of an area with a radius of 80 km. This area is considered low risk, with measures such as public awareness and community reinforcement implemented to ensure personnel safety (Jones et al., 2007). Since then, the NNR has published a description of the three zones in their Position Paper Emergency Planning Technical Basis for New Nuclear Installations (NNR, 2012). It states that in determining the extent of an emergency planning zone, site characteristics, source term analysis, design details, radiological impact evaluation, and risk insights are important. Once considered, the following zones should be demarcated (NNR, 2012: 13):

- "An exclusion zone (EZ) is a radius determined for the purpose of evacuating persons in the event of a nuclear accident. Within the boundaries of that zone or within any erven intersecting with that zone, there must be no members of the public resident, no uncontrolled recreational activities, no commercial activities, or institutions which are not directly linked to the operation of nuclear installations situated within this zone, or for which an authorisation has not been granted;

- An overall Emergency Planning Zone (EPZ) of such size that emergency or remedial measures must be considered where the potential exists that any members of the public may receive more than an annual effective dose of 1 mSv owing to the source term;

- A Long-Term Protective Action Planning Zone (LPZ), where preparations for effective implementation of protective actions to reduce the risk of stochastic health effects from long-term exposure to deposition and ingestion must be developed in advance, consistent with international standards."

A comparison of previous nuclear disaster incidents was conducted for further analysis. Historical events, such as the Chernobyl nuclear power plant disaster that took place in April 1986, left two Chernobyl workers dead due to the disaster that occurred on the day of the incident. A further 28 people died within a few weeks of the incident due to radiation exposure and acute radiation syndrome (WNA, 2024c). Decades later, the Fukushima Daiichi nuclear disaster occurred on 11 March 2011. There were no deaths or cases of radiation sickness from the incident; however, it caused over 100,000 people to evacuate their homes as a preventative measure.

Methodology

This study focuses on secondary source data from Statistics South Africa and geospatial data, such as shapefiles, for analysis. Analysis of such secondary source data is made possible by applications/software, namely ArcMap 10.8.1. The research area was first selected, and its boundaries were determined and described using buffering.

In GIS, a buffer is a reclassification that occurs at a certain distance. It refers to a designated zone or area established around or within a geographic feature, such as a point, line, or polygon (Dempsey, 2023). Buffering is a crucial component of GIS,[3] also referred to as buffer analysis, which shows and identifies features that are within a user-specified distance of a given feature or location. To carry out this function, a zone (i.e. buffer) is created around a map feature to evaluate and identify what is located there. This can be compared to drawing a safety zone around a feature, where all entities are subject to its influence or effects. The buffer can be used to determine possible effect zones, assess a feature's proximity, or clarify spatial patterns. As such, they can be variable or uniform in their dimensions (Dempsey, 2023).

3 GIS is "a system designed to capture, store, manipulate, analyse, manage, and present all types of geographical data" (University of Wisconsin–Madison, 2025).

Using ArcMap 10.81, a shape file of the towns in the Western Cape around the specified municipality that falls within our buffers and hospitals within the specified buffer zones was inserted, and the shape file data were obtained from the university database. For the attribute data for places, the point of interest, which is Koeberg, was selected and exported. Three buffer zones at several distances were created using multiple ring buffers under the proximity tool around Koeberg to determine the distribution of residential areas and hospitals within the research area. Several areas that fell within the high, medium, and low-risk zones were calculated. Data collected from Statistics South Africa (2024) was used to calculate the number of people affected in the specified zones. To determine the zone distances around the Koeberg guidelines, the IAEA and the NNR were followed.

Results

The maps generated in this study provided a visual representation of the data used. This study used EPZs from the NNR. The results, showing different buffer zones from Koeberg at different radii, are shown in Figure 9.3. The first zone is the Precautionary Action Zone (PAZ), which consists of an area of 5 km in radius; the second zone is the Urgent Protective Action Zone (UPZ), which consists of an area of 16 km in radius; and the last zone is the Long-Term Protective Action Planning Zone (LPZ), which consists of an area of 80 km in radius.

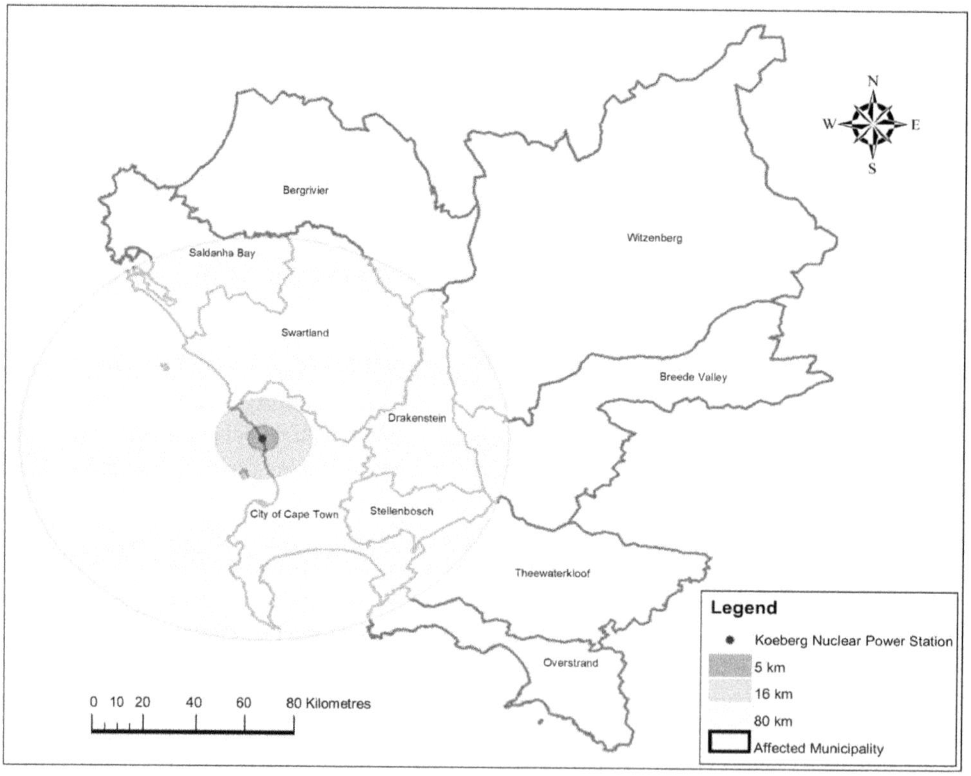

Figure 9.3: Map of buffer zones and municipalities that fall within the emergency planning zones in the Western Cape. The red buffer zone denotes a high-risk area, the orange buffer zone denotes a medium-risk area, and the yellow buffer zone denotes a low-risk area. Ten municipalities were affected.
(*Source:* Authors' own work using ArcMap 10.8.1)

As indicated in Figure 9.3, ten municipalities were affected. Within these municipalities, there are various typologies of human settlements, classified in a study by Van Huyssteen et al. (2018). A City of Cape Town census document (2013) classifies, for example, Melbosstrand as a census suburb, and describes a sub-place as the second (lowest) tier in the place name category, which includes a suburb, section, or zone of an (apartheid) township, smallholdings, village, subvillage, ward, or informal settlement. The census suburb Melkbosstrand encompasses the following sub-places: Duynefontein, Melkbosstrand SP, and Van Riebeeckstad (City of Cape Town, 2013). Thus, for the simplification of the findings of this research, the various towns, census suburbs, and sub-places will be referred to as human settlements.

Figure 9.4 depicts the different human settlements that fall within the three emergency planning zones. Only human settlements fall within the high-risk zone, which is the 5 km zone. Within the 16 km zone, there are 12 human settlements, and within the 80 km zone, there are 503 human settlements (among several towns and suburbs – almost the entire Cape Town metropole, Swartland, Drakenstein, and Stellenbosch regions, and half of the Saldanha Bay Municipal region).

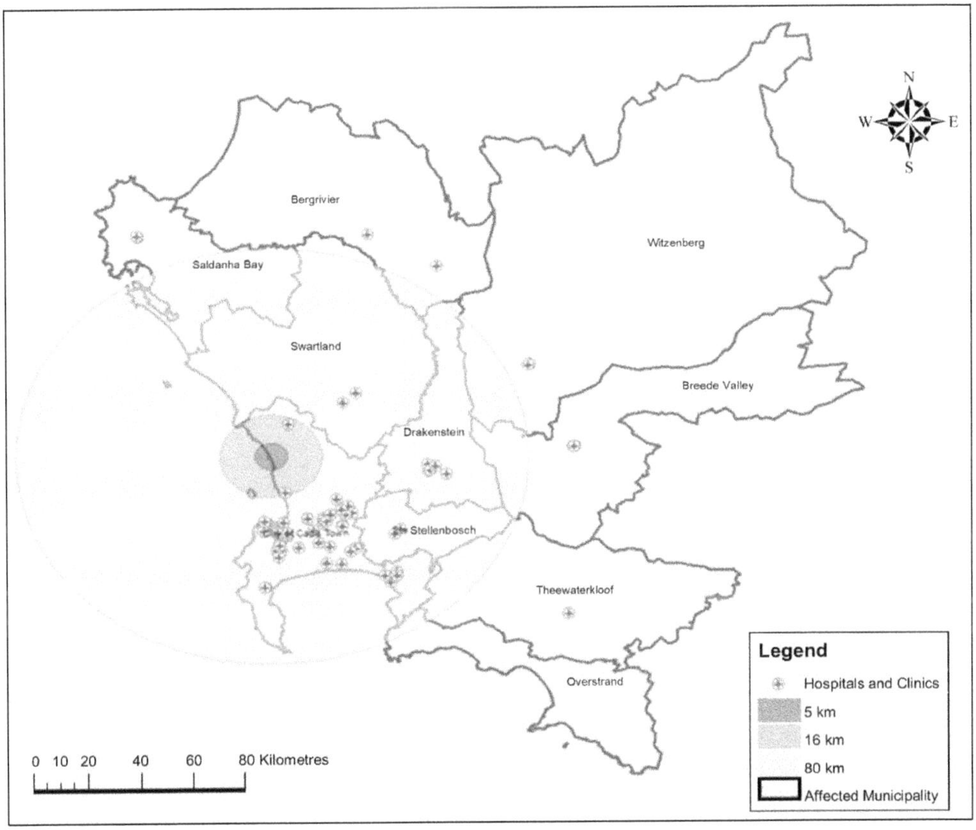

Figure 9.4: Map of towns/census suburbs/sub-places (or human settlements) within the emergency planning zones. The red colour dots indicate towns that are at a higher risk of radiation exposure, the orange colour dots denote all the towns that are at a medium risk of exposure to radiation, and the green colour denotes all the towns that are at a lower radiation exposure.
(*Source:* Authors' own work using ArcGIS 10.8.1)

Public health is defined as a community's capacity to prevent and respond to acute and chronic health and safety challenges. This study focused on public health facilities located in the Western Cape within the affected municipalities. Figure 9.5 shows the distribution of hospitals and medical clinics that should be functionally capable of providing various degrees of medical assistance in the case of a nuclear disaster. The GIS analysis indicated that only two hospitals were within the 16 km zone (UPZ), whereas 60 medical facilities (hospitals and clinics) were located within the 80 km radius (LPZ). Additionally, outside the 80 km zone but still within the affected municipality, six nearby hospitals were available. There are no hospitals within the 5 km zone.

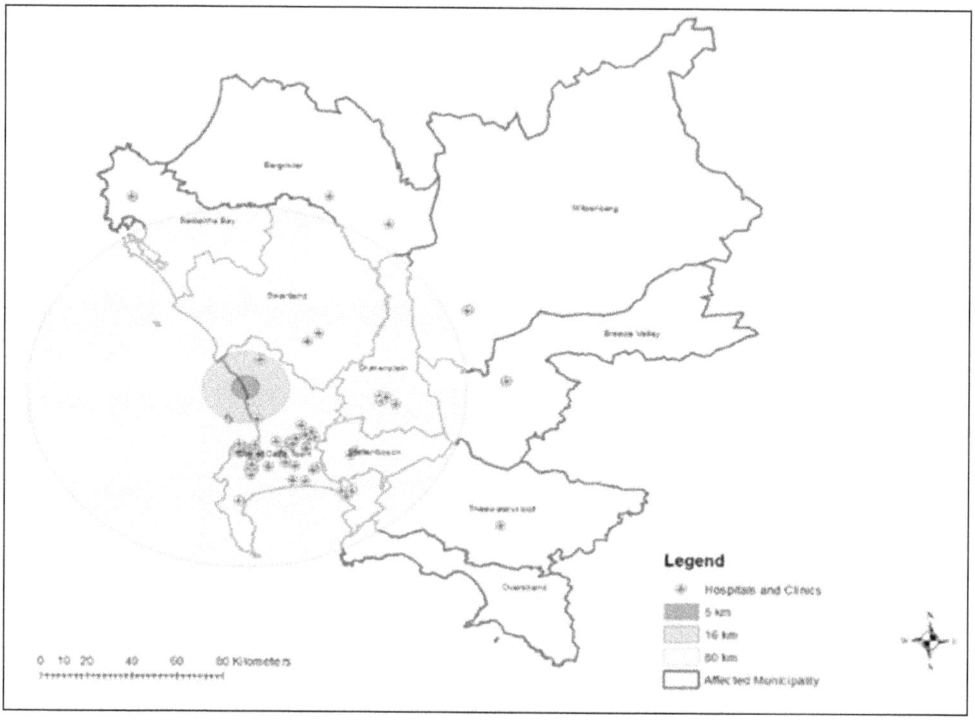

Figure 9.5: Map of hospitals within the emergency planning zones.
(*Source:* Authors' own work using ArcGIS 10.8.1)

Statistical analysis

Table 9.1 shows the population that falls within each human settlement, as well as the different human settlements that fall within the high- and moderate-risk emergency planning zones and their respective population sizes.

The 5 km EPZ consists of approximately 3,387 people, and the 16 km EPZ consists of 209,775 people (Table 9.2). The 80 km EPZ consists of approximately 5,207,512 people.

Table 9.1: Affected municipalities and their population

Municipality	**Population**
Overstrand	132 495
Theewaterskloof	139 563
Stellenbosch	175 411
City of Cape Town	4 772 846
Swartland	148 331
Breede Valley	212 682
Saldanha Bay	154 695
Drakenstein	276 800
Bergrivier	70 276
Witzenberg	103 765
TOTAL	**6 186 864**

Source: Statistics South Africa (2024).

Table 9.2: Populations within the high-risk (5 km) and moderate-risk zones (16 km)

Human setteements	**Distance**		**Population**
	Within 5 km	**Within 16 km**	
Duynefontein	Yes	-	1 453
Melkbosstrand	-	No	11 303
Atlantis	-	No	67 491
Avondale	-	No	3 246
Blouberg Rise	-	No	106 222
Philadelphia	-	No	570
Protea Park	-	No	12 388
Robinvale	-	No	3 497
Van Riebeeckstrand	Yes	-	1 934
Wesfleur	-	No	1 671
TOTAL PEOPLE	**3 387**	**208 322**	**209 775**

Source: Statistics South Africa (2024).

To estimate the affected population within the 80 km zone, we included municipalities (Table 9.3) that were completely affected and those that were half-affected. For the half-affected municipalities, we considered half of their total population. Our analysis indicates that municipalities such as the City of Cape Town, Stellenbosch, Drakenstein, and Swartland are completely affected, while Saldanha is partially affected.

Table 9.3: Affected municipalities within the emergency planning zones and their populations

Municipality	Population
Stellenbosch	175 411
City of Cape Town	4 772 846
Swartland	148 331
Saldanha Bay / 2	154 695 / 2 = 77 347
Drakenstein	276 800
TOTAL	**5 417 287**

The total population within the 80 km zone is calculated as follows (1):

$$T\ (80\ km) = (80\ km) = P\ (80) - P\ (5\ km\ \&\ 16\ km)$$

where:

T (80 km) refers to the total population within the 80 km EPZ.
P (n = 80) is the total population within the affected EPZ.
P (5 km and 16 km) is the total population within the 5 km and 16 km EPZs.

The total population within the affected EPZ was calculated as 5,207,512. To create a comparative context for nuclear disasters, let us consider some prominent examples dating back to 1986. The context will only be shaped by focusing on nuclear energy production disasters, with no consideration given to nuclear armament testing and/or nuclear devices used for kinetic effects.

Comparing the impact of nuclear power plant disasters: Chernobyl vs. Fukushima Daiichi

The nuclear disasters at Chernobyl in 1986 and Fukushima Daiichi in 2011 remain landmark events in the history of nuclear energy (WNA, 2024c). This highlights the catastrophic implications of nuclear energy disasters on human survival (health and settlement at the very least), the environment, and societal infrastructure. Although both incidents share the commonality of being nuclear accidents, their geographic settings, approaches to providing healthcare, and effects on the impacted populations vary greatly from one another. This thorough examination explores various facets to clarify the complex effects of each catastrophe.

The Chernobyl nuclear disaster (1986)

The Chernobyl Nuclear Power Plant is located in a rural landscape approximately 130 km north of Kiev (Ukraine) and approximately 20 km south of the border with Belarus (WNA, 2024c) (Figure 9.6). This area of Ukraine is described as a Belarusian-type woodland with a low population density. Approximately 3 km from the reactor, in the new city of Pripyat, there were 49,000 inhabitants. The old town of Chornobyl, which had a population of 12,500, is approximately 15 km southeast of the complex. Within a 30 km radius of the power plant, the total population was between 115,000 and 13,000 at the time of the accident (WNA, 2024c). It is thus a predominantly rural setting characterised by small villages and agricultural land.

Figure 9:6: Geographic location of Chernobyl.
(*Source:* Adapted from WNA, 2024c)

The reactor explosion on 26 April 1986 released a catastrophic number of radioactive particles into the atmosphere, which were dispersed over large areas owing to prevailing winds (Goldberg, 2012). The steam explosion and subsequent fires released at least 5% of the radioactive material from the reactor core into the atmosphere, causing contamination in various regions of Europe. On the night of the incident, two workers at the Chernobyl plant lost their lives due to the explosion, and an additional 28 died within weeks from acute radiation sickness. According to the United Nations Scientific Committee on the Effects of Atomic Radiation (UNSCEAR), aside from approximately 5,000 cases of thyroid cancer (leading to 15 deaths), there is no significant evidence of a widespread public health impact due to radiation exposure two decades after the disaster. In total, around 350,000 individuals were evacuated as a result of the event, with ongoing efforts to resettle those who were displaced (WNA, 2024c)

The immediate area surrounding the plant became known as the Chernobyl Exclusion Zone (CEZ), initially covering a 10 km radius that was changed to a 30 km radius and encompassing parts of Ukraine, Belarus, and Russia (Goldberg, 2012; WNA, 2024c). The predominantly rural nature of the region complicated the monitoring of radioactive contamination and hindered effective evacuation strategies, as many residents lived in isolated villages and were dispersed over a vast area. Although the CEZ was established for safety, it has since become a haunting reminder of the disaster's impact. The landscape, once filled with human activity, became desolate, with abandoned towns like Pripyat, previously home to 49,000 residents, now silent and overgrown (Ager et al., 2019). The geographic context of Chernobyl contributed to the long-term challenges of managing contaminated land and providing healthcare to affected populations, as evacuation and monitoring required vast resources and coordination across national borders.

The Fukushima Daiichi nuclear disaster (2011)

Fukushima Daiichi is situated on the eastern coast of Japan, in a region renowned for its high population density and agricultural productivity (Figure 9.7). Following a very strong earthquake (named the Great East Japan Earthquake), a 15-metre-high tsunami disrupted the power supply and cooling systems of three of the six Fukushima Daiichi reactors, leading to a nuclear disaster that ensued on 11 March 2011 (WNA, 2024a; NIDS, 2012: 237). Within the first three days, the cores of all three reactors disintegrated, resulting in a level 7 nuclear incident, as per the International Nuclear and Radiological Event Scale. This classification is based on radioactive releases from days 4–6, which amounted to approximately 940 PBq (I-131 equivalent) (WNA, 2024a).

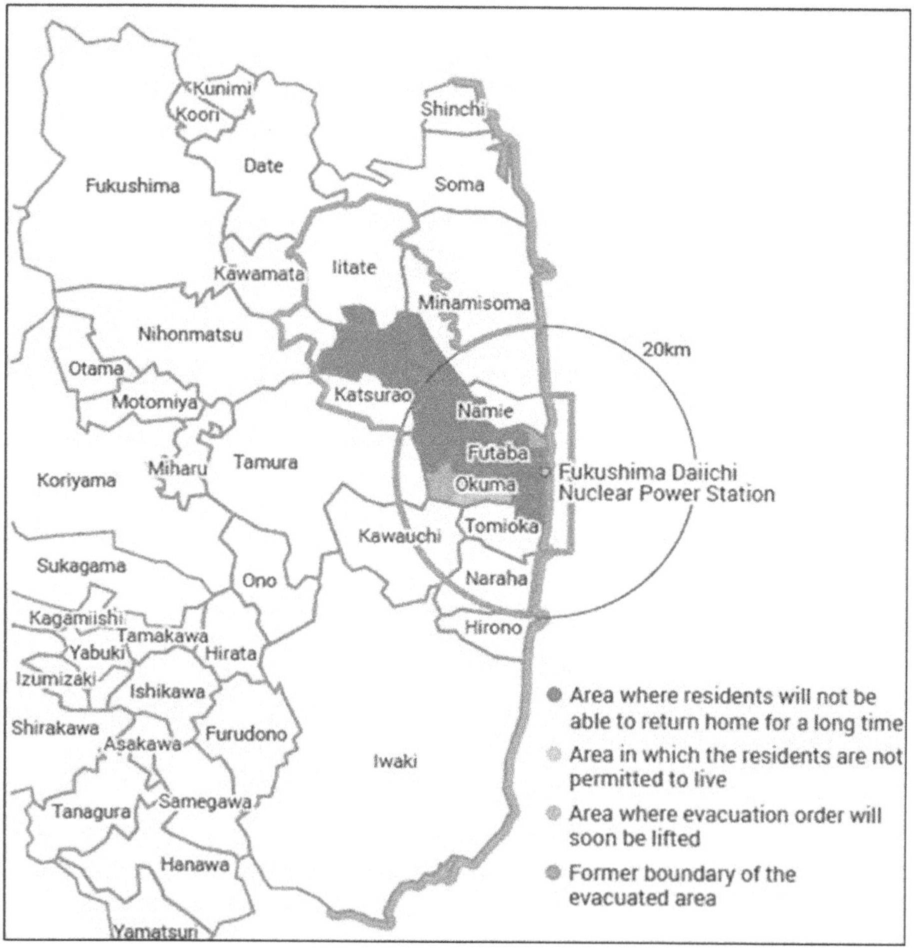

Figure 9.7: Evacuation and restricted areas surrounding Fukushima Daiichi Nuclear Power Plant, Japan (2011).
(*Source:* Adapted from WNA, 2024a)

Unlike the rural setting of Chernobyl, Fukushima is characterised by its proximity to populous urban areas, including Fukushima City and Minamisoma, which are home to a significant number of residents. This dense population made the evacuation process more urgent and complex, as the government had to swiftly relocate over 160,000 individuals within days. The geographical layout of Fukushima, with its mountainous terrain and proximity to the Pacific Ocean, posed additional challenges for emergency responders during the crisis. The tsunami caused extensive damage to infrastructure and disrupted communication channels, making it difficult to coordinate evacuation efforts and provide timely information to residents (Koshimura & Shuto, 2015). The high population density in Fukushima also meant that the immediate health and safety risks from radiation exposure were felt more acutely than the relatively dispersed population around Chernobyl.

The damage caused by the accident resulted in the decommissioning of all four Fukushima Daiichi reactors, which had a combined net capacity of 2719 MWe. After two weeks, reactors 1–3 became stable with the addition of water, and by July, they were cooled using recycled water from a newly-built treatment plant. In mid-December, the reactors were officially declared to be in a cold shutdown state. Beyond cooling, the main challenge was to prevent the release of radioactive materials, especially from the contaminated water that leaked from the three affected units. This issue became prominent in the media in August 2013. Although no deaths or radiation sickness cases were reported from the accident, over 100,000 people were evacuated as a precaution. However, government concerns have delayed the return of many evacuees to their homes. Official records indicate that 2,313 disaster-related deaths occurred among those evacuated from Fukushima Prefecture, in addition to approximately 19,500 people who died due to the earthquake and tsunami (WNA, 2024a).

Healthcare responses and medical infrastructure – Chernobyl and Fukushima compared

The Chernobyl disaster healthcare response was significantly hampered by the political and social contexts of the Soviet Union. Initially, the Soviet government attempted to downplay the severity of the incident, resulting in delayed evacuations and a lack of coordinated medical response. First responders, including firefighters and plant workers, rushed to the site without adequate protective gear or training to deal with radiation exposure. The casualties included firefighters who responded to the initial fires on the roof of the turbine building. Although the fires were extinguished within a few hours, radiation exposure on the first day led to 28 deaths by the end of July 1986, including the six firefighters. The radiation doses received by the firefighters and power plant workers were sufficiently high to cause acute radiation syndrome (ARS), which occurs when an individual is exposed to over 700 milligrays (mGy) in a short period (usually minutes). The common symptoms of ARS include nausea, vomiting, gastrointestinal issues, headaches, burns, and fever. Whole-body radiation doses between 4,000 mGy and 5,000 mGy within a brief period would result in a 50% fatality rate, while doses between 8,000 and 10,000 mGy are fatal. The doses received by the firefighters who died were estimated to be as high as 20,000 mGy (WNA, 2024b). The rising number of people suffering from ARS in the days and months to come were systematically hospitalised in ill-equipped facilities to handle such cases (Yablokov et al., 2010).

As the number of evacuees surged, hospitals in the surrounding areas were overwhelmed. The centralised Soviet healthcare system struggled to provide adequate care, particularly for those suffering from long-term radiation-related illnesses (typically caused by exposure to the radionuclide caesium-137, whereas exposure to iodine-131 resulted in shorter-term effects [WNA, 2024b]). A marked increase in thyroid cancer, particularly among children exposed to radioactive iodine, became evident in the years following the disaster (UNSCEAR, 2008). Despite acknowledging these health issues, the Soviet government lacked the resources and infrastructure to effectively address the ongoing healthcare needs of the affected population.

In addition to the physical health impacts, the psychological consequences of the disaster were also profound. The lack of mental health support for survivors, coupled with the stigma associated with radiation exposure, contributed to widespread anxiety and depression, collectively referred to as the Chernobyl Syndrome (Yablokov et al., 2010). Many individuals experienced social isolation and economic hardship because of their displacement and the long-term effects of the disaster.

In contrast to Chernobyl, the response to the Fukushima disaster benefited from Japan's relatively advanced emergency preparedness systems. The Japanese government quickly declared a nuclear emergency, allowing for the rapid implementation of evacuation plans and mobilisation of resources (Yamamoto, 2013). The Japan Self-Defense Forces played a crucial role in rescue operations, providing medical care and logistical support to affected communities (NIDS, 2012).

Following the disaster, the Japanese government established the Fukushima Health Management Survey (FRIPS, 2023) to monitor the health of evacuees and provide free medical monitoring and care to residents at risk of radiation exposure (Yasumura et al., 2012). This initiative represents a proactive approach to healthcare, enabling more structured health monitoring and assessment of long-term impacts on the population. The survey facilitated a focus on physical health and mental wellbeing, recognising the psychological toll of the disaster on affected individuals, as well as the monitoring of the "health effects of long-term low-dose radiation exposure" (Yasumura et al., 2022: S3).

Despite these efforts, challenges remain. Many evacuees faced barriers to accessing healthcare services due to relocation and the stigma associated with radiation exposure. Mental health issues, including anxiety and depression, have become prevalent among evacuees, with some studies indicating increased rates of suicide in affected populations (Yamamoto, 2013). The Japanese government's attempts to provide support were commendable; however, the emotional distress experienced by many individuals was compounded by uncertainty regarding their future and the disaster's long-term consequences.

Impact on affected populations – Chernobyl and Fukushima compared

The impact of the Chernobyl disaster on the affected population was profound and multifaceted, with long-term consequences. Over 350,000 people were evacuated from the exclusion zone and surrounding areas (UNSCEAR, 2008). The mass relocation resulted in significant social disruption, as communities were disbanded and individuals were forced to abandon their homes, often with little time to prepare. The psychological impact of this dislocation was severe, with many evacuees grappling with feelings of loss, grief, and anxiety about their future (Fairlie, 2018).

The long-term health consequences of radiation exposure became evident in the years after the disaster. The UNSCEAR 2000 report stated that, aside from the rise in thyroid cancer, there was no indication of a significant public health effect linked to radiation exposure 14 years post-accident. There is no scientific proof of an increase in overall cancer rates, mortality, or non-cancerous conditions associated with radiation exposure. The UNSCEAR (2008) reported that studies have shown a marked increase in thyroid cancer, particularly among children who were exposed to radioactive iodine released during the explosion. However, attributing specific health outcomes directly to radiation exposure remains complex because of the multitude of factors involved, including lifestyle, diet, and healthcare access. Moreover, the lack of systematic long-term health monitoring and follow-up exacerbates the challenges faced by survivors in accessing appropriate care.

The social stigma associated with radiation exposure further complicates the lives of many survivors. Discrimination in employment and social interactions often results in isolation, affecting the mental health and social status of affected individuals. The consequences of the disaster extended beyond immediate health effects, disrupting family structures and community cohesion for generations to come.

The impact on the population affected by the Fukushima disaster was similarly significant but differed in its dynamics because of Japan's sociopolitical context. Over 160,000 residents were evacuated from their homes, with the government implementing a swift and organised evacuation process. Many evacuees were placed in temporary shelters that often lacked adequate facilities and privacy, leading to heightened stress and anxiety during an already traumatic period.

The psychological impact of displacement was profound, with many individuals experiencing anxiety, depression, and a sense of hopelessness about returning home. Studies have indicated that mental health issues have become prevalent among evacuees, with some reports suggesting higher rates of suicide and social withdrawal than pre-disaster levels (Yamamoto, 2013). The trauma of the event, compounded by the uncertainty surrounding radiation exposure and its long-term effects, created a challenging environment for those affected by the incident.

However, the establishment of health consultations and screening programmes allowed for a more structured approach to address the health needs of evacuees. Ongoing health monitoring in Fukushima aims to assess potential radiation-related illnesses and provide support to individuals at risk (Yasumura et al., 2012). Although the long-term health effects of radiation exposure remain a topic of ongoing study and debate, the proactive measures taken by the Japanese government represent an important step in addressing the needs of the affected population.

Despite these efforts, the fear associated with radiation exposure continues to affect residents' quality of life. Many individuals remain anxious about their long-term health and the potential consequences of living in contaminated areas. The perception of radiation as a lingering threat influences social interactions and community cohesion, further complicating the recovery process for affected populations.

Discussion

EPZs around nuclear power plants are crucial for public safety during crises. Considering Koeberg's EPZs, the events of Chernobyl and Fukushima are very relevant. At Chernobyl, the safety zones had to be expanded considerably because of the rapid spread of radiation owing to European climatic conditions. This makes the lessons learned from the Chernobyl disaster the most relevant.

Lessons learned from Chernobyl regarding prevailing winds and climatic conditions should have a significant impact on determining the optimal size and direction (vectors) of Koeberg EPZs. This is particularly relevant to the west coast region of the Western Cape because of the known windy climatic conditions during virtually all the seasons.

The most significant health threat from nuclear incidents is the release of radioactive materials into the atmosphere, see (amongst others) the World Health Organisation webpage (WHO, 2023) and the U.S. Nuclear Regulatory Commission (USNRC, 2023 and 2025). In the case of Chernobyl, it was caesium 137. In the case of Koeberg, it would be uranium 235. Radiation contamination from these radioactive isotopes can cause acute radiation sickness, which might start with symptoms like feeling sick and exhausted and, in severe cases, could lead to organ failure and, depending on the level of exposure, death. People who are not in the immediate disaster area, which is within an 80 km radius (Figure 9.7), also face serious long-term health risks. Radioactive particles can travel hundreds of kilometres, especially if the climatic conditions of a particular area are characterised by prevailing winds. This not only poses an immediate danger to humans but also has secondary implications for food and fresh water. This implies that the community may experience higher rates of cancer, birth defects, and other health issues for years to come.

Those living in the PAZ, which consists of an area 5 km in radius, and the UPZ, which consists of an area 16 km in radius, should be permanently evacuated. Based on current estimates, a disaster at Koeberg would affect approximately 209,775 people living within a 16 km radius. These individuals must be evacuated. The number of people who need to be evacuated depends on how many people survive the initial incident.

Beyond the direct impact on human life, a nuclear plant explosion would have catastrophic consequences for the terra, fauna, and flora within these zones. Radioactive contamination of soil, air, and water sources devastates local ecosystems. The long half-life of many radionuclides means that certain areas may remain uninhabitable for decades or even centuries. To illustrate the point, uranium 235 has an estimated half-life of 700 million years, depending on the level of enrichment (Australian Government, 2024). Agricultural and livestock production would be severely disrupted, creating food insecurity and economic ruin for affected communities and South Africa at large.

The cleanup and decontamination processes are arduous, expensive, and pose ongoing health risks to workers. Populations within the ingestion pathway are more affected by this process. An estimated 1,386,182 people will suffer from radioactive contamination of soil, air, and water sources, and farmers from this area will suffer the most as the environment will not be conducive to farming.

Damage to critical infrastructure compounds the human and environmental toll. Power grids, transportation networks, communication systems, and other vital services are crippled, hampering emergency response and recovery efforts. The costs of repairing and replacing damaged infrastructure, not to mention long-term economic disruption, would be staggering. Entire cities or regions might have to be abandoned, halting economic development in the region, with severe implications for the economic development of South Africa due to the strain on the fiscus resulting from containment and emergency measures.

Based on the statistical analysis, approximately 5,417,287 people living in the Western Cape will be affected by the impact of the nuclear disaster, with approximately 209,775 at high risk of being exposed to strong radiation with less reaction time. The distributed nature of medical care facilities (Figure 9.7) could have a positive impact on possible care for the injured because, with evacuation, injured people will still have access to care within the 80 km radius and beyond. However, the density of medical care facilities closes to the PAZ (5 km) and UPZ (16 km) will have a severe impact on the ability to care for the injured because of the possible contamination of the facilities within a very short period of time, depending on the prevailing winds during that season. The prevailing wind in the Western Cape during summer is south-southeast. In winter, it is northwest with rain (South Africa Online, 2025). Thus, for example, a north-westerly wind pattern will push a nuclear plume into the Cape Town Metropolitan area with devastating effects for the population, critical infrastructure, government, and the economy at large. If this pattern also produces rain, it will probably increase freshwater contamination. If it is a typical summer weather pattern, a nuclear plume will be pushed up the coast to less densely populated areas. Critical economic infrastructure in the Saldanha Bay Municipality will be severely affected, with a negative impact on the South African economy.

Conclusion

This study examined the risks of a possible nuclear explosion at the Koeberg Nuclear Power Station. This shows that living close to the plant is risky, whereas the dangers lessen with increasing distance. Alarmingly, approximately 209,775 people in the Western Cape face serious threats if a major nuclear incident occurs. Our findings highlight the urgent need for residents within a 5 km and 16 km radius to consider relocating to safer areas, as their proximity to the plant heightens their exposure to radiation during a nuclear emergency. These results highlight the importance of policymakers planning for such scenarios. They must create strategies that boost community safety and resilience against possible nuclear threats.

If you live within 16 km of the plant, it is important to stay informed about emergency plans and alerts from local authorities and the plant itself. Sign up for community alert systems to receive timely updates. Prepare an emergency kit with essentials such as water, non-perishable food, a flashlight, batteries, a battery-powered radio, and first-aid supplies. If health officials recommend it, potassium iodide tablets should be included. Familiarise yourself with the evacuation routes and procedures. It is advisable to practice these drills with your family so that everyone knows what to do in the event of an emergency.

References

Ager, A.A., Lasko, R., Myroniuk, V., Zibtsev, S., Day, M.A., Usenia, U., Bogomolov, V., Kovalets, I. & Evers, C.R. 2019. The wildfire problem in areas contaminated by the Chernobyl disaster. *Science of the Total Environment*, 696. https://doi.org/10.1016/j.scitotenv.2019.133954

Australian Government. 2024. Uranium-235. *ANSTO*, 26 August. https://www.ansto.gov.au/media/2237/download?inline [28 March 2025].

City of Cape Town. 2013. 2011 Census Suburb: Melkbosstrand. https://resource.capetown.gov.za/documentcentre/Documents/Maps%20and%20statistics/2011_Census_CT_Suburb_Melkbosstrand_Profile.pdf [30 March 2025].

Dempsey, C. 2023. Buffers in GIS. *Geography Realm*. https://www.geographyrealm.com/buffers-in-gis/ [28 March 2025].

EIA [U.S. Energy Information Administration]. 2022. Nuclear explained: Nuclear power and the environment. https://www.eia.gov/energyexplained/nuclear/nuclear-power-and-the-environment.php [22 March 2025].

Eskom. 2025. Koeberg Nuclear Power Station. https://www.eskom.co.za/eskom-divisions/gx/koeberg-nuclear-power-station/ [23 March 2025].

Fairlie, I. 2018. Evacuations after severe nuclear accidents. https://www.ianfairlie.org/news/evacuations-severe-nuclear-accidents/ [28 March 2025].

FRIPS [Fukushima Reconstruction Information Portal Site]. 2023. An overview of Fukushima Health Management Survey. https://www.pref.fukushima.lg.jp/site/portal-english/en-4-3-1.html [5 August 2024].

Goldberg, K. 2012. A question of benefit versus risk: Pilgrim Nuclear Power Station. B.A. thesis. Worcester, MA: Worcester Polytechnic Institute.

IAEA [International Atomic Energy Agency]. 2007. Arrangements for preparedness for a nuclear or radiological emergency. Safety guide No. GS-G-2.1. https://www-pub.iaea.org/MTCD/Publications/PDF/Pub1265web.pdf [28 March 2025].

Jones, J., Naude, S., van Wyngaardt, G. & Marks, A. 2007. Koeberg nuclear emergency plan: Traffic evacuation model. Proceedings of the 26th Southern African Transport Conference (SATC 2007), 9–12 July 2007, Pretoria, South Africa.

KAA [Koeberg Alert Alliance]. 2024. Civil society responds to NNR decision to extend Koeberg's unit one licence. https://koebergalert.org/2024/07/22/civil-society-responds-to-nnr-decision-to-extend-koebergs-unit-one-licence/ [26 March 2025].

Koshimura, S. & Shuto, N. 2015. Response to the 2011 Great East Japan Earthquake and Tsunami disaster. *Philosophical Transactions of the Royal Society A*, 373(2053). https://doi.org/10.1098/rsta.2014.0373

Latlong.net. 2025. Cape Town, South Africa. https://www.latlong.net/place/cape-town-south-africa-854.html [15 March 2025].

NIDS [The National Institute for Defense Studies]. 2012. *The East Asian Strategic Review.* Chapter 7: Japan: Toward the establishment of a dynamic defense force. Tokyo: The Japan Times. 235–267. https://www.nids.mod.go.jp/english/publication/east-asian/pdf/2012/east-asian_e2012_07.pdf

NNR [National Nuclear Regulator]. 2012. Position paper: Emergency planning technical basis for new nuclear installations (PP-0015). https://nnr.co.za/wp-content/uploads/2020/10/PP-0015-EPTB-Rev-0-Final.pdf [28 March 2025].

NNR [National Nuclear Regulator]. 2022. Regulatory framework. https://nnr.co.za/about/acts-and-regulations/ [28 March 2025].

Ramdass, N. 2024. Koeberg Unit 1 licence extended to 2044. *Creamer Media's Engineering News*, 23 August. https://www.engineeringnews.co.za/article/koeberg-unit-1-licence-extended-to-2044-2024-08-12 [28 March 2025].

Schwikowski, M. 2017. South Africa's controversial nuclear power plans. *Deutsche Welle*, 4 July. https://www.dw.com/en/south-africas-controversial-nuclear-power-plans/a-38338085 [28 March 2025].

South Africa Online. 2025. Western Cape climate. https://southafrica.co.za/western-cape-climate.html [28 March 2025].

Statistics South Africa. 2024. Statistics by theme: People. https://www.statssa.gov.za/?page_id=595 [3 December 2024].

University of Wisconsin–Madison. 2025. Mapping and Geographic Information Systems (GIS): What is GIS? https://researchguides.library.wisc.edu/GIS [10 April 2024].

UNSCEAR [United Nations Scientific Committee on the Effects of Atomic Radiation]. 2008. Sources and effects of ionizing radiation: UNSCEAR 2008 Report, Volume I. https://www.unscear.org/unscear/en/publications/2008_1.html [28 March 2025].

USNRC [U.S. Nuclear Regulatory Commission]. 2023. Emergency preparedness & response. "[3 https://www.nrc.gov/about-nrc/emerg-preparedness.html [3 April 2024].

USNRC [U.S. Nuclear Regulatory Commission]. 2025. About NRC. https://www.nrc.gov/about-nrc.html [24 January 2025].

Van Huyssteen, E., Green, C., Sogoni, Z., Maritz, J. & McKelly, D. 2018. South African Functional Town Typology (CSIR 2018 v2). http://stepsa.org/socio_econ.html#Indicator [28 March 2025].

WHO [World Health Organization]. 2023. Radiation emergencies. https://www.who.int/health-topics/radiation-emergencies [2 April 2024].

WNA [World Nuclear Association]. 2024a. Fukushima Daiichi accident. https://world-nuclear.org/information-library/safety-and-security/safety-of-plants/fukushima-daiichi-accident [29 April 2024].

WNA [World Nuclear Association]. 2024b. Safety of nuclear power reactors. https://world-nuclear.org/information-library/safety-and-security/safety-of-plants/safety-of-nuclear-power-reactors [23 August 2024].

WNA [World Nuclear Association]. 2024c. Chernobyl accident 1986. https://world-nuclear.org/information-library/safety-and-security/safety-of-plants/chernobyl-accident [3 December 2024].

Yablokov, A.V., Nesterenko, V.B., Nesterenko, A.V. & Sherman-Nevinger, J.D. 2010. *Chernobyl: Consequences of the Catastrophe for People and the Environment*. Boston. MA: Blackwell.

Yamamoto, L.G. 2013. Risks and management of radiation exposure. *Pediatric Emergency Care*, 29(9): 1016–1026. https://doi.org/10.1097/PEC.0b013e3182a380b8

Yasumura, S., Hosoya, M., Yamashita, S., Kamiya, K., Abe, M., Akashi, M., Kodama, K. & Ozasa, K. 2012. Study protocol for the Fukushima health management survey. *Journal of Epidemiology, 22(5): 375–383*. https://doi.org/10.2188/jea.JE20120105

Yasumura, S., Ohira, T., Ishikawa, T., Shimura, H., Sakai, A., Maeda, M., Miura, I., Fujimori, K., Ohto, H. & Kamiya, K. 2022. Achievements and current status of the Fukushima health management survey. *Journal of Epidemiology, 32(*Suppl_XII): S3–S10. https://doi.org/10.2188/jea.JE20210390

CHAPTER 10

Enhancing utility management through GIS and GPS integration: A case of the Saldanha Military Area

S. Nyembe and I. Henrico

Abstract

This study addresses utility infrastructure management challenges at the South African Military Academy (SAMA), focusing on outdated floor plans and inaccurate records encountered during a 2017 refurbishment. Success depended on leveraging the institutional knowledge of the facilities manager. Under a Memorandum of Understanding dated 5 October 2017, SAMA became responsible for the Faculty of Military Science's bulk utility maintenance. This prompted the development of an interactive, cost-effective system for efficient utility location within the Saldanha Military Area (SMA). The primary objective was to create a geospatial database that visually represents the precise locations of existing utilities, above and below ground, using GIS software and GPS data collection. These include water valves, pipelines, sewerage systems, fire hydrants, electrical substations, and kiosks. The aim was to produce a geospatial PDF with ArcMap 10.8.1, enhancing service delivery to the Faculty of Military Science and providing crucial information to the SMA stakeholders. The results show that the geodatabase significantly improved utility location accessibility and efficiency, benefiting SAMA and military stakeholders. This study underscores the pivotal role of geospatial technologies in utility infrastructure management.

Keywords: geospatial database, geospatial PDF, GIS, GPS, reticulation plans, Saldanha Military Area, service delivery, South African Military Academy, utilities

Introduction

Rapid access to geospatial information is paramount in the era of heightened need for swift and effective responses. The advent of geographical information systems (GIS) has transformed the world in various ways that support analytical models in location science (Goodchild, 2009). Rapid technological advances in GIS continue to maximise the availability of geographically referenced information (Bozkaya et al., 2019). In addition, studies suggest various disciplines, including military bases, over the past years, have shown interest in the utilisation of GIS for managing their utility infrastructure (Bernhardsen, 2002; Crawford, 2012; Geilhausen & Otto, 2012 Dangermond & Goodchild, 2020). The efficiency of geospatial techniques in capturing and mapping utilities has proven to be significant.

Rahman et al. (2020) explore a recurring thematic literature concern around GIS as an effective tool utilised and managed by current utility infrastructure and future expansion. Several in-depth studies indicate that over 80% of water and sewerage systems' data are georeferenced, which provides a cross-cultural experience and increases the inclusivity of GIS in data capture and digitisation (Shamsi, 2005; Crawford, 2012). In this regard, the duo argued that water and sewerage system data significantly help locate assets while maximising the input of generating characteristics through attribute tables. A parallel study by Meyers (1999) implies that when GIS and Global Positioning Systems (GPS) are mutually integrated, they enable organisations to optimise their capacity to manage utilities by improving the accuracy of collected data, minimising data entry costs, and providing excellent management of existing utility networks. Meyers's findings imply that GIS effectively creates digital maps of utility networks that execute tracking, analysis, and management of networks with unique potential in the provision of precise geospatial coordinates for utility features.

Emergency preparedness, response, and mitigation necessitate the rapid acquisition of the most accurate and updated geospatial content to ensure the widespread adoption of GIS in utility management. However, the rate at which the latter justification is manifested and readily accepted by the military is low. Okpuvwie & Toko Mouhamadou (2023) argue that to effect the change, military command should prioritize investing in the training of its personnel in field of various technologies. A similar study by Pincevičius et al. (2005) showed that the military utilises a digital base map in the creation of various maps without clustering unwanted data, thereby aiding in the viewing of geospatial information within their jurisdiction areas. Khan et al. (2021) affirm that military organisations depend on facility management through GIS when undertaking their missions, thereby ensuring the acquisition of relevant resources and infrastructure needed for successful operations. To sustain civil-military understanding, Rahman Jumat et al. (2012) examine solutions aimed at improving facility management in the military, including the implementation of an integrated facility management system, the provision of a predictive maintenance component for capturing, and the digitisation and mapping of utilities within the Saldanha Military Area (SMA).

Andrew (2020) argued that pinpointing the exact location of existing utilities when it comes to installations, repairs, and planning is challenging. Therefore, geospatial products are needed to support the management and administration of utilities within the SMA. To address the challenges experienced at SMA, a handheld GPS and GIS will be used to capture and map utilities. These services will enhance the South African Military Academy's (SAMA) service delivery model for the faculty and contribute to the immediate stakeholders. This study explores the use of geospatial technology to capture and map utilities in SMA. How can geospatial technologies improve the management and administration of bulk utilities within the SMA?

This study aims to design and develop a geospatial database capable of accurately representing the physical locations of utilities, above and below ground. Subsequently, this database will be used to create a GeoPDF document for enhanced data visualisation and accessibility. While this study focused on the development of a GeoPDF for utility mapping, ongoing efforts are being made to integrate the Saldanha Campus into Stellenbosch University's existing campus map. This ensures that the facilities and utilities of the Faculty of Military Science are accurately represented in the university's digital infrastructure.

An overview of GIS application in utility management

Utility companies are increasingly adopting GIS technology to manage their assets and infrastructure (Mennecke, 1997; Chen, 2007; Pick, 2008). With GIS mapping, utilities can efficiently manage, analyse, and visualise data related to their infrastructure, such as power lines, pipelines, and water distribution networks. Meyers (1999) stated that GIS technology offers significant benefits to utilities by enhancing their operations, maintenance, and planning. For instance, GIS mapping can assist utilities in identifying and ranking maintenance needs based on asset locations and conditions. By leveraging GIS data, utilities can optimise the deployment of crews and equipment to respond to emergencies and outages. Furthermore, GIS mapping provides utilities with a tool to identify potential risks and vulnerabilities in their infrastructure, enabling them to implement mitigation strategies to avoid breakdowns or losses. The implementation of GIS technology can facilitate more effective decision-making (Hughes, 2006; Crawford, 2012).

Rayburn's (2004) study found that GIS may improve the effectiveness and precision of water service providers. This study demonstrated how the use of GIS technology simplifies asset management and improves customer service. Anozie (2019) illustrated the value of utilising GIS for mapping subsurface water pipes with current tools such as GPS and GIS. Because of its ability to provide information on the physical locations of distribution infrastructure, this type of utility mapping is crucial in routine maintenance planning. Mwaniki and Odera (2012) conducted a study that focused on creating a geospatial database that included spatial and non-spatial data from ILRI sites to manage space. The study's findings demonstrated that GIS is a practical tool for managing space facilities because it simplifies the storage, updating, modification, querying, and sharing of information on a single platform.

Ufoegbune et al. (2016) used GIS to analyse data and create maps of the water supply system in Ota, demonstrating the significant role of GIS in planning and improving water supply in the area. Ajwaliya and Patel (2014) aimed to design and create a GIS-based utility management system for the DOS Housing Colony that would provide a thorough overview of the area's utilities, including water supply, sewage, electricity, and gas. After testing, it was discovered that the system worked well for controlling colony utilities. Planning is essential to prevent delays in utility detection. Maintaining utility data across several maps is challenging, especially in emergencies where time is limited. Owing to outdated maps, contractors frequently spend a lot of time in the field, adding extra costs and delays. Likewise, maintaining utility data across multiple maps may be difficult in an emergency. Correct knowledge of the existing subsurface infrastructure is essential for ensuring correct utility design, installation, and excavation. Failure to do so may result in fatalities and severe loss. Studies by Momin and Kolekar (2017), Jamil et al. (2012), and Cazzaniga et al. (2013) all emphasise the necessity of having accurate 3D mapping and GIS applications, as well as the relevance of having a centralised system for presenting immediate information to engineers and departments.

The safety and wellbeing of workers is a significant concern that is often overlooked in the utility mapping industry. Surveyors who deal with utility mapping must be aware of the risks and threats that these utilities may present (Kolera & Bernold, 2006). They must adhere to the requirements set forth by the Occupational Safety and Health Act (DOSH, 2008). Serious repercussions, such as fines or jail time for the company and the employee, may result from failing to take essential safeguards. This makes it imperative for land surveyors to prioritise the safety of workers, as it may be a matter of life or death.

There are, without a doubt, challenges to using GIS mapping for utility management. According to Buyuksalih et al. (2017), a critical obstacle is the high expense associated with acquiring and maintaining the necessary technology, as well as the requirement for workers to undergo specialised training. In addition, environmental conditions, such as weather and topography, may affect data accuracy and make it difficult to collect reliable data. Despite these difficulties, GIS mapping for utility management offers several advantages, and as it becomes more available and economical, it is predicted to be used more frequently in the industry in the future. GIS can be used to map several utilities, including water supply systems, sewage systems, gas pipelines, electrical distribution networks, and telecommunications networks (Ajwaliya & Patel, 2014). This technology can help detect crucial infrastructure assets, including valves, pumps, and transformers, to improve maintenance and repair schedules. In addition, GIS can offer valuable knowledge in areas with high utility demand, such as industrial zones and heavily residential areas, easing infrastructure construction and ensuring effective utility distribution (Li et al., 2015).

It is imperative for an organisation to choose a methodology that suits the specific needs of its department and outlines the steps necessary to carry out the project. However, it is not always necessary to follow every step of the methodology, as some GIS projects may only require a few specific steps (Crawford, 2012). Rahman et al. (2020) employed geospatial techniques to create a GIS database for infrastructure management at the University of Kebangsaan Malaysia (UKM). The methodology used in this study included four steps: data gathering, data preparation and processing, geodatabase development, and web-GIS application development. The system was successful in providing users with a new data-sharing platform and approach that enhanced operations and decision-making. In related work by Shadin and Tahar (2015), fire hydrants were mapped using a mobile GIS application in Bandung, Indonesia. This study found that mobile GIS is a useful tool for precisely mapping fire hydrants and enhancing the safety of urban areas. Cannistra's (1999) approach was also helpful in determining how to convert the data.

This study investigated the water supply network of a section of Tarkwa Municipality in Southwestern Ghana using GIS techniques. The analogue maps of the main and lateral pipelines were scanned and georeferenced to the same projection, and the digital maps of the study region were re-projected to a specific projection (WGS84 UTM zone 30 North). A geodatabase model was built, the pipes and node valves were manually digitised, and attribute tables for each were created. An analogue map of the main and lateral pipelines, shape files of the roads, contour lines, and cultural elements were used as resources. Using a handheld GPS receiver, the precise locations of the valves and fire hydrants were obtained. A geodatabase was created, a geometric network was constructed, spatial and attribute queries were run in GIS, map layouts were produced from spatial analysis, and graphs were plotted using attribute data. According to the study, more than half of the Tarkwa network needed to be replaced because it was outdated. Detailed geographically referenced data on pipelines and the cost of replacement as they approach their intended lifespan were supplied by the research (Pindiga et al., 2015).

Materials and methods

Study area

The Saldanha Military Area (SMA) is located on the west coast of South Africa, approximately 140 km northwest of Cape Town (Figure 10.1). It is a strategically important military area that serves as a training base for the South African National Defence Force (SANDF). This area is situated near the town of Saldanha and overlooks Saldanha Bay, which provides access to the Atlantic Ocean and is a key location for the South African Navy (SAN). It has an average elevation of 27 m above sea level. The military area covers approximately 10 km^2.

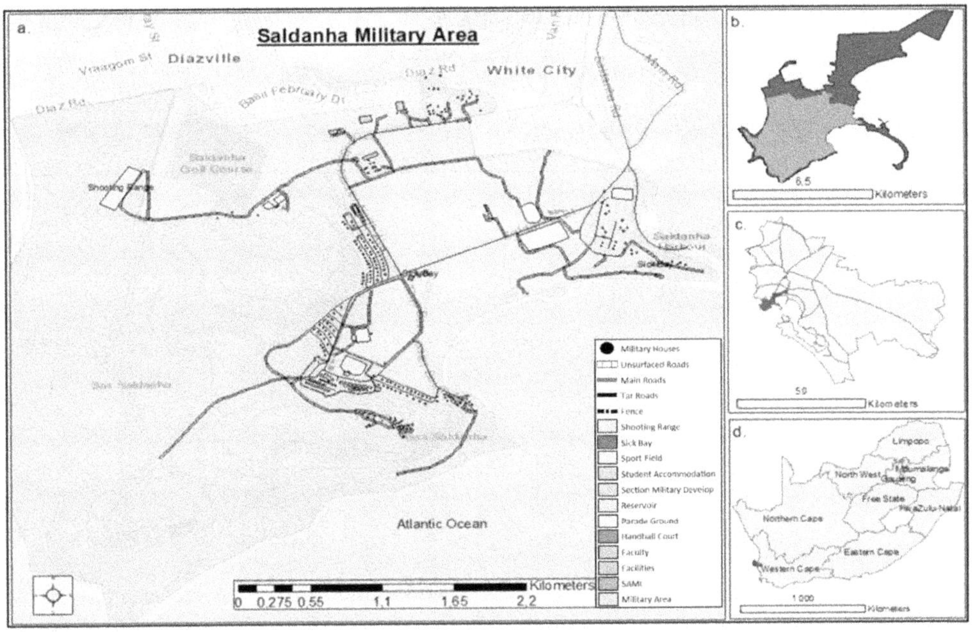

Figure 10.1: (a) Illustration of the study area (Saldanha Military Area); (b) Location of the SMA within Saldanha Bay; (c) Location of Saldanha Bay within Saldanha Bay Municipality (SBM); (d) Location of SBM within the Western Cape province of the Republic of South Africa.

(*Source:* Data courtesy of Municipal Demarcation Board, 2016; map created by the author using Esri ArcMap 10.8.1)

Bezuidenhout (2014) states that the SMA is home to the South African Military Academy (SAMA), Faculty of Military Science, SAS Saldanha, and Military Sick Bay. The abovementioned stakeholders use bulk services in the area. The area is characterised by a Mediterranean climate, with most rainfall occurring in winter and hot conditions in summer (Schulze, 1965). The military area is surrounded by open ocean areas to the west and south in the form of the Atlantic Ocean, with the cold Benguela current flowing northward along the coast. The study area was purposefully selected because of the lack of prior research on this specific topic in the SMA.

Methodology

This study used quantitative data to precisely locate the utilities in the study area using latitude and longitude coordinates. Various utilities, such as water valves, kiosks, pipelines, electrical substations, and fire hydrants, were recorded and their quantities were measured. To ensure accuracy, data were collected through in-situ measurements of the physical locations of utilities. A handheld GPS was used to capture the spatial locations of utilities by employing their latitude and longitude coordinates (x, y) on site. Attribute data for each utility were also recorded in the field. The collected data were projected onto the ArcGIS 10.8.1 interface for further analysis. To create a geospatial database for the study, we used ArcGIS 10.8.1 software, which offers tools for conversion, shape file creation, and analysis. We opted for the Geodatabase (gdb) format because of its capability to store large datasets that demand significant storage space. A Geospatial database (gdb) is a file structure that stores geometry, spatial reference system, attributes, and behavioural rules for data, as defined by Esri (2023). The flowchart of the methodology is shown in Figure 10.2.

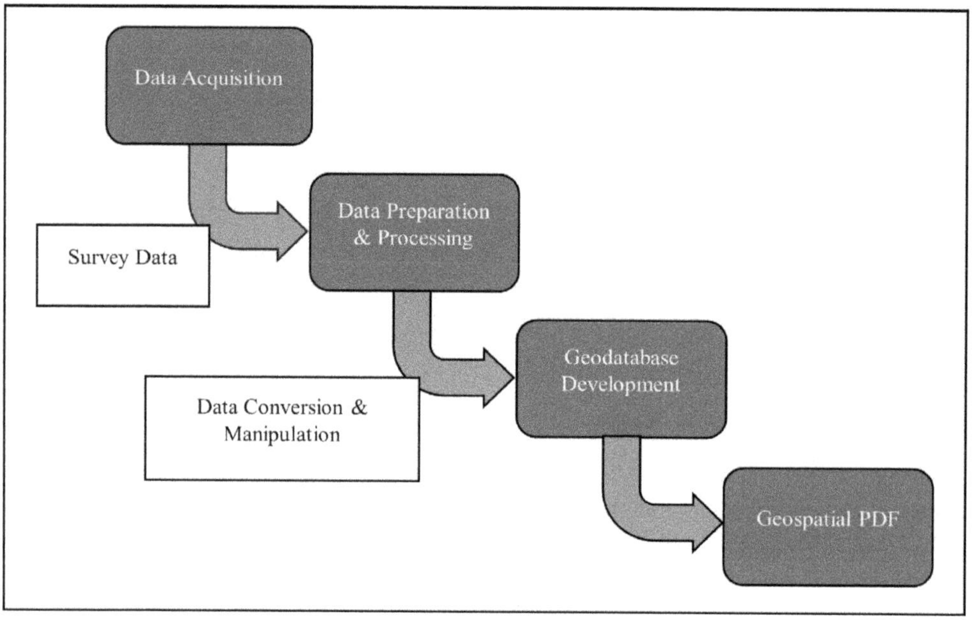

Figure 10.2: Methodological workflow.

Data acquisition

Accurate utility mapping is crucial for the effective analysis, planning, and optimisation of utility flow studies. Therefore, it is fundamental to have a GIS-based utility map that provides precise geo-coordinates and utility configuration information. Successful implementation of GIS demands an appropriate GPS survey which facilitates the creation of a digital map for the distribution of utilities within the SMA. Before conducting field observations and gathering data, the survey team conducted a reconnaissance mission. This mission provided the researcher and his team with a first-hand understanding and mental picture of the project site and study area. Based on the reconnaissance conducted, the team determined the survey approaches suitable for fast-tracking data gathering and techniques to be used for the study. In addition, the recce helped determine and select the necessary equipment for the survey.

Enhancing utility operations with GPS technology

The GPS is a satellite-based system that uses a constellation of 24 operational satellites orbiting the Earth at an altitude of 12,000 m above sea level, transmitting information to users worldwide (Saini et al., 2016). GPS receivers receive data from several satellites and use the information encoded in the signals to triangulate a geographic position on Earth (Zeiler, 1992). The significance of GPS for utilities lies in its ability to provide precise, up-to-date geographic and attribute data for a specific utility, thus making it the most effective method. A handheld Garmin GPSMAP 65s was used to gather data (Figure 10.3) with a precision of 1.8 m.

Figure 10.3: An illustration of the handheld GPSMAP 65s device (left) sourced from the Garmin website, and an in-situ photo of field measurements using the device (right) taken by the authors. (*Sources:* Garmin, 2024; photo credit: Authors' own image)

Data preparation and processing

In this phase, all spatial information from the GPX file was converted into a compatible GIS format. The GPS receiver transferred the data to the workstation via a USB cable. These data were then imported into the GIS database for validity and accuracy and visualised on a base map. This includes the coordinates of all utilities captured and the attributes of each feature. GIS encompasses two distinct data types: vector and raster. In the context of this project, vector data represent the precise locations of water valves, fire hydrants, building footprints as polygons, and pipelines and roads as lines. In addition, the processes that were conducted included the creation of shape files, georeferencing, and digitisation. The digitised and created shapefiles of features were all georeferenced in the Universal Transverse Mercator (UTM) projection system WGS84 and UTM zone 34 S. Digitisation (on-screen) is the process of converting geographic features on a map into a digital format. The x- and y-coordinates of point, line, and polygon features were recorded and stored as spatial data.

Geodatabase development

A geodatabase is a collection of different geographic datasets within a common file system that can be used to store, query, and manipulate spatial data. This encompasses all the feature datasets and feature classes based on the user's requirements, which are georeferenced and already converted into GIS format, such as shapefiles.

Geospatial PDF

The most fundamental stage of this study was the creation of an interactive PDF. To achieve this, a composite map of all utilities was first created using ArcMap. The export option was selected from the menu bar, and the map was saved as a PDF. However, this was not an ordinary export; the trick lay in the advanced tab, where the layers and attributes were set to export PDF layers and feature attributes. Concluded by checking the box to export Map Georeferenced information was checked, and therefore, once the PDF was opened, an object data tool could be selected to perform some analysis.

GeoPDF is widely used to present GIS mapping data using a standard PDF viewer. The geospatial functionality of a GeoPDF includes scalable map display, layer visibility control, access to attribute data, coordinate queries, and spatial measurements (Cokinos et al., 2008; Geilhausen & Otto, 2012). GeoPDF has provided people in the field with the ability to receive geospatially referenced products without using sophisticated software. Figure 10.4 demonstrates the steps employed to create a GeoPDF. The target group for this product is the Faculty of Military Science. Four group layers were created according to the requirements. The focus of this product is the Saldanha Military Area (SMA) and its associated areas.

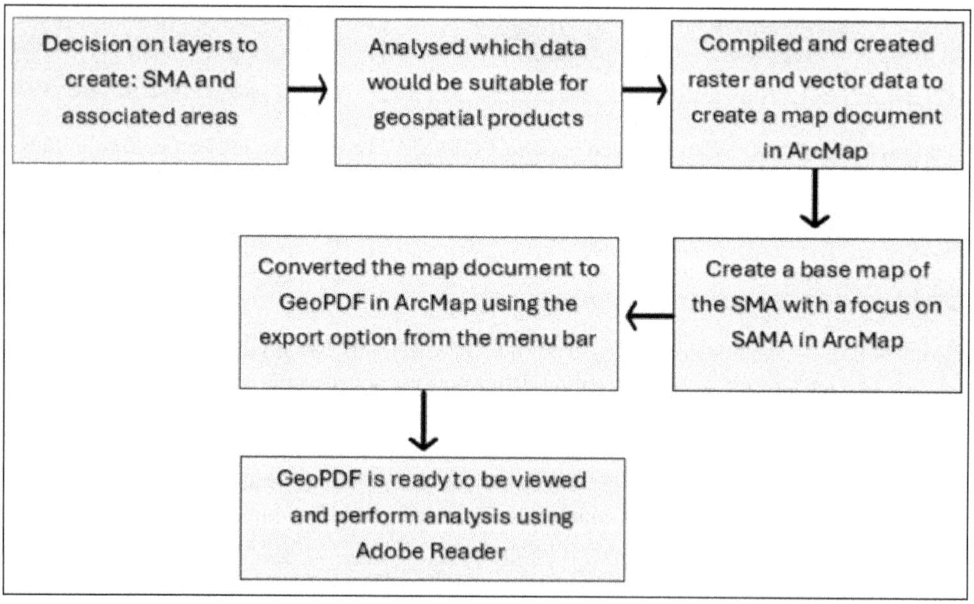

Figure 10.4: Process of creating GeoPDF.

Results and discussions

The geodatabase was designed with all the necessary details for the feature classes. The file geodatabase consisted of four feature datasets (Figure 10.5): (1) fresh water supply, (2) grey water, (3) firefighting, and (4) electrical reticulation.

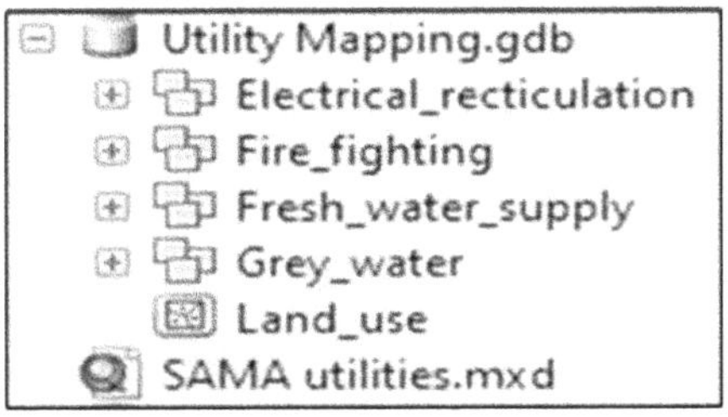

Figure 10.5: Geodatabase design.

The GPS coordinates of all utilities, as demonstrated on the composite map, were captured. These include water valves, fire hydrants, kiosks, electrical substations, sewers, and pipelines. A plethora of queries were raised. Among the queries is the identification of the water valves within the study area, as demonstrated in Figure 10.6. To locate underground utilities such as water valves and sewerage, all that is required is the x- and y-coordinates of the utility in question to be entered into the GPS and then navigated. Although they are subsurface utilities, they are visible from the surface, making them easier to locate. It should be noted that the main purpose of displaying the attribute in the snap is solely to demonstrate how the spatial query works, and this can be done for all mapped features with attribute data.

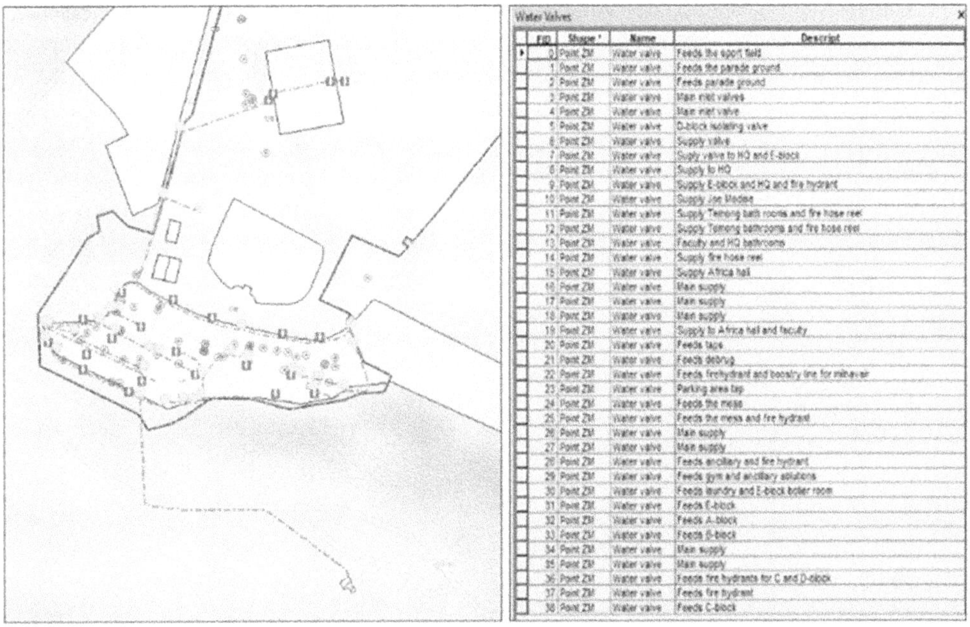

FID	Shape *	Name	Descript
0	Point ZM	Water valve	Feeds the sport field
1	Point ZM	Water valve	Feeds the parade ground
2	Point ZM	Water valve	Feeds parade ground
3	Point ZM	Water valve	Main inlet valves
4	Point ZM	Water valve	Main inlet valve
5	Point ZM	Water valve	D-block isolating valve
6	Point ZM	Water valve	Supply valve
7	Point ZM	Water valve	Supply valve to HQ and E-block
8	Point ZM	Water valve	Supply to HQ
9	Point ZM	Water valve	Supply E-block and HQ and fire hydrant
10	Point ZM	Water valve	Supply Joe Modise
11	Point ZM	Water valve	Supply Temong bath rooms and fire hose reel
12	Point ZM	Water valve	Supply Temong bathrooms and fire hose reel
13	Point ZM	Water valve	Faculty and HQ bathrooms
14	Point ZM	Water valve	Supply fire hose reel
15	Point ZM	Water valve	Supply Africa hall
16	Point ZM	Water valve	Main supply
17	Point ZM	Water valve	Main supply
18	Point ZM	Water valve	Main supply
19	Point ZM	Water valve	Supply to Africa hall and faculty
20	Point ZM	Water valve	Feeds taps
21	Point ZM	Water valve	Feeds [illegible]
22	Point ZM	Water valve	Feeds firehydrant and [illegible] line for [illegible]
23	Point ZM	Water valve	Parking area tap
24	Point ZM	Water valve	Feeds the mess
25	Point ZM	Water valve	Feeds the mess and fire hydrant
26	Point ZM	Water valve	Main supply
27	Point ZM	Water valve	Main supply
28	Point ZM	Water valve	Feeds ancillary and fire hydrant
29	Point ZM	Water valve	Feeds gym and ancillary ablutions
30	Point ZM	Water valve	Feeds laundry and E-block boiler room
31	Point ZM	Water valve	Feeds E-block
32	Point ZM	Water valve	Feeds A-block
33	Point ZM	Water valve	Feeds B-block
34	Point ZM	Water valve	Main supply
35	Point ZM	Water valve	Main supply
36	Point ZM	Water valve	Feeds fire hydrants for C and D-block
37	Point ZM	Water valve	Feeds fire hydrant
38	Point ZM	Water valve	Feeds C-block

Figure 10.6: Query identifying water valves.
(*Source:* Map created by the author using Esri ArcMap 10.8.1)

Figure 10.7 illustrates the composite map for all utilities mapped in the study area. This type of map shows how utilities are distributed within the study area.

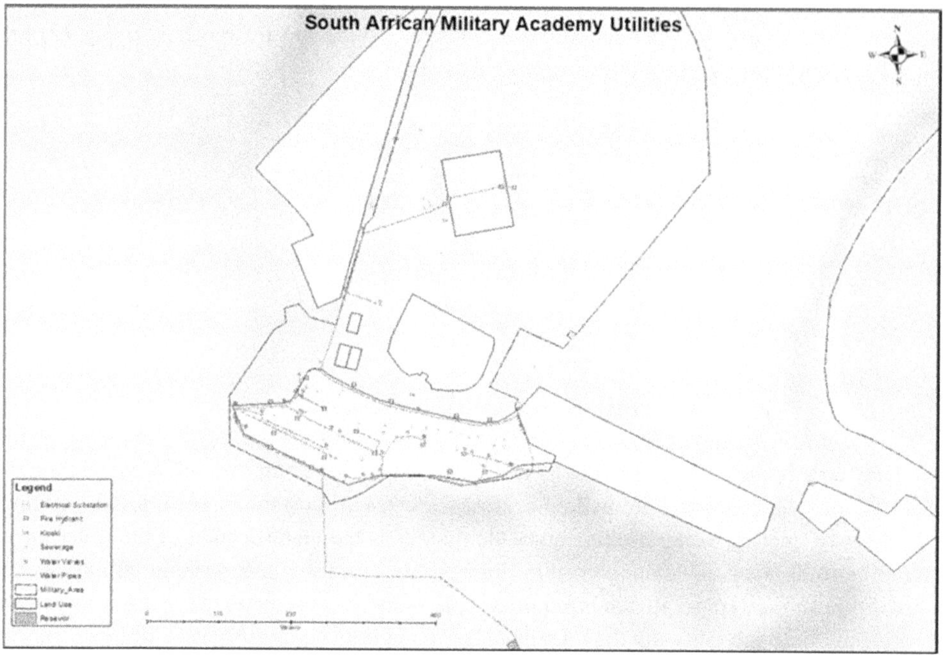

Figure 10.7: Composite map.
(*Source:* Map created by the author using Esri ArcMap 10.8.1)

The National Intelligence Agency initially designed a GeoPDF to provide military commanders with a technologically advanced map. In particular, it was developed to provide mapping information efficiently to commanders in the field as well as the community (Cervantes, 2009). The product allows the end user to access analysis tools such as object data, geolocation, and measuring tools. The data contained in the GeoPDF are geospatially referenced and allow users to turn layers on and off. The product can be viewed using a PDF viewer, such as Adobe Acrobat Reader. The screenshots below (Figure 10.8) demonstrate the analysis tools that can be accessed in the PDF viewer.

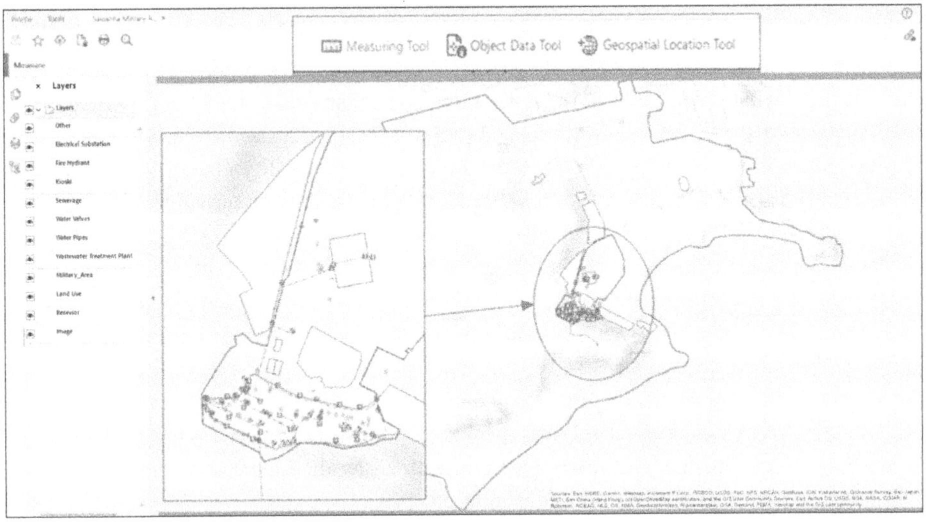

Figure 10.8: GeoPDF analysis tools.
(*Source:* Map created by the author using Esri ArcMap 10.8.1)

The results indicate the unlimited capabilities of GIS integrated with GPS in managing spatial data of SMA utilities. The GeoPDF provides the institution with the ability to locate, measure, and analyse different datasets within the study area. Facility management can access data from computers and smartphones at any time. Layers were created for each feature dataset in ArcMap, which consisted of data from the geodatabase. The interesting part about this product is that when new facilities are replaced, geodatabases and composite maps can be updated effectively to accommodate the new features, thereby making a GeoPDF interactive and reliable. Documents and data associated with the SMA are linked to the geodatabase and can be easily accessed as downloadable files or shared via direct hyperlinks. To fully utilise its interactive features, the document should be opened in Adobe Acrobat Reader, which allows users to access embedded geospatial data, toggle map layers, and perform basic spatial analyses.

This study demonstrated the capability of GIS in utility mapping with the integration of GIS and handheld GPS. Utilities are crucial for every organisation and should be managed properly to avoid accidents that may result in monetary losses. The SMA does not have a map showing the distribution of utilities and their layout. Consequently, SMA requires a digital map showing the locations of these utilities. Reticulation plans play a critical role when repair and replacement are required. Facility managers should maintain and continuously check the serviceability of these utilities. For facility managers, it is important to have this kind of artefact that simply shows the exact location and features instead of spending time in the field trying to locate these features. It should be noted that for any construction to begin with their work, in any location, knowledge of the features should be presented and made available to them. Therefore, geospatial products allow the Faculty of Military Science and associated stakeholders to visualise and manage their utilities. The products offer users an accurate variety of information linked to each feature of the product. Information pertaining to bulk utilities at the SMA can be accessed from the geodatabase that was created to specifically address the current challenge. The mapped data can, therefore, be updated at any time should the need arise, such as future revamping. Moreover, this study demonstrates the use of GIS in utility mapping as an effective and efficient tool.

The geospatial product (GeoPDF) is easy to use because it requires no prior GIS experience. In the table of contents, the layers are labelled, making it easier for users to obtain information related to a specific feature. The table of contents also provides clear and concise symbols, which are all unique. Updates can be performed on the geodatabase (mxd. document), which contains spatial and non-spatial data of all utilities. Therefore, these data can be exported as GeoPDFs to integrate new information pertaining to the study area. If a facility manager needs to replace and repair utilities, a geodatabase is readily available to incorporate new data. In addition, the layers in the table of contents were labelled to allow users to choose which feature to visualise and how to attribute information about a specific feature. During the study, the main challenge encountered was pinpointing the exact location of the sewerage lines, which made it difficult to determine the direction of flow. In addition, there was insufficient equipment to identify underground utilities. Moreover, a significant number of these drains were found to be sealed with concrete, further complicating the task; hence, they were labelled 'unknown' in the attribute table. As a result, physical inspection with specialised tools is necessary for proper identification and marking. In conclusion, this study was constrained by the visual accessibility of utilities, which restricted the ability to plot underground utilities.

While the GeoPDF provides a static but interactive solution for visualising utility infrastructure, a more dynamic and integrated approach is being undertaken by updating Stellenbosch University's existing campus map (Figure 10.9) to include the Saldanha Campus. This update ensures that the Faculty of Military Science's facilities and utilities are accessible within the university's official mapping system, filling a critical gap in digital campus representation. By incorporating this data into the layer list of the campus map, facility managers and stakeholders will gain improved accessibility to spatial information, enabling more effective decision-making and facility management.

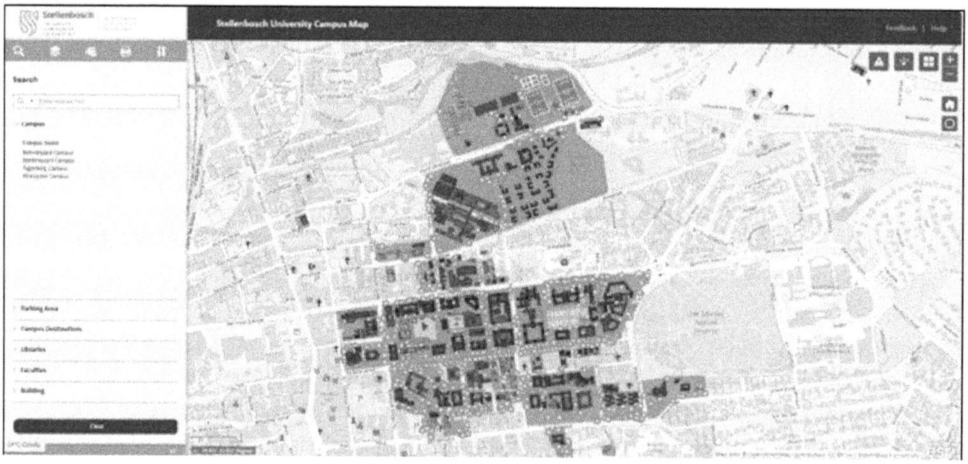

Figure 10.9: Stellenbosch University Campus Map interface, showcasing the structured layout of buildings, faculties, parking areas, and key campus destinations. This interactive GIS-based system allows users to search for facilities and navigate the campus efficiently.

Conclusion

This project successfully demonstrated the role of GIS and GPS technology in enhancing utility management at the Saldanha Military Academy. By adhering to survey protocols and conducting detailed reconnaissance, this study ensured precise spatial mapping of essential utilities, addressing the long-standing challenge of outdated or missing reticulation plans. The integration of field-collected GPS data with GIS software provided a structured and accurate geodatabase, enabling improved visualisation and accessibility of infrastructure information.

The findings highlight the practical advantages of using GeoPDFs as an intuitive tool for non-GIS users, offering a static yet interactive method for visualising and retrieving spatial data. However, recognising the need for a more scalable and integrated approach, the geodatabase developed in this study will contribute to updating Stellenbosch University's existing campus map, ensuring that the Saldanha Campus is fully incorporated into the university's broader digital mapping system. This update will not only enhance access to facility and utility data but also support long-term planning and infrastructure management within the Faculty of Military Science.

This study reinforces the enduring importance of GIS in facility and utility management, particularly in institutional and military environments, where the accuracy and accessibility of spatial data are critical. As technology advances, continued updates to the geodatabase, along with the potential incorporation of remote sensing techniques such as Ground Penetrating Radar (GPR), will further improve underground infrastructure mapping and maintenance. By integrating GIS-based solutions into existing university systems, this study lays the foundation for more efficient, data-driven decision-making in infrastructure planning and management.

References

Ajwaliya, R.J. & Patel, S. 2014. Design and development of GIS-based utility management system at DOS Housing Colony, Vikramnagar, Ahmedabad. *International Journal of Engineering Research and Applications*, 4(9): 56–73.

Andrew, L. 2020. The vulnerability of vital systems: How "critical infrastructure" became a security problem. In: M.A. Dunn & K.S. Kristensen (eds.). *Securing 'the Homeland': Critical Infrastructure, Risk and (In)Security*. Oxfordshire: Routledge. 17–39. https://doi.org/10.4324/9780203926529-2

Anozie, M.C. 2019. Application of Geographic Information System (GIS) in utility management: A case study of New Haven water distribution management. Master's dissertation. Enugu: Godfrey Okoye University.

Bernhardsen, T. 2002. *Geographic Information Systems: An Introduction*. New York, NY: John Wiley & Sons.

Bezuidenhout, J. 2014. The background radiation and exposure levels at various South African west coast military units. *Scientia Militaria: South African Journal of Military Studies*, 42(2): 164–176. https://doi.org/10.5787/42-2-1098

Bozkaya, B., Bruno, G. & Giannikos, I. 2019. Location and geographic information systems. In: G. Laporte, S. Nickel, & F. Saldanha da Gama (eds.). *Location Science*. Cham: Springer. 559–590. https://doi.org/10.1007/978-3-030-32177-2_19

Buyuksalih, I., Bayburt, S.G., Baskaraca, A., Karim, H. & Rahman, A. 2017. 3D modelling and visualization based on the Unity game engine: Advantages and challenges. *ISPRS Annals of the Photogrammetry, Remote Sensing and Spatial Information Sciences*, IV-4/W4: 161–166. https://doi.org/10.5194/isprs-annals-IV-4-W4-161-2017

Cannistra, J.R. 1999. Converting utility data for a GIS. *American Water Works Association Journal*, 91(2): 55–64. https://doi.org/10.1002/j.1551-8833.1999.tb08585.x

Cazzaniga, N.E., Carrion, D., Migliaccio, F. & Barzaghi, R. 2013. A shared database of underground utility lines for 3D mapping and GIS applications. *ISPRS International Archives of the Photogrammetry, Remote Sensing and Spatial Information Sciences*, 1: 105–108. https://doi.org/10.5194/isprsarchives-XL-4-W1-105-2013

Cervantes, D. 2009. Using GIS to create an interactive GeoPDF mapbook for the Big Island of Hawaii. Master's dissertation. Maryville, MO: Northwest Missouri State University.

Chen, R.J.C. 2007. Geographic information systems (GIS) applications in retail tourism and teaching curriculum. *Journal of Retailing and Consumer Services*, 14(4): 289–295. https://doi.org/10.1016/j.jretconser.2006.07.004

Cokinos, J.S., Egorov, V.I., Vivian, J.R., Haney, J.W. & Shields, G.W. 2008. A versatile alternative for distribution of geospatial data. https://www.searchanddiscovery.com/documents/2009/40387cokinos/

Crawford, D. 2012. Implementing a utility geographic information system for water, sewer, and electric. Master's dissertation. Maryville, MA: Northwest Missouri State University.

Dangermond, J. & Goodchild, M.F. 2020. Building geospatial infrastructure. *Geo-Spatial Information Science*, 23(1): 1–9. https://doi.org/10.1080/10095020.2019.1698274

DOSH [Department of Occupational Safety and Health]. 2008. *Occupational Safety and Health Act and Regulations 1994*. Kuala Lumpur: MDC Publishers Sdn Bhd.

Esri [Environmental Systems Research Institute]. 2023. Geodatabase definition. GIS Dictionary. https://support.esri.com/en-us/gis-dictionary/geodatabase [3 April 2023].

Geilhausen, M. & Otto, J.C. 2012. Smart geomorphological map browsing: A tale about geomorphological maps and the internet. *EGU General Assembly Conference Abstracts*.

Goodchild, M.F. 2009. Geographic information systems and science: Today and tomorrow. *Annals of GIS*, 15(1): 3–9. https://doi.org/10.1080/19475680903250715

Hughes, J. 2006. GIS combines geography and information for effective utility management. *American Water Works Association Journal*, 32(12): 10–11. https://doi.org/10.1002/j.1551-8701.2006.tb01904.x

Jamil, H., Nomanbhoy, Z. & Mohd Yusoff, M.Y. 2012. Underground utility mapping and its challenges in Malaysia. https://www.fig.net/resources/proceedings/fig_proceedings/fig2012/papers/ts05j/TS05J_jamil_nomanbhoy_et_al_5636.pdf [25 September 2023].

Khan, S.Z., Mohsin, M. & Iqbal, W. 2021. On GPS spoofing of aerial platforms: A review of threats, challenges, methodologies, and future research directions. *PeerJ Computer Science*, 7. https://doi.org/10.7717/peerj-cs.507

Kolera, B.T. & Bernold, L.E. 2006. Intelligent utility locating tool for excavators. *Journal of Construction Engineering and Management*, 132(9): 919–927. https://doi.org/10.1061/(ASCE)0733-9364(2006)132:9(919)

Li, S., Cai, H. and Kamat, V.R., 2015. Uncertainty-aware geospatial system for mapping and visualizing underground utilities. *Automation in Construction*, 53, pp.105-119.

Mennecke, B.E. 1997. Understanding the role of geographic information technologies in business: Applications and research directions. *Journal of Geographic Information and Decision Analysis*, 1(1): 44–68.

Meyers, J.R. 1999. GIS in the utilities. In: P.A. Longley, M.F. Goodchild, D.J. Maguire & D.W. Rhind (eds.). *Geographical Information Systems: Management Issues and Applications*. Volume 2. New York, NY: John Wiley & Sons. 801–818.

Momin, A.A.S. & Kolekar, S. 2017. Mapping of the assets and utilities: A vision for the development of smart cities in India. *International Journal of Applied Engineering Research*, 12(24): 15378–15383.

Mwaniki, M.W. & Odera, P.A. 2012. Application of GIS in facility space management: A case study of ILRI. *International Journal of Science and Research (IJSR)*, 1: 1–8.

Okpuvwie, E.J. & Toko Mouhamadou, I. 2023. Application of geospatial technologies in military operations. *Socialscientia: Journal of Social Sciences and Humanities*, 8(2). https://journals.aphriapub.com/index.php/SS/article/view/2119

Pick, J.B. 2008. *Geo-business: GIS in the Digital Organization*. New York, NY: John Wiley & Sons.

Pincevičius, A., Baušys, R. & Jankauskas, P. 2005. Application of geographical information systems in modelling of military operations. *Aviation*, 9(3): 36–42. https://doi.org/10.3846/16487788.2005.9635909

Pindiga, A.M., Sani, M.J. & Garba, T. 2015. Mapping of water distribution network using GIS technology in Bauchi Metropolis, Nigeria. *Journal of Environment and Earth Science*, 5(13): 88–94.

Rahman, M.A., Maulud, K.A., Bahri, M.S., Hussain, M.S., Oon, A.R., Suhatdi, S., Hashim, C.C. & Mohd, F.A. 2020. Development of GIS database for infrastructure management: Power distribution network system. *IOP Conference Series: Earth and Environmental Science*, 540(1). https://doi.org/10.1088/1755-1315/540/1/012067

Rahman Jumat A., Coffey, V. & Skitmore, M. 2012. Stakeholder service delivery expectations of military facilities management. *Built Environment Project and Asset Management*, 2(2): 146–166. https://doi.org/10.1108/20441241211280891

Rayburn, S.R. 2004. The use of GIS in water utility asset management: Four Seasons Subdivision of Lufkin, Texas. Master's dissertation. Nacogdoches, TX: Stephen F. Austin State University.

Saini, D., Merotha, S. & Sharma, A. 2016. Empirical study of global positioning system analyzing issues and challenges. *International Journal of Innovative Research in Science and Engineering*, 2(4): 1–9.

Schulze, B.R. 1965. *Climate of South Africa*. Part 8. General Survey. Pretoria: Government Printer & Weather Bureau.

Shadin, M.S. & Tahar, K.N. 2015. The implementation of mobile GIS for fire hydrant mapping. *2015 International Conference on Space Science and Communication (IconSpace)*. 65–70. https://doi.org/10.1109/IconSpace.2015.7283782

Shamsi, U.M. 2005. *GIS Applications for Water, Wastewater, and Stormwater Systems*. Boca Raton: CRC Press. https://doi.org/10.1201/9781420039252

Ufoegbune, G.C., Atanley, P.A., Eruola, A.O., Makinde, A.A. & Ojekunle, Z.O. 2016. Geographical information system (GIS) application for planning and improvement of public water supply in Ota, Ogun State, Nigeria. *Journal of Applied Sciences and Environmental Management*, 20(4): 1105–1111. https://doi.org/10.4314/jasem.v20i4.26

Zeiler, M.N. 1992. Integrating GIS and GPS for mapping and analysis of electric distribution circuits. *Proceedings 1992 Rural Electric Power Conference*.

CHAPTER 11

Maritime surveillance in Saint Helena Bay using a multisource approach

J.M. Mametja and D. Putter

Abstract

This study focuses on the integration of the Sentinel-1 Synthetic Aperture Radar satellite with the Automatic Identification System to detect and analyse illegal, unreported, and unregulated activities conducted within Saint Helena Bay, South Africa. Such activities threaten the marine ecosystem and the national security of South Africa. The activities include illegal drug trade and the fishing of endangered species, such as lobsters and abalone. This highlights the need for enhanced and enforced surveillance methods. The proposed method integrates the use of the Automatic Identification System with the Sentinel-1 satellite, which provides a tool that enhances the traditional way of monitoring marine activities. The Sentinel-1 satellite can provide high-resolution imagery regardless of weather conditions or time of day. This complements the Automatic Identification System by tracking vessels even when the transponder is manipulated or turned off. The integration will allow security authorities to detect suspicious vessel movements and patterns, which can indicate potential illegal activities. This will lead to more effective enforcement and protection of South Africa's marine resources. However, there are still challenges that need to be investigated to improve the systems and surveillance methods.

Keywords: Automatic Identification System, illegal, Sentinel-1, Sentinel Application Platform, Synthetic Aperture Radar, unreported and unregulated

Introduction

"Protection of coastal waters against illegal activities is essential for the maintenance of economic security and sovereign integrity." (Reutech Radar Systems, 2025)

South Africa is experiencing an increase in coastal activities, such as fishing and shipping, which are essential for economic improvement. This is also driven artificially by the instability caused by conflicts close to the Horn of Africa (Martin, 2024). There has been a recent flurry of exploration for oil and gas on the west coast of South Africa and Namibia. This will increase shipping traffic owing to the very large oil and gas fields opened just north of the South African maritime border with Namibia (Koning, 2024). However, this importance is overlooked, and state and non-state actors tend to exploit this through the illegal harvesting of maritime resources and shipping. These illegal activities are not unique to the South African coastline but are a global phenomenon that threatens marine ecosystems and national security. These activities include the shipping of narcotics, people, and endangered and protected fauna and flora, as well as illegal fishing of endangered species such as lobsters and abalone (Reva & Walker, 2020; UNODC, 2014; Brick et al., 2014). A 2014 UNODC report states that "[w]hile some 500 million containers move across the globe each year, accounting for nearly 90 per cent of global trade, less than two per cent of these are ever physically inspected. Although the majority transport legal cargo, freight containers are constantly targeted for the shipment of drugs, counterfeit goods, weapons, and other illicit products, which are smuggled alongside legal commodities" (UNODC, 2014). Ten years later, with significant increases in worldwide trade and transport, this threat has grown substantially.

In South Africa, the development of Marine Protected Areas (MPAs) was introduced to combat, or at least mitigate the negative effects of these threats, and to increase the visibility and awareness of maritime security. MPAs are "sites that have been identified and selected to protect ecosystems, serve as a sanctum for commercially harvested fish species to recover, and provide some benefits to humans" (Keebine, 2023). Currently, there are 42 MPAs covering a maritime area of 57,736 km^2, an increase from 0.4% of the South African marine real estate to 5.4% thereof (Keebine, 2023). However, the development of MPAs highlights challenges, as they do not cover the entire South African coastal area (Figure 11.1).

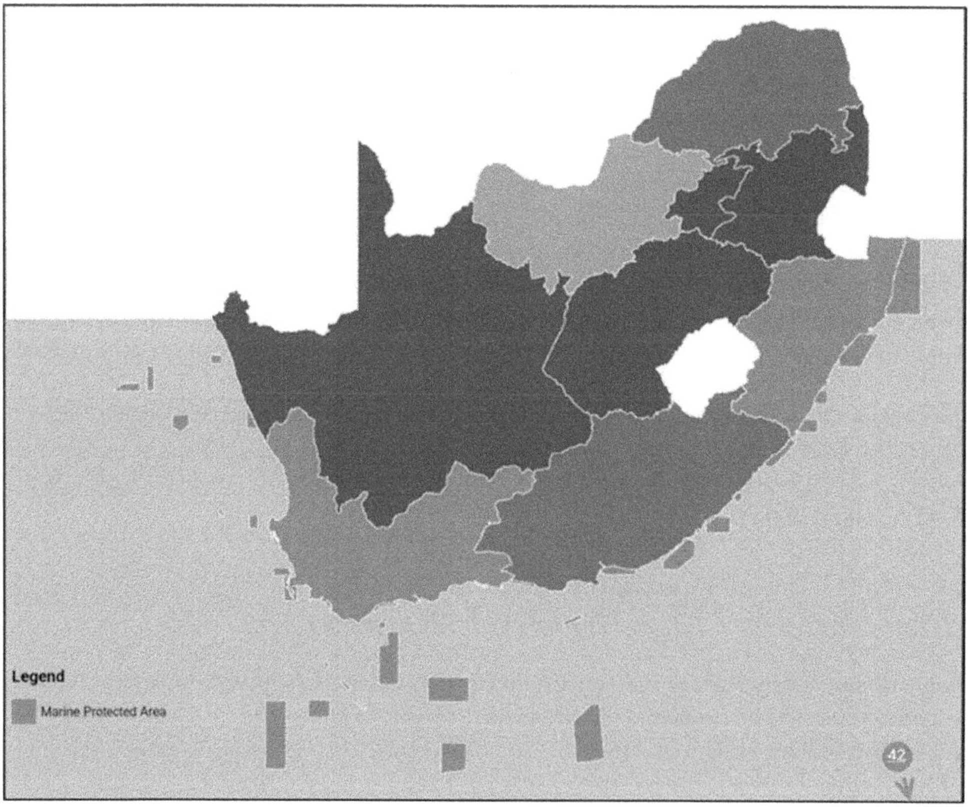

Figure 11.1: South African Marine Protected Areas.
(*Source:* Keebine, 2023)

Patrolling such a large area with widely distributed MPAs across the South African maritime real estate (Figure 11.1) is problematic and challenging. Thus, there is a need for different policing and enforcement strategies and enhanced surveillance methods at sea (Fielding, 2021).

The increase in illegal marine activities around the South African coast and in the oceans highlights the need for advanced surveillance. Traditionally, surveillance methods used were patrolling vessels and the monitoring of the Automatic Identification System (AIS). However, monitoring vessel movement becomes difficult when the AIS is shut down (Iacarella et al., 2023). The AIS is a short-range automatic tracking system that shares information between ships and shores for traffic purposes. The AIS enables authorities to identify specific vessels and their activities within or near a nation's Exclusive Economic Zone. When AIS data are fused with existing radar systems, authorities can more easily differentiate between vessels. AIS data can be automatically processed to create normalised activity patterns for individual vessels, which, when breached, create an alert, thus highlighting potential threats for more efficient use of security assets. The AIS improves maritime domain awareness and allows for heightened security and control (NATO, 2021).

The introduction of radar satellite systems, especially Sentinel-1 (ESA, 2025a) and TerraSAR-X (ESA, 2025b), has revolutionised the traditional sea surveillance *modus operandi*. These satellites use Synthetic Aperture Radar (SAR) technology for detection and data collection with a mission life of approximately seven years. Specifically, the Sentinel systems use Sentinel-1C satellites since their launch on 5 December 2024 (Sentinel-1B was terminated in 2022 [ESA, 2025a]). Radar satellites are used to complement AIS and mitigate its limitations. Radar was selected for its capability to detect vessels that intentionally switch off their transponder or location. This is called the dark target. "A dark target at sea can be an object adrift, someone who has accidentally turned off their Automatic Identification System (AIS) transponder, or someone who intentionally wants to hide. These can all threaten critical infrastructure assets at sea, whether it is accidental or malicious" (Kongsberg, 2024). Radar satellites can also be used in any weather conditions. It provides more relief to surveillance efforts as it reduces the need for physical patrolling during low visibility conditions (such as nighttime) because the radar can provide data during the day and night (Grover et al., 2018).

Sentinel-1 is the focus of this chapter because of the extensive coverage it provides. This, along with the fact that its data are publicly available and have a high resolution compared to other satellites, makes it a very cost-effective sensor (Velotto et al., 2016).

The integration of Sentinel-1 and AIS data provides a rigorous and holistic surface picture in support of Maritime Domain Awareness (MDA). MDA is defined as "the effective understanding of anything associated with the maritime domain that could impact security, safety, the economy or the marine environment" (IMO, 2025a). MDA consists of at least the following three components that focus on threat and safety within the maritime space: Maritime Situational Awareness, Maritime Threat Awareness, and Maritime Response (IMO, 2025a). This would allow the identification of ships that power down their AIS while underway to avoid satellite detection (Grover et al., 2018). This study explores the integration of AIS and Sentinel-1 Synthetic Aperture Radar (SAR) data to monitor maritime activities and identify ships suspected of conducting illegal activities. This study focused on the Saint Helena Bay area. Maps were created using Esri ArcGIS® software. ArcGIS® and ArcMap™ are the intellectual properties of Esri and are used herein under licence (ArcGIS, 2025).

Geographic area of interest

This study focuses on Saint Helena Bay, hereafter referred to as 'the Bay'. The Bay has a rich history stretching back to the landing of Vasco da Gama in 1497, who named the Bay "Bahia de Santa Elena" (SAHO, 2025). The Bay was selected for its unique marine biodiversity and socio-economic significance. The Bay is located, as a pinpoint geo-location, at the geographical coordinates 32°45'30" S, 18°01'40" E. It is situated along the southwestern coastline of South Africa, within the Western Cape province, and is approximately 150 km north of Cape Town. Figure 11.2 illustrates the geographic area, where A represents the provincial map of South Africa, and B focuses on the Bay.

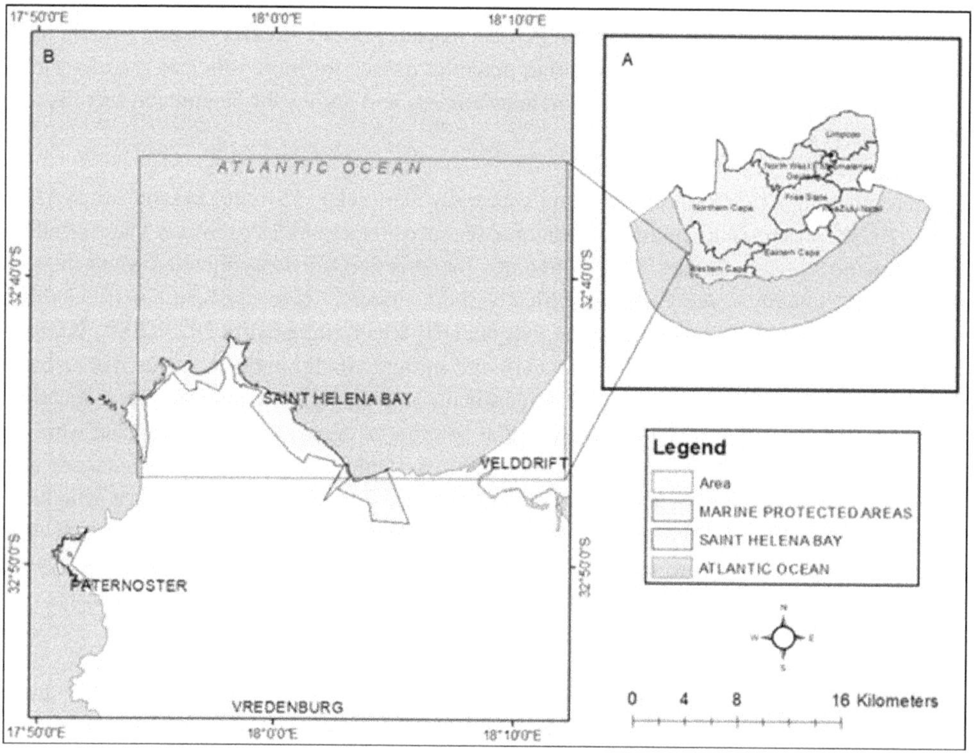

Figure 11.2: Saint Helena Bay.
(*Source:* Authors' own work using ArcMap 10.8.1)

The Bay is bordered by the towns of Vredenburg to the south, Paternoster to the southwest, the South Atlantic Ocean to the west and north, and Velddrif to the east (Schultz, 2015; Ward et al., 2022).

More granularly, the Bay stretches from Dwarskersbos in the north past Shelley Point Peninsula to Cape St Martin in the west. It comprises a total of 18 small bays. It also incorporates the neighbouring towns of Stompneusbaai and Britannia Bay (SAHO, 2025). The Report on the Economic and Socio-Economic State and Growth Prospects of the 12 Proclaimed Fishing Harbours in the Western Cape (Department of the Premier, 2012) describes Saint Helena Bay as a notable destination due to its Mediterranean climate and natural beauty, making it a perfect destination for tourism and thus socio-economically important for the Western Cape economy. Biccard et al. (2020) note that the residents of the Bay experience a Cape Mediterranean climate, characterised by hot, dry summers and mild and wet winters.

The Bay is of considerable significance owing to its diverse marine ecosystems and economic role as a hub for fishing activities. According to Burger et al. (2020), the Bay benefits from the nutrient-rich Benguela current, rendering the harbour, situated on the southwestern perimeter of the Bay, an active commercial fishing port that processes the largest volume of small pelagic fish in South Africa. The harbour also supports vibrant small-scale fishing activities, targeting marine species such as snoek, abalone, and an endangered type of rock lobster. However, due to inadequate harbour management and protection measures, the area is vulnerable to exploitation. A significant number of individuals engage in the harvesting of marine species due to poor socio-economic opportunities in the area (Schultz, 2015), positioning the area as a hotspot for illegal fishing and marine poaching activities. This geographic area of interest extends over approximately 800 km^2.

Methodology

The methodology of this study delineates the comprehensive approach undertaken to analyse data sourced from the AIS and Sentinel-1 satellite imagery, which specifically focused on monitoring illegal, unreported, or unregulated fishing activities within the geographic area of interest (i.e. Saint Helena Bay). This method combines qualitative approaches to data collection with quantitative approaches to data analysis. Furthermore, the investigation integrates multiple data streams, i.e. AIS and Sentinel-1 data, where AIS data represent the real-time positional information of vessels and are critical for maritime traffic pattern analysis. Sentinel-1, along with its SAR technology, offers the high-resolution imagery required for vessel detection under most environmental conditions. A robust analytical framework was created by combining these datasets, enabling maritime activity monitoring and the identification of potential illegal activities.

Data sources

Sentinel-1 data were downloaded from the Copernicus website, and the AIS data were extracted from an open-source website.[1]

Sentinel-1 SAR Data

Sentinel-1 is a near-polar sun-synchronous satellite developed by the European Space Agency (ESA) as part of the Space Component of Copernicus. Sentinel-1 comprises two satellites (A and B), each carrying a C-band SAR sensor, enabling it to capture in dual polarisation (VH + VV). Both satellites provide freely available data at relatively short revisit times of up to 6 days (Grover et al., 2018; Torres et al., 2017). The open and free data policy of Sentinel-1 provides the ability to download Ground Range Data (GRD) images captured with only interferometric wide-swath (IW) mode. IW is the mode used by the satellite to capture images that cover a 250 km swath at a high spatial resolution of 5×20 m, with an orbital period of 98.6 min (Mdakane et al., 2023; Grover et al., 2018). The image was captured on 12 January 2024 at 17:33:49 when the satellite

1 Global Fishing Watch: https://globalfishingwatch.org/

passed from north to south. Table 11.1 highlights the specifications of the Sentinel-1A satellite and the date acquired for this study.

Table 11.1: Characteristics of the satellite

Specifications	**Descriptions**
Mission	Sentinel-1A
Product type	Ground Range Detected with high resolution
Acquisition mode	IW (Interferometric Wide Swath Mode)
Acquisition time	12 Jan 2024 - 17:33:49
Pass	Ascending
Polarisation	Dual (VH + VV)
Band	C Band
Resolution	5 m x 20 m
Swath width	250 km
Orbit	Near-Polar Sun-Synchronous @ 693 km, 12 days repeat cycle
Orbital period	98.6 minutes
Launch weight	2,300 kg
Battery capacity	≥ 324 Ah
Science data/storage capacity	1,410 Gigabits (End-of-Life)

Automatic Identification System (AIS) data

This study incorporated the use of an AIS based on satellite (Sat-AIS) and terrestrial AIS (t-AIS) data. Iacarella et al. (2023) claim that Sat-AIS provides data with wide coverage on the open sea but contains some temporal gaps due to AIS messages received late after the satellite passed the area. In contrast, t-AIS is used to collect data on ground satellites, which are spatially restricted depending on the extent of land-based stations. However, various companies provide high-quality AIS data that are not restricted but are expensive to acquire. The AIS data used in this study were collected from a company group called Global Fishing Watch. The company provides publicly available AIS data on human activities in the ocean, including fishing activities.

Method

This study followed a quantitative approach to data analysis. A multisource vessel detection system that integrates the use of SAR data and AIS for enhanced monitoring of vessel activities within the Bay was explored. The methodology includes the pre-processing of Sentinel-1 data, that is, the detection of vessel data and the analysis of such data.

Preprocessing

The Sentinel Application Platform (SNAP) software enables users to process and analyse SAR imagery. SNAP software allows users to observe the Earth and is particularly preferred for monitoring ship activities. This software enables, with the pre-processing of Sentinel-1 images, the detection and identification of activities at sea. In other words, it is perfectly suitable for MDA functions.

A series of seven preprocessing steps was implemented to reduce the noise and distortions within the imagery for better analysis. These steps were conducted within the SNAP software's Graph Builder. This allows users to create a workflow diagram with input data (read), nodes, and output (write). A workflow (Figure 11.3) was constructed on the SNAP platform to facilitate image pre-processing. Initially, a subset of the geographic area of interest was created to reduce the processing, storage, and time of the image (Moskolaï et al., 2022). The original image that was captured lacked accuracy because it was affected by the ascending movement of the satellite and required geocoding. By applying the orbit file, the geocoding of the image is improved by assigning the correct geographic coordinates to the image. However, this process does not correct all errors in the image. The next process, Border Noise Removal (BNR), was applied to remove noise (or distortion) from the edges of the image. This noise could be due to atmospheric interaction with satellite radiation during image acquisition. Once BNR was achieved, calibration was performed to correct the image angle (Filipponi, 2019).

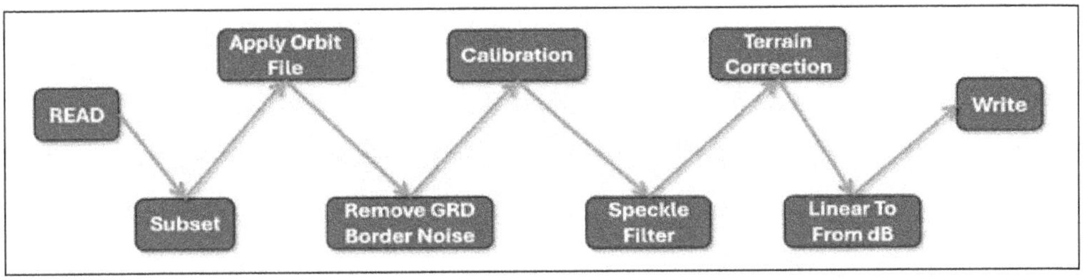

Figure 11.3: Workflow diagram depicting pre-processing steps.

Radar images often have noise called speckle (also referred to as the salt-and-pepper effect). This type of noise is common. "Noises degrade image quality, causing information loss and unsatisfying visual effects. Salt-and-pepper noise is one of the most popular types of noise that affect image quality. In RGB colour image, salt and pepper noise changes the number of occurrences of colours combination depending on the noise ratio" (Azzeh et al., 2018:252). Speckle filtering processes were applied to remove such noise. This process improves the visual quality of SAR images. However, this does not correct the coordinates of the image that were incorrectly misplaced during the acquisition of the image, making the analysis of the image difficult.

The next process entailed georeferencing the image according to the World Geodetic System 1984 (WGS84). "WGS84 consists of a reference ellipsoid, a standard coordinate system, altitude data, and a geoid. The error of WGS84 is believed to be less than 2 cm to the centre mass" (GIS Geography, 2025), and this process corrects the distortions which occurred during the acquisition of the image. After correcting the coordinates and georeferencing of the image, the final step in pre-processing was converting the pixel values of the image to decibels (dB). Generally,

SAR images captured by radar satellites have high pixel values, and a decibel transformation is performed to improve the visualisation of the image for a more precise analysis. This conversion is often achieved through a logarithmic transformation that adjusts radar backscatter values to a more standardised and interpretable scale (Madaan & Kaur, 2019; Moskolaï et al., 2022).

Ships typically appear brighter with a high backscattering value in SAR images, whereas darker areas represent smooth sea surfaces. After completing image pre-processing, the image was ready for ship detection, which was carried out through the ocean object detection process within the SNAP software. An ocean object detection tool was employed to identify objects that backscatter on the sea surface, such as ships and other buoyant platforms. The object detection process was completed in three steps. The first step was land and sea masking, which was used to detect any metal, such as bridges and ships, by covering the land areas to minimise the detection of terrestrial features that may appear brighter in SAR images. The second step was pre-screening, which uses the Constant False Alarm Rate detector to assist in the fast detection of ships on the sea surface as captured by the image. The last step was the discrimination process, which was applied to eliminate the false detection of any brighter object on the sea surface with less than 30 m, which is considered a false detection (Mdakane et al., 2023; Bioresita et al., 2018).

Data analysis

Sentinel-1 GRD images and AIS layers were created and merged for analysis to detect ships suspected of conducting illegal activities within the geographic area of interest, which provided an enhanced maritime surveillance method. From the AIS layer, linear interpolation was applied to estimate the ship path depicted as a straight line before merging the two layers. A line was drawn, assuming that the ships travelled in a straight line. This line was drawn by connecting each ship track. This process was performed so that the ship path could sync with the timestamps of Sentinel-1 because the AIS data were recorded continuously while the Sentinel-1 image was captured at fixed intervals.

After linear interpolation, an association was established between the two layers. This involved creating a buffer of 50 m around each ship detected in the Sentinel-1 image to identify intersections and non-intersections with interpolated ship tracks. Inconsistencies in ships reporting their locations could be identified by a thorough analysis of both layers (Sentinel-1 and AIS datasets), which could send a signal of suspected illegal activity (Mdakane et al., 2023). This includes ships that intentionally do not share their location to mask activities or ships that do not carry their AIS, which is prohibited by law (IMO, 2025b). It is mandated under SOLAS regulation V/19 that ships should be equipped with an AIS transponder for automatic reporting in areas where it is mandatory to exchange information between ships and the shore for traffic management purposes (IMO, 2025b). This greatly enhances safety at sea, owing to the increased visibility of vessels underway and static, which reduces the probability of collisions. Failure to share such information is illegal (except for certain categories of vessels under national regulations, such as naval vessels), and it is an indication of suspicious behaviour, especially if it is detected in MPAs.

Results

This study merged the use of publicly available data from the Sentinel-1 satellite with AIS data for enhanced monitoring of human activities within the geographic area of interest. The Sentinel-1 image followed a pre-processing of imagery, which included steps ranging from creating subset images to correcting errors and adjusting the coordinate system, which significantly improved the quality and usability of the SAR imagery. Subsequent object detection operations facilitated the accurate differentiation between ships and a smooth sea surface, allowing for precise ship identification within the geographic area of interest. The detected ships are visually represented with red buffers in Figure 11.4, and this data was further enriched with AIS data through linear interpolation to depict vessel tracks, which provided a full understanding of vessel movements and activities.

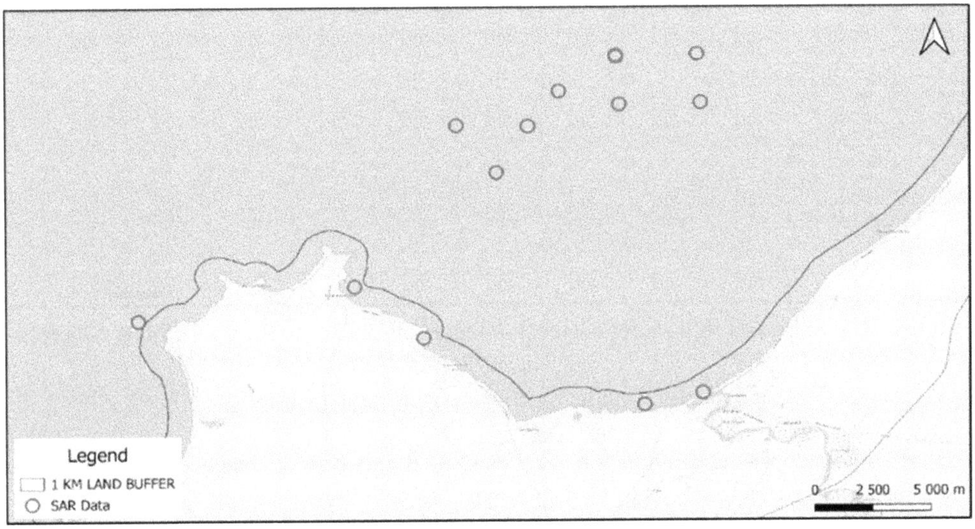

Figure 11.4: Visual representation of detected ships with red buffers in Saint Helena Bay. (*Source:* Authors' own work using ArcGIS 10.8.1)

The buffer zone in Figure 11.4 was created 1 km from the land to avoid the detection of small ships less than 30 m in length. Generally, these ships conduct small-scale fishing or are stationary along the coast. These ships were excluded from the analysis for various reasons. The focus of this study was on large ships with lengths greater than 30 m to illustrate the usefulness of the methodology. Ships near the coastal lines were also excluded because they were small ships that conducted legal fishing on a small scale or were stationary and could lead to false alarms if included. These vessels constitute a sample that can be studied using this methodology in the future.

The Bay is known for its commercial shipping activity as a waiting area for large bulk carriers that enter the Saldanha Harbour to load iron ore. Thus, there is constant movement of shipping conforming to the specifications of the study (i.e. ships larger than 30 m). The results, as depicted in Figure 11.4, show that eight ships were detected at the time of data collection within the geographic area of interest using the SAR satellite. Four other ships that were detected near the coastline were excluded, depicted within the 1 km land buffer because they represented either a false alarm or were small ships conducting small-scale fishing, along with one ship that was outside the geographic area of interest. On the other hand, AIS data (Figure 11.5), showing recorded timestamps of 10 vessels that were continuously sharing their locations, are displayed in points and lines connecting the points to show the vessel tracks.

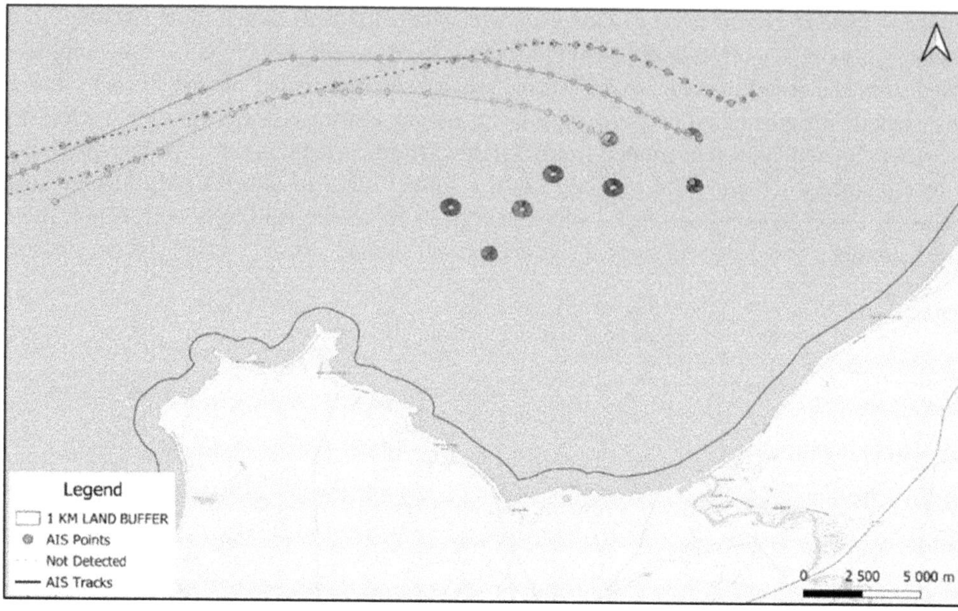

Figure 11.5: Ships and ship tracks detected in Saint Helena Bay at the time of data collection using SAR data.
(*Source:* Authors' own work using ArcGIS 10.8.1)

Figure 11.6 illustrates the combination of AIS and SAR data from 12 January 2024, which was used for a comparative analysis to identify suspicious activities of vessels larger than 30 m in length within the geographic area of interest.

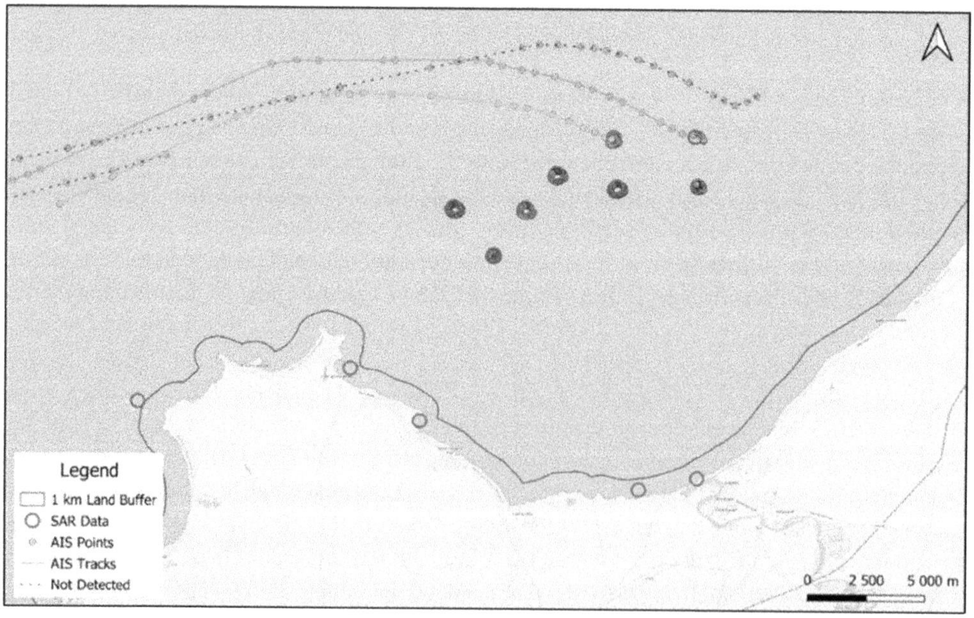

Figure 11.6: Ships and ship tracks detected in Saint Helena Bay at the time of data collection using SAR and AIS data.
(*Source:* Authors' own work using ArcGIS 10.8.1)

An effective comparative analysis of both datasets was done on separate software called QGIS. The results revealed a 100% match rate between the SAR and AIS data for the ships detected by SAR, as depicted in Figure 11.6. However, only two vessels recorded in the AIS data were not captured by the satellite.

Discussion

This study addresses a significant knowledge gap regarding the security situation in the waters of Saint Helena Bay, where illegal fishing activities are prevalent. In response to this issue, there is a robust requirement to develop an enhanced monitoring method for maritime activity. This study proposes the integration of SAR and AIS datasets as an effective means of monitoring the area in the spirit of MDA. The results indicated a 100% match rate between the SAR and AIS datasets for detected vessels, with only two vessels recorded by AIS that were not captured by satellite imagery. This represents 20% (two out of ten) of the vessels identified by AIS that went undetected by SAR. These findings demonstrate that the combined use of SAR and AIS data provides a more comprehensive understanding of maritime activities, ensuring accurate monitoring and revealing that no vessels are expected to be engaged in illegal activities within the geographic area of interest.

Previous studies of Kurekin et al. (2019) and Galdelli et al. (2021) have shown that merging these databases may provide a better understanding of ship movements, even in adverse weather conditions or when the AIS transponder does not transmit the signal. Suspicious activities are detected when ships operating within restricted areas manipulate their AIS information by spoofing their identity. This is done by reporting activities that do not match observed behaviour. For example, a ship claiming to be a cargo vessel but showing commercial fishing-related movement patterns can be construed as an illegal activity in progress, which can then be flagged for further investigation by the authorities. Other ships power down their AIS transponders to avoid detection during illegal activities. Other activities include ships that show unusual behaviour, such as long static periods in a fishing zone or MPA. This constitutes suspicious behaviour which should be flagged for investigation. These examples ground the requirement for enhanced MDA in the interest of the country in question, in this case, South Africa. Ghana has adopted the implementation of integrated maritime systems, which resulted in a 91% success rate during the test phase. This integration proved to be effective, as it detected eight out of ten ships in the study area. However, the results were not 100% accurate because of some limitations of AIS and SAR in identifying a ship (Kurekin, et al., 2019: 17).

AIS data have limitations because they rely on ships that carry AIS transponders. These transponders can also be turned off for various reasons, such as hiding activities or technical issues. It is also mentioned that some ship identities can be manipulated for malicious actions. This shows the need for other technologies to fill the gap in traditional methods used for marine surveillance and provide a more accurate picture of sea activities (Reggiannini et al., 2024). According to Achiri et al. (2018), the combination of AIS and SAR is an effective maritime surveillance method for identifying illegal, unreported, and unregulated fishing activities. By cross-referencing detected ships with AIS data, we could identify ships that did not report their location, which is a common tactic used in illegal fishing activities. Evans (2018) argued that illegal fishing costs South Africa millions of Rands, harms marine ecosystems, and leaves local fishermen with empty nets. Many of these illegal activities are conducted in South Africa, and resources are exported to Hong Kong. Reports made by marine traffic stated that since 2000, poaching syndicates have smuggled more than 96 million individual abalones to Hong Kong, which is the epicentre of the trade.

Conclusion

Illegal fishing poses a threat to maritime security and requires serious attention. If this goes unchecked, it could result in geopolitical instability and tensions between countries. The introduction of SAR satellites with the inclusion of AIS data in maritime security could help combat the threat of illegal fishing. This study shows how the integration of SAR with AIS data can improve maritime surveillance capabilities when tracking vessel movements and activities within the area of interest (Saint Helena Bay). The findings revealed that the integration of the two technologies makes it easier to detect and monitor ships at sea.

The use of Sentinel-1 complements the limitation of AIS in cases when ships turn off their transponder to avoid detection. Sentinel-1 can still identify such vessels. However, one challenge with Sentinel-1 is that it does not provide real-time imagery, which means that some illegal activities might go unnoticed. This delay could affect the effectiveness of using Sentinel-1 alone for monitoring illegal activities at sea. This highlights the need to consider more than one satellite or the use of commercial satellites that could provide real-time monitoring of sea activities. Therefore, it is important that future research refine the method to include more than one satellite in monitoring activities and provide an automatic identification of ships, as manually merging the datasets is time-consuming. This could also include the addition of optical satellites which provide the true colour of the image.

References

Achiri, L., Guida, R. & Iervolino, P. 2018. SAR and AIS fusion for maritime surveillance. *2018 IEEE 4th International Forum on Research and Technology for Society and Industry (RTSI)*. 1–4. https://doi.org/10.1109/RTSI.2018.8548517

ArcGIS. 2025. ArcGIS Online – Powered by ESRI. https://www.arcgis.com/index.html [23 January 2025].

Azzeh, J., Zahran, B. & Alqadi, Z. 2018. Salt and pepper noise: Effects and removal. *JOIV: International Journal on Informatics Visualization*, 2(4): 252–256. https://doi.org/10.30630/joiv.2.4.151

Biccard, A., Hutchings, K., Wright, A.G., Mostert, B.M., Sedick, S. & Clark, B.M. 2020. *The State of St Helena Bay 2020*. Report No. AEC 1908/1. Prepared by Anchor Environmental Consultants (Pty) Ltd for the St Helena Bay Water Quality Trust.

Bioresita, F., Pribadi, C.B. & Firdaus, H.S. 2018. Ship detection in Madura Strait and Lamong Gulf using Sentinel-1 SAR data. *Proceedings of the 3rd International Conference on Science and Technology*. 13–23. https://doi.org/10.29037/digitalpress.11224

Brick, K., Muchapondwa, E. & Visser, M. 2014. Abalone poaching, methamphetamine use, and criminal activity in South Africa: Implications for resource management. https://www.researchgate.net/publication/255608360_Abalone_poaching_methamphetamine_use_criminal_activity_in_South_Africa_and_the_associated_implications_for_resource_management_Draft [20 August 2024].

Burger, J.M., Moloney, C.L., Walker, D.R., Parrott, R.G. & Fawcett, S.E. 2020. Drivers of short-term variability in phytoplankton production in an embayment of the southern Benguela upwelling system. *Journal of Marine Systems*, 208. https://doi.org/10.1016/j.jmarsys.2020.103370.

Department of the Premier. 2012. Report on the economic and socio-economic state and growth prospects of the 12 proclaimed fishing harbours in the Western Cape: Final consolidated report. https://cisp.cachefly.net/assets/articles/attachments/42643_harbour_project_final_consol_doc_nov12.pdf [30 March 2025].

ESA [European Space Agency]. 2025a. Sentinel-1: Radar vision for Copernicus. https://www.esa.int/Applications/Observing_the_Earth/Copernicus/Sentinel-1 [23 January 2025].

ESA [European Space Agency]. 2025b. TerraSAR-X and TanDEM-X. https://earth.esa.int/eogateway/missions/terrasar-x-and-tandem-x [23 January 2025].

Evans, J. 2018. SA has lost 96 million abalone to poaching since 2000 - new report. *Africa Wild*. https://africawild-forum.com/viewtopic.php?t=8843 [20 February 2024].

Fielding, P. 2021. Marine and coastal areas under protection: Republic of South Africa. In: *Western Indian Ocean Marine Protected Areas Outlook: Towards Achievement of the Global Biodiversity Framework Targets*. United Nations Environment Programme/Nairobi Convention Secretariat. 131–166.

Filipponi, F. 2019. Sentinel-1 GRD preprocessing workflow. *MDPI Proceedings*, 18(1). https://doi.org/10.3390/ECRS-3-06201

Galdelli, A., Mancini, A., Ferrà, C. & Tassetti, A.N. 2021. A synergic integration of AIS data and SAR imagery to monitor fisheries and detect suspicious activities. *Sensors*, 21(8). https://doi.org/10.3390/s21082756

GIS Geography. 2025. World Geodetic System (WGS84). https://gisgeography.com/wgs84-world-geodetic-system/ [23 January 2025].

Grover, A., Kumar, S. & Kumar, A. 2018. Ship detection using Sentinel-1 SAR data. *ISPRS Annals of the Photogrammetry, Remote Sensing and Spatial Information Sciences*, 4: 317–324. https://doi.org/10.5194/isprs-annals-IV-5-317-2018

Iacarella, J.C., Burke, L., Clyde, G., Wicks, A., Clavelle, T., Dunham, A., Rubidge, E. & Woods, P. 2023. Application of AIS- and flyover-based methods to monitor illegal and legal fishing in Canada's Pacific marine conservation areas. *Conservation Science and Practice*, 5(6). https://doi.org/10.1111/csp2.12926

IMO [International Maritime Organization]. 2025a. Maritime Domain Awareness. https://www.imo.org/en/OurWork/Security/Pages/Maritime-Domain-Awareness.aspx [23 January 2025].

IMO [International Maritime Organization]. 2025b. AIS transponders. https://www.imo.org/en/OurWork/Safety/Pages/AIS.aspx [23 January 2025].

Keebine, G. 2023. Marine protected areas. *South African Environmental Observation Network*, 20 January. https://sarva.saeon.ac.za/marine-protected-areas/ [23 January 2025].

Kongsberg. 2024. Vessel detection from space. https://www.kongsberg.com/what-we-do/protecting-sea-infrastructure/vessel-detection-from-space/ [23 January 2025].

Koning, T. 2024. The Orange Basin, Deepwater Namibia – What's going on with its resources, reserves and future production of natural gas? *Africa Oil + Gas Report*, 1 September. https://africaoilgasreport.com/2024/09/oil-patch-sub-sahara/the-orange-basin-deepwater-namibia-whats-going-on-with-its-resources-reserves-and-future-production-of-natural-gas [23 January 2025].

Kurekin, A.A., Loveday, B.R., Clements, O., Quartly, G.D., Miller, P.I., Wiafe, G. & Adu Agyekum, K. 2019. Operational monitoring of illegal fishing in Ghana through exploitation of satellite earth observation and AIS data. *Remote Sensing*, 11(3): 1–28. https://doi.org/10.3390/rs11030293

Madaan, E.S. & Kaur, S. 2019. Pre-processing of synthetic aperture radar Sentinel-1 images for agricultural land. *International Journal of Control and Automation*, 12(5): 443–459.

Martin, G. 2024. SAMSA reports a growing number of incidents at sea as maritime traffic around SA increases. *Protection Web*, 6 November. https://www.protectionweb.co.za/maritime-security/samsa-reports-a-growing-number-of-incidents-at-sea-as-maritime-traffic-around-sa-increases/ [23 January 2025].

Mdakane, L.W., Sibolla, B.H. & Haupt, S.A. 2023. Maritime domain awareness in South Africa: A multisource approach using remote sensing and AIS data. *Proceedings of the 2023 IEEE International Symposium on Geoscience and Remote Sensing (IGARSS)*. 1473–1478. https://doi.org/10.5194/isprs-archives-XLVIII-1-W2-2023-1473-2023

Moskolaï, W.R., Abdou, W., Dipanda, A. & Kolyang. 2022. A workflow for collecting and preprocessing Sentinel-1 images for time series prediction suitable for deep learning algorithms. *Geomatics*, 2(4): 435–456. https://doi.org/10.3390/geomatics2040024

NATO [North Atlantic Treaty Organisation]. 2021. AIS (Automatic Identification System) overview. https://shipping.nato.int/nsc/operations/news/2021/ais-automatic-identification-system-overview [23 January 2025].

Reggiannini, M., Salerno, E., Bacciu, C., D'Errico, A., Lo Duca, A., Marchetti, A., Martinelli, M., Mercurio, C., Mistretta, A., Righi, M. & Tampucci, M. 2024. Remote sensing for maritime traffic understanding. *Remote Sensing*, 16(3). https://doi.org/10.3390/rs16030557

Reutech Radar Systems. 2025. Security solutions. https://www.reutechradar.com/solutions/security-solutions/ [23 January 2025].

Reva, D. & Walker, T. 2020. Tech innovations are key to South Africa's maritime security. *ISS Today*, 14 September. https://issafrica.org/iss-today/tech-innovations-are-key-to-south-africas-maritime-security [23 January 2025].

SAHO [South African History Online]. 2025. Saint Helena Bay. https://www.sahistory.org.za/place/saint-helena-bay [26 January 2025].

Schultz, O.J. 2015. Defiance and obedience: Regulatory compliance among artisanal fishers in St Helena Bay. *Marine Policy*, 60: 331–337. https://doi.org/10.1016/j.marpol.2015.06.007

Torres, R., Navas-Traver, I., Bibby, D., Lokas, S., Snoeij, P., Rommen, B., Osborne, S., Ceba-Vega, F., Potin, P. & Geudtner, D. 2017. Sentinel-1 SAR system and mission. *2017 IEEE Radar Conference (RadarConf)*. 1582–1585. https://doi.org/10.1109/RADAR.2017.7944460

UNODC [United Nations Office on Drugs and Crime]. 2014. Crime in a box: The international trafficking of clandestine goods in containers. https://www.unodc.org/islamicrepublicofiran/en/crime-in-a-box_-the-international-trafficking-of-clandestine-goods-in-containers.html [23 January 2025].

Velotto, D., Bentes, C., Tings, B. & Lehner, S. 2016. First comparison of Sentinel-1 and TerraSAR-X data in the framework of maritime targets detection: South Italy case. *IEEE Journal of Oceanic Engineering*, 41(4): 993–1006. https://doi.org/10.1109/JOE.2016.2520216

Ward, C., Gammage, L. & Jarre, A. 2022. Community profiles for Western Cape marine fisheries. The University of Cape Town. https://doi.org/10.25375/uct.19383602.v1

SECTION 4

Sustainable and Renewable Energy Solutions

The final section addresses the growing need for sustainable energy solutions within military and national planning frameworks. The studies focus on site selection, resource modelling, and operational applications of renewable energy in support of energy security and environmental stewardship.

Chapter 12 employs a Multi-Criteria Decision-Making (MCDM) approach, combining the Analytical Hierarchy Process (AHP) with GIS, to identify optimal locations for photovoltaic (PV) solar farms in the Saldanha Military Area. The analysis integrates terrain, climate, and proximity variables to support a cost-effective and resilient energy infrastructure.

Chapter 13 models the wave energy potential of North Bay, Saldanha, using oceanographic datasets (e.g. wave height and bathymetry) and tools such as SWAN and MATLAB. This chapter identifies high-energy zones suitable for small-scale energy conversion, contributing to localised marine energy planning.

Chapter 14 concludes the volume with a study of military exercise planning, evaluating how GIS tools supported reconnaissance and logistical coordination during the Exercise Trans Enduro. The authors demonstrate how open-source spatial datasets and mobile GPS mapping facilitated the successful execution of complex field operations by the cadets and staff.

Together, these chapters underscore the strategic and environmental value of spatial tools in advancing energy resilience, supporting operational planning, and reinforcing sustainable defence infrastructure.

CHAPTER 12

Photovoltaic solar farm site selection using the Analytical Hierarchy Process and geographic information systems: A case in the Saldanha Military Area

K.T. Kortman and S. Henrico

Abstract

South Africa's reliance on coal-powered generators, coupled with the inadequate replacement of aging infrastructure, has led to load shedding by Eskom, resulting in frequent power outages. These power outages, lasting between 2 and 10 hours per day (stages 2 to 6) depending on the stage of load shedding implemented by Eskom, aim to protect the national grid from collapse due to an inability to meet the current energy demand. Institutions such as the South African Military Academy rely on microgrids powered by diesel generators to sustain their operations during these periods of power outages. However, the use of diesel generators is costly and environmentally unfriendly. A photovoltaic (PV) solar farm offers a sustainable alternative, but selecting an optimal site is a complex Multi-Criteria Decision-Making (MCDM) process. This study aims to identify a suitable site for constructing a PV solar farm in the Saldanha Military Area (SMA) using the Analytical Hierarchy Process (AHP) and geographic information systems (GIS). The key selection criteria include climatic factors (solar radiation, surface temperature), orographic characteristics (slope, aspect), environmental considerations (land use), and economic factors (proximity to roads and substations). Protected areas and water bodies were excluded from the constraints. Geospatial data from various sources were analysed using GIS tools to visualise and assess the land suitability. The results demonstrate that integrating GIS and AHP is an effective and cost-efficient approach for PV solar farm site selection. This method provides a valuable framework for spatial decision support and can be applied to similar studies in South Africa and other countries.

Keywords: Analytical Hierarchy Process, geographic information system, load shedding, PV solar farm, renewable energy, Saldanha Military Area

Introduction

The increase in the global population and energy consumption has led to an increase in global energy demand (Manieniyan et al., 2009). According to the IEA's World Energy Outlook report (IEA, 2022), over 65% of the world's energy production comes from fossil fuels (oil, gas, and coal), and the other 35% comes from renewable fuels (biofuel, hydroelectricity, solar and wind energy, and nuclear energy). Fossil fuels are limited and non-renewable; therefore, with the rapid increase in energy demand, the world is facing accelerated depletion of fossil fuel reserves. The uneven global distribution of these resources has triggered political and economic conflicts (Pretorius et al., 2015). Moreover, fossil fuels significantly contribute to greenhouse gas emissions, exacerbating global warming (Böhm & Peterson, 2021). Growing awareness of these environmental and geopolitical challenges has prompted many countries to invest in renewable energy, which is cleaner, cheaper, and more sustainable for meeting future energy demand. Among renewable energy options, solar energy is recognised as one of the most reliable and environmentally-friendly solutions (Akkas et al., 2017).

South Africa has faced a severe energy crisis since the mid-2000s owing to ageing and deteriorating infrastructure. Eskom, the national electricity supplier, has struggled to meet current energy demands, resulting in frequent load shedding with stages ranging from 1 to 6 to prevent the collapse of the national grid (Eskom, 2013). Approximately 85% of South Africa's primary energy is generated from coal-operated power stations (Pretorius et al., 2015). This ongoing crisis has significantly impacted the economy and key sectors, including military infrastructure, which relies on diesel-powered microgrids for uninterrupted operations during load shedding. However, microgrids are costly, environmentally harmful, and inefficient (Goldberg, 2015).

Many countries have adopted solar-powered microgrids to address the energy demands of critical institutions. Selecting an ideal location for a PV solar farm requires consideration of factors such as proximity to main roads, distance from transmission lines, solar radiation, surface temperature, and slope steepness (Akinci & Özalp, 2022). Site selection is a complex process that can be facilitated by integrating Multiple-Criteria Decision Analysis (MCDA) with geographic information systems (GIS) to define, evaluate, and prioritise criteria (Tariq et al., 2020). MCDA tools, such as the Analytical Hierarchy Process (AHP) and Analytical Network Process (ANP), help identify optimal locations (Uyan, 2013). This study focuses on the AHP and GIS, as AHP is recognised for its accuracy and simplicity (Zoghi et al., 2017).

Literature review

Various MCDA methods have been used over the years (Aruldoss et al., 2013; Rodríguez Mañay et al., 2022). These methods include studies such as Sozen et al. (2015), where the authors applied a different approach to identify a suitable location for PV solar farms using Data Envelopment Analysis (DEA) and the Technique for Order of Preference by Similarity to Ideal Solution (TOPSIS). DEA is a performance measurement technique that evaluates the efficiency of decision-making units (e.g. solar sites) by comparing inputs and outputs, identifying the most efficient units, and highlighting inefficiencies. Similarly, TOPSIS tolerates the comparison of objective weights. TOPSIS selects the best alternative by minimising the distance from the positive ideal solution and maximising the distance from the negative ideal solution (Sozen et al., 2015).

Furthermore, several studies have used the Fuzzy Analytical Hierarchy Process (Fuzzy AHP) to determine solar site suitability. For instance, Kengpol et al. (2012) used Fuzzy AHP, incorporating decision analysis to select a site in Thailand. Similarly, Asakereh et al. (2014) used Fuzzy AHP and GIS to identify suitable sites for solar PV farms in Iran. Fuzzy AHP extends the standard AHP by integrating fuzzy logic to handle uncertainty and imprecision in decision-making, particularly when the criteria weights or judgements are subjective and require a range of values. However, this methodology is complex and requires extensive calculations, increasing the risk of human error.

A study conducted by Lee et al. (2017) investigated the feasibility of finding a suitable solar site by incorporating Interpretive Structural Modelling (ISM), Fuzzy Analytical Network Process (Fuzzy ANP), and VIKOR. ISM is initially used to identify relationships between criteria and sub-criteria, forming a decision-making network. Then, the Fuzzy ANP solves the network and calculates the weight of each sub-criterion. Finally, VIKOR is used to rank PV solar plant locations (Lee et al., 2017). These methods require specialised software and are often complex to implement.

The AHP is a widely used MCDA method for analysing and solving complex problems, with applications in planning, resource allocation and management, and site selection (Xin et al., 2019; Ahmad & Tahar, 2014). Shrestha et al. (2022) conducted a study to identify power plant installation sites. GIS have been used by many organisations to solve complex problems, as it offers tools to analyse, manipulate, manage, and store spatial data for diverse applications (Chang, 2008). Over time, scholars have combined AHP with GIS to identify suitable locations for constructing nuclear plants, wind farms, and PV solar farms. For example, Yushchenko et al. (2018) and Tahri et al. (2015) integrated AHP and GIS to investigate land suitability for constructing PV solar farms in West Africa and Morocco. Other MCDA methods, such as TOPSIS, ELECTRE, and VIKOR, have been applied. ELECTRE is an MCDA method that compares alternatives based on multiple criteria and eliminates less favourable options using pairwise comparisons and thresholds. Akkas et al. (2017) used these methods to select suitable sites for constructing PV solar plants in Turkey. As the demand for geospatial tools has increased, many organisations have adopted GIS-based approaches to determine land suitability for PV solar farms. In another example, Tunç et al. (2019) compared AHP weights with survey weights and concluded that both methods yielded similar results.

Several studies have shown that integrating GIS with AHP is effective for land suitability analyses. For instance, a study in Nepal utilised criteria such as solar irradiation, slope, and distance from roads and urban areas to determine suitable locations. Similarly, Shrestha et al. (2022) investigated land suitability for PV solar farms using criteria such as solar radiation, annual mean temperature, distance from roads, distance from transmission infrastructure, distance from urban areas, slope, aspect, and land use. The results highlighted areas with high suitability percentages. Doorga et al. (2019) conducted a case study in Mauritius using predetermined criteria to determine land suitability and obtained similar results. Notably, there is no standardised criterion for land suitability analysis for constructing PV solar farms, as site-specific factors influence the selection process. These factors play a critical role in determining feasibility and examining land suitability.

This study adopted a methodology similar to what was applied in Nepal and Thailand, combining GIS and AHP to identify suitable sites for PV solar farm development. The criteria included annual solar irradiation, annual surface temperature, slope, aspect, distance from roads, distance from miniature substations, and land use/land cover. Table 12.1 summarises the literature that integrates AHP and GIS for PV solar farm site selection.

Table 12.1: Summary of literature for combining AHP and GIS for PV solar farmland suitability

MCDA	Location	Reference
AHP	Mauritius	Doorga, Rughooputh & Boojhawon (2019)
	West Africa	Yushchenko, de Bono, Cjatenoux, Patel & Ray (2018)
	South Eastern Europe	Sánchez-Lozano, Teruel-Solano, Soto-Elvira & Socorro Garcia-Cascales (2013)
	Tanzania	Aly, Jensen & Pedersen (2017)
	Morocco	Tahri, Hakdaoui & Maanan (2015)

Study area, data, and methodology for PV solar farm site selection

Study area

The study area is located in the Saldanha Military Area (SMA), situated at 33°02'53" S and 17°54'53" E (Axelson, 1977) on the West Coast of the Western Cape, South Africa (Figure 12.1). This area is home to the largest and deepest natural harbour in the Southern Hemisphere (Henrico, 2021) and is approximately 155 km northwest of Cape Town. The SMA hosts the South African Navy training unit, SAS Saldanha, and the South African Military Academy (SAMA). Saldanha has a Mediterranean climate with most of its rainfall occurring in winter; however, the annual rainfall is relatively low. Saldanha has an average high temperature of 24 °C and an average low of 11 °C, with an average wind speed of 23 kph. Saldanha has an average elevation of 27 m above sea level (Rossouw, 2008). The SMA is adjacent to the Saldanha Provincial Nature Reserve, which is 876 hectares with various fynbos and animal life (Bezuidenhout, 2012). Most of the land within the SMA is reserved for the provincial nature reserve, while the military uses it for residential, office, and training purposes. This justifies the need for an independent microgrid amidst the current energy crisis and forms the basis for selecting the study area.

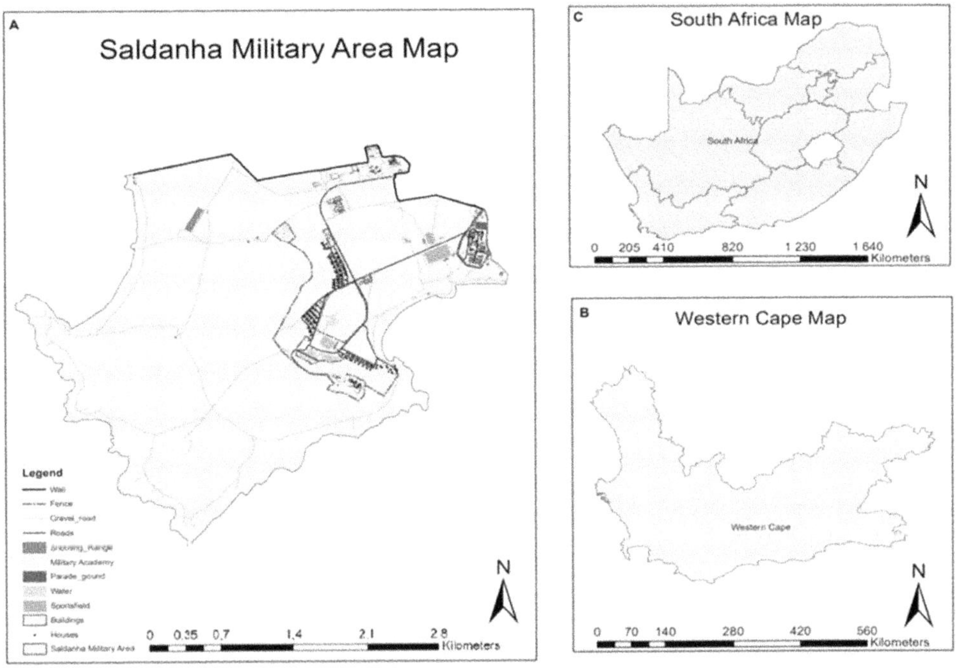

Figure 12.1: Illustration of study area. Saldanha military area (A) located in the Western Cape province (B) in South Africa (C).
(*Source:* Authors' own work using ArcGIS 10.8.1)

Methodology

Selecting a PV solar farm site is a complex process that requires considering various criteria before reaching a conclusion (Wiguna et al., 2016). To overcome this complexity, researchers have opted to use a Multi-Criteria Decision-Making (MCDM) method because it provides strong decision-making when the selection of alternatives is complex (Aruldoss et al., 2013). Identifying a suitable location involves evaluating a wide range of climatic, geographic, economic, and environmental factors. Each criterion consists of sub-criteria that further influence the decision-making process. This study was divided into two parts. The first part identifies and defines the main and sub-criteria, while the second part applies the AHP to calculate the criteria weights. These weights were then used in a weighted overlay analysis to conduct a land suitability assessment. Figure 12.2 illustrates the adapted methodology flow (Shrestha et al., 2022).

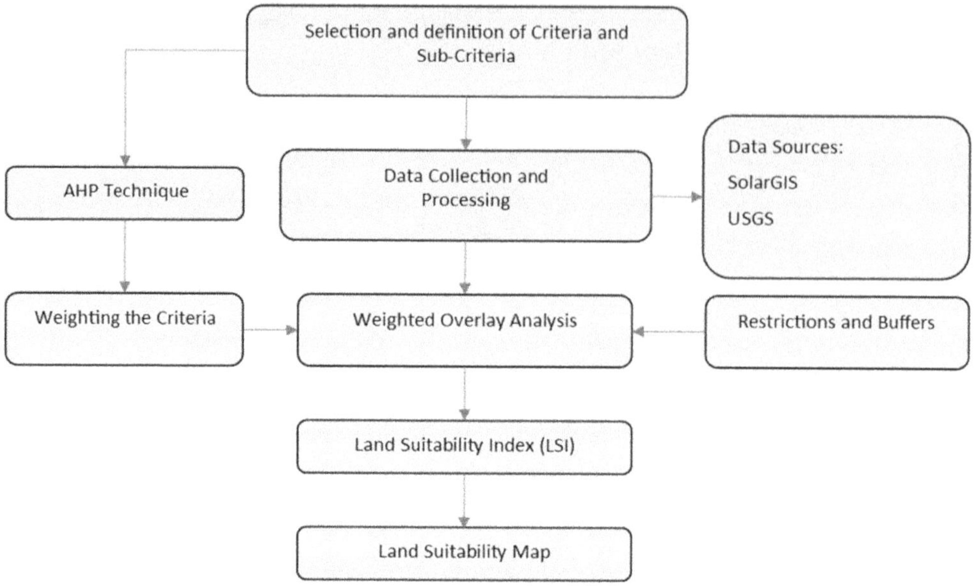

Figure 12.2: Methodology flowchart.
(*Source:* Adapted from Shrestha et al., 2022)

Criteria selection

Numerous scholars have conducted studies to establish criteria for PV solar farm site selection. Across these studies, four common criteria have been identified: climatic, geographic, economic, and environmental factors (Shrestha et al., 2022; Wiguna et al., 2016; Lee et al., 2017). For this study, only two restrictions were identified: protected areas and waterbodies, along with the criteria and sub-criteria assessed, as shown in Table 12.2.

Table 12.2: Criteria description and restrictions

Criteria	**Sub-criteria**	**Description**
Climatic	Annual solar irradiation (kWh/m^2)	Sites with higher solar radiation potential are preferred as they yield more energy.
	Annual surface temperature (°C)	Solar energy production is heavily reliant on incoming solar irradiation. However, surface temperature has a significant impact on energy production. Areas with high surface temperatures tend to have less energy production compared to areas with low surface temperatures (Dubey et al., 2013).

Geographic	Slope (°)	Land should meet slope requirements,; otherwise, excavations or filling operations might be needed (Colak et al., 2020). Sites with a gentler slope are preferred as they receive more solar irradiation.
	Aspect	The direction in which slopes are facing is important (Colak et al., 2020) as it dictates the duration, amount, and intensity of solar irradiation received by a particular area. Slopes facing south tend to absorb more solar irradiation.
Economic	Distance from roads (m)	Sites near roads are preferred as they cut the cost of having to build new roads for accessing the site (Colak et al., 2020).
	Distance from miniature substations (m)	Sites near miniature substations are preferred as they reduce the cost of adding new transmission infrastructure (Colak et al., 2020).
Environmental	Land use/cover	Generating a land use/cover map helps the researcher to identify what different areas within the area of study are used for, and to also map the vegetation cover of the area. This will also assist when identifying protected/restricted areas (Colak et al., 2020).
	Restrictions	
	Protected areas	Protected areas like nature reserves, heritage sites, etc.
	Water bodies	Land covered by water.

Geographic information systems (GIS) processing

GIS are software used to analyse and interpret geospatial data to help solve complex problems (Malczewski & Rinner, 2015). Integrating GIS with the MCDM method helps to enhance the selection of criteria by considering various subjective and contradictory parameters (Wiguna et al., 2016). As stated earlier, GIS-AHP, as one of the MCDM methods, is widely used in spatial analysis to help find suitable sites for a variety of applications, including renewable energy, landfills, nuclear site selection, and urban planning (Yushchenko et al., 2018). These methods have been proven efficient in the PV solar farm site sections (Doorga et al., 2019; Sánchez-Lozano et al., 2013). A Land Surface Temperature (LST) map was created to find an optimal site for constructing the solar farm. This was done using the ArcGIS raster calculator to extract data from Landsat 9 for the period of 1 January 2022 to 1 January 2023. This period was chosen because it was the most recent data available for the study area. According to Hizam et al. (2017), the following steps were used to calculate the surface temperature from the Landsat image:

1. Top of Atmospheric (TOA) spectral radiance TOA=M_L * Q_{cal} + A_L, where M_L = Band-specific multiplicative rescaling factor from the metadata. Q_{cal} = Band 10, and A_L= Band-specific additive rescaling factor from the metadata.
2. Conversion of TOA to Brightness Temperature (BT) BT = (K_2/(In(K_1/L)) - 273.15 where K_1 and K_2 are Band-specific thermal conversion constants from the metadata. L = TOA, and 273.25 = absolute zero to obtain results in °C.
3. NDVI calculation NDVI = (Band 5 – Band 4)/(Band 5 + Band 4).
4. Proportion of vegetation (P_v) calculation P_V= Square((NDVI-$NDVI_{min}$)/($NDVI_{max}$-$NDVI_{min}$)).
5. Emissivity (ε) calculation: ε = 0.004 * Pv + 0.986.
6. LST calculation: LST = (BT/(1+(0.0015 * BT/1.4388) * In(ε))).

Usage of the Analytical Hierarchy Process (AHP)

After creating all necessary maps using ArcGIS (Figures 12.3–12.10), the next step in the process is to use the AHP to assign weights to the identified criteria (see Table 12.2). The AHP is a structured method used to analyse complex decisions based on pair-wise comparison methods to assign weights to each criterion (Shrestha et al., 2022). This process helps simplify complex problems by turning them into manageable parts, then each part is compared with the other to determine its importance. The pairwise comparison uses a matrix A = [Cij] to calculate the weight of each criterion using Saaty's scale, ranging from value 1-9 (Ruiz-Fernández et al., 2020; Shrestha et al., 2022). Table 12.3 shows Saaty's scale used in this study.

Table 12.3: Saaty's scale

Scale	Importance	Description
1	Equal importance	Both criteria have equal influence.
3	Moderate importance	One criterion is slightly favoured over the other.
5	Strong importance	One criterion is strongly favoured over the other.
7	Very strong importance	One criterion is very strongly favoured over the other.
9	Extreme importance	One criterion is extremely favoured over the other.
2,4,6,8	Intermediate values	Between the degrees of importance.

(*Source:* Shrestha et al., 2022)

The criteria and sub-criteria weights are calculated using a pairwise decision matrix (1) based on the level of influence each criterion has on the other. The formation of a pairwise decision matrix (Shrestha et al., 2022) depends on the following steps:

$$A = \begin{vmatrix} C_{11}\, C_{12}\, C_{1n} \\ C_{21}\, C_{22}\, C_{2n} \\ * \qquad * \\ C_{n1}\, C_{n2}\, C_{nn} \end{vmatrix}$$

where:

A is the pairwise comparison matrix, where each element represents the relative importance or preference of one criterion (or alternative) compared to another; and

Ci is the element of the matrix, where:

Cij represents the relative importance of criterion *i* compared to criterion *j*.

Cii=1 indicates that the diagonal elements are always 1 because a criterion compared to itself is equally important.

Cji=1 means the matrix is reciprocal, therefore if *i* is twice as important as *j*, then *j* is half as important as *i*.3

n is the number of criteria or alternatives being compared.

According to Shrestha et al. (2022), the formation of a pairwise decision matrix depends on the following steps:

1. Pairwise decision matrix *A* of size *N*N* is created if *Cij* is the entry in the ith row and j^{th} column, and the product of C_{ij} and C_{ji} must be unity.
2. A pairwise decision matrix is prepared by dividing each entry of the column by the total sum of the elements of the same column.
3. The average of each entry of the row provides the relative weight of each criterion.
4. To ensure consistency in the calculated weights, a consistency ratio factor (CR) is calculated (2). CR is the ratio of the consistency index (CI) to the random index (RI) (Shrestha et al., 2022).

$$CR = \begin{vmatrix} \alpha_{max}\text{-}N \\ N\text{-}1 \end{vmatrix}$$

where:

αmax is the largest (principal) eigenvalue of the comparison matrix. This value is derived from the pairwise comparison matrix.

N is the number of criteria or alternatives being compared in the decision-making process.

N-1 is the degree of freedom for the consistency calculation, representing the maximum permissible inconsistency.

According to Shrestha et al. (2022), if CI < 0.10, the results are considered satisfactory. However, if CI > 0.10, the consistency ratio (CR) must be recalculated to eliminate inconsistencies within the pairwise comparison matrix.

Data collection

This study was a desktop-based analysis, with raw data gathered from various sources. Solar irradiance data were obtained from the SolarGIS website (SolarGIS, 2023). Surface temperature, slope, and aspect maps were generated using Landsat 9 imagery spanning 1 January 2022 to 1 January 2023. Data for road networks and miniature substations were created using ArcGIS digitising and analysis tools, including the buffering function. Land use/land cover data were sourced from Esri's Sentinel-2 (2022) Land Cover Explorer and processed using ArcGIS (Esri, 2022).

Results and discussion

Results from the Analytical Hierarchy Process (AHP)

This study aimed to identify a suitable location for constructing a community-scale PV solar farm within the SMA. To achieve this objective, the influence of seven sub-criteria on solar energy production was analysed using the AHP method. A pairwise comparison matrix was employed for each sub-criterion to assess its relative importance against the others and to determine the corresponding weights of the criteria. The pairwise comparison matrix (Table 12.4) was calculated using Equation 1, which was applied to derive the weights of the identified sub-criteria. Values 1 to 7 on the x- and y-axes of the matrix represent the following sub-criteria:

1 = Annual solar irradiation

2 = Annual surface temperature

3 = Slope

4 = Aspect

5 = Land use/cover

6 = Distance from roads

7 = Distance from miniature substations

Table 12.4: Sub-criteria pairwise comparison matrix

Sub-criteria	**1**	**2**	**3**	**4**	**5**	**6**	**7**
1	1	3	5	7	5	6	5
2	0.33	1	2	3	6	6	4
3	0.20	0.50	1	2	5	4	3
4	0.14	0.33	0.50	1	4	3	0.50
5	0.20	0.17	0.20	0.25	1	2	0.50
6	0.27	0.17	0.25	0.33	0.50	1	0.33
7	0.20	0.25	0.33	2	2	3	1

Table 12.5 shows the calculated weight of each criterion and the consistency ratio (CR) by calculating the ratio between the consistency index (CI) and randomness index (RI) using Equation 2. The calculated CR was 0.066, which shows consistency in the pairwise comparison matrix. Annual solar irradiation had the highest weight, as it had the greatest influence on solar energy production, and land use ranked last, as it did not have a significant influence on solar energy production.

Table 12.5: Sub-criteria weights and consistency ratio

Sub-criteria	**Weight (%)**	**Rank**
Annual surface temperature	21.6	2
Annual solar radiation	41.2	1
Aspect	7.7	5
Distance from roads	3.3	7
Distance from the miniature substation	7.9	4
Land use/cover	4.0	6
Slope	14.1	3
Consistency Ratio (CR) = 0.066		

After analysing the weights of the sub-criteria against each other, it was found that solar irradiation had the most influence on solar energy production, followed by surface temperature. As stated previously, land use/land cover had the least influence on solar energy production. Solar irradiation yielded the highest weight, which corresponds well with other studies (Colak et al., 2020; Zoghi et al., 2017).

Results from the GIS analysis

In this section, we discuss the results obtained when analysing the data using ArcGIS. The first map that was analysed was the aspect map. This map (Figure 12.3) shows the aspect of the study area, which was generated from Landsat 9 data using the ArcGIS Aspect analysis tool. This indicates the compass direction in which the terrain surface is facing. This is important as it dictates the duration, amount, and intensity of solar irradiation received by the solar panels. In the Southern Hemisphere, slopes facing true north tend to absorb more solar irradiation (Williams, 2022).

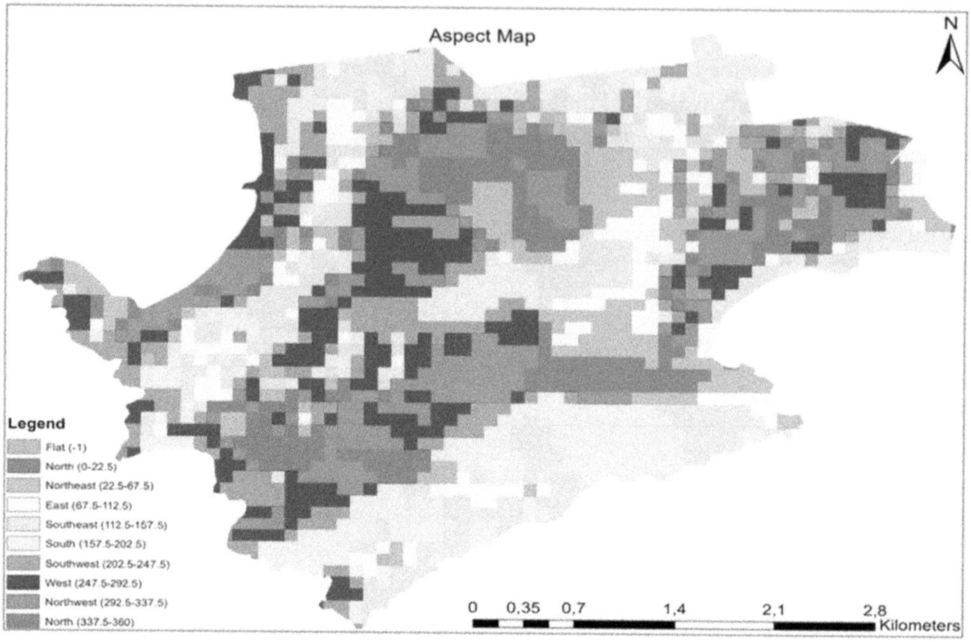

Figure 12.3: Aspect map.
(*Source:* Authors' own work using ArcGIS 10.8.1)

The next GIS analysis was a slope analysis (Figure 12.4), aimed at identifying the angle of steepness of the terrain within the study area. This analysis is crucial in determining the suitability of land for the construction of a community-scale PV solar farm, as the slope of the terrain can significantly affect the construction costs and the efficiency of the solar farm.

The slope map was generated using Landsat 9 imagery in combination with the ArcGIS Slope analysis tool. The resulting map was then reclassified into five equal classes, each representing a different range of slope values. These classes were visually represented by a colour gradient, with each colour indicating varying levels of terrain steepness.

Areas with slopes between 0° and 5° are considered the most favourable for PV solar installation, as they generally present low construction costs. These areas are relatively flat, making it easier and less costly to develop solar farm infrastructure. In contrast, steeper slopes can complicate construction efforts, increase costs, and reduce the available space for solar panels to be installed.

Figure 12.4: Slope map.
(*Source:* Authors' own work using ArcGIS 10.8.1)

Figures 12.5 and 12.6 illustrate the digitised road networks and locations of miniature substations within the SMA, both of which are essential factors in determining land suitability for a community-scale PV solar farm.

Figure 12.5 presents the buffered distance from roads generated using ArcGIS. The analysis began by creating a shapefile of the road network, followed by the digitisation of road locations within the study area. To assess accessibility, multiple ring buffer analysis was conducted, creating proximity zones at 200 m intervals from the roads. The resulting map used a colour gradient from green to red, where green areas represented locations closest to roads and red areas indicated those farthest away.

Areas closer to roads are generally preferred for infrastructure development because of the lower economic costs associated with transportation and site accessibility. Constructing a PV solar farm in areas near existing roads eliminates the need for extensive road development, thereby reducing construction costs and logistical challenges.

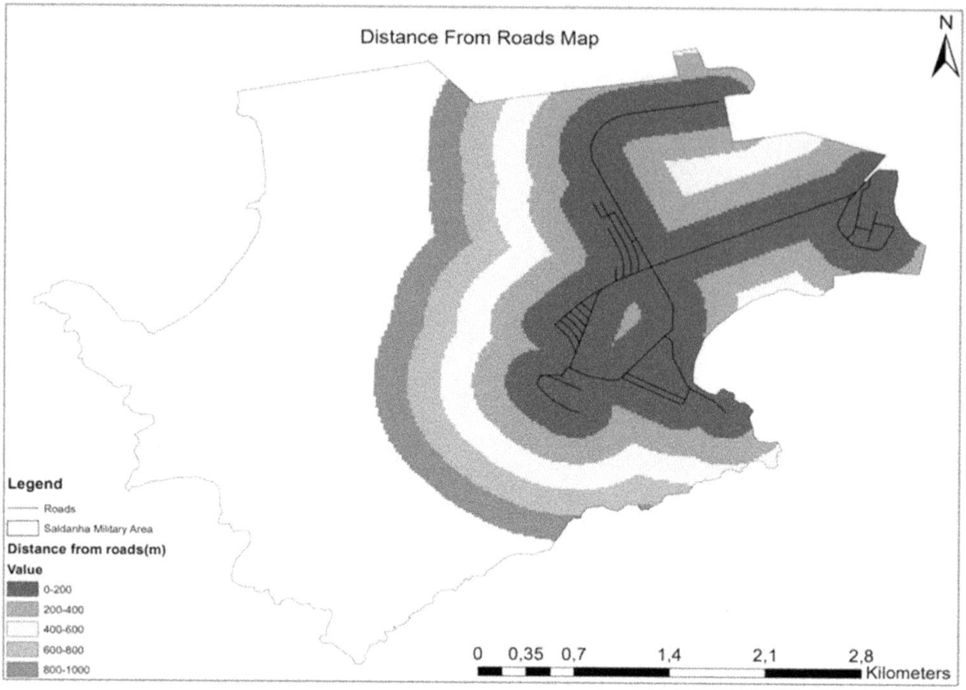

Figure 12.5: Distance from roads map.
(*Source:* Authors' own work using ArcGIS 10.8.1)

Similarly, Figure 12.6 shows the buffered distance from the miniature substations, highlighting their role in energy distribution. A shapefile of the miniature substations was created, and their locations were digitised within the study area. A multiple-ring buffer analysis was then applied, generating proximity zones at 200 m intervals. The colour gradient follows the same pattern as the road buffer analysis, with green areas indicating close proximity to substations and red areas representing the least favourable locations owing to their greater distance.

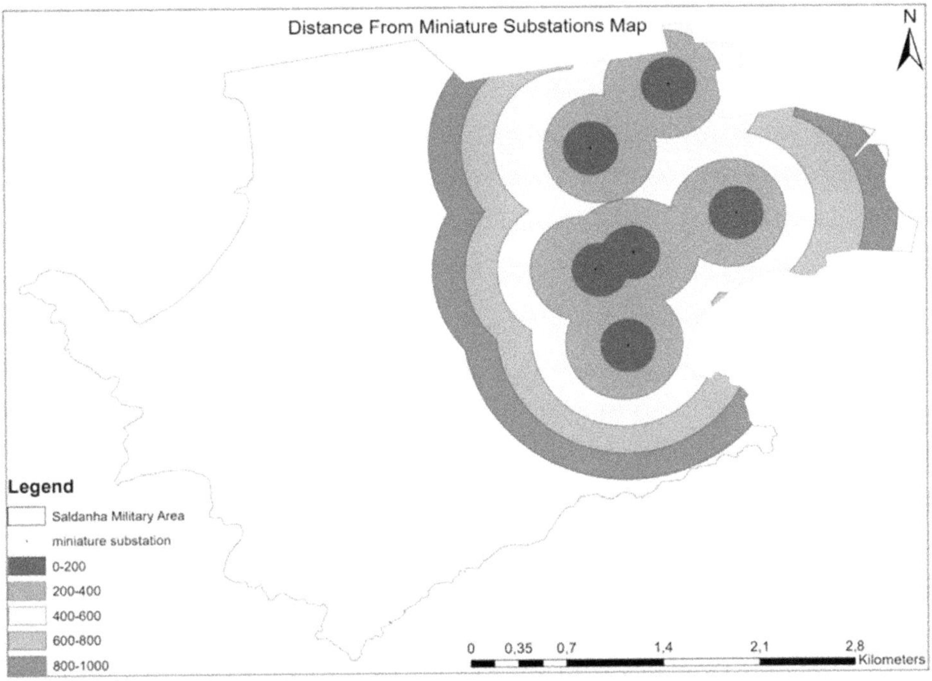

Figure 12.6: Distance from miniature substations map.
(*Source:* Authors' own work using ArcGIS 10.8.1)

The proximity of a PV solar farm to road networks and electrical substations is crucial for ensuring cost-effectiveness and operational efficiency. Sites closer to roads facilitate easier transportation of materials, maintenance, and construction, whereas proximity to substations minimises the need for additional infrastructure to connect the solar farm to the grid. By integrating these factors into the land suitability analysis, this study provides a comprehensive framework for identifying the most economically viable and logistically feasible locations for a PV solar farm within the SMA.

The next step was to analyse the climatic factors, namely, land surface temperature (LST) and the solar irradiation map. Figure 12.7 shows the LST map generated using ArcGIS and Landsat 9 by calculating the Top of Atmospheric (TOA) spectral radiance. The TOA was then converted to Brightness Temperature (BT). After BT, NDVI was calculated as it plays a major role in calculating the proportion of vegetation (P_v) and emissivity (ε) within the study area. Finally, the LST map is calculated by the equation: $LST = (BT/(1+(0.0015 * BT/1.4388) * In(\varepsilon)))$. The LST map illustrates variations in the surface temperature using a gradient colour scheme. The areas represented by dark and light blue shades indicate lower surface temperatures, making them more suitable for solar panel efficiency. Conversely, the regions depicted in yellow and orange shades exhibited relatively higher surface temperatures, which could negatively affect the PV solar performance. Because solar panel efficiency is inversely proportional to surface temperature, elevated temperatures can lead to increased resistance in PV cells, reducing the overall energy output (Dubey et al., 2013). Consequently, identifying locations where temperature conditions remain within the optimal range is essential for ensuring the long-term efficiency and sustainability of solar energy production.

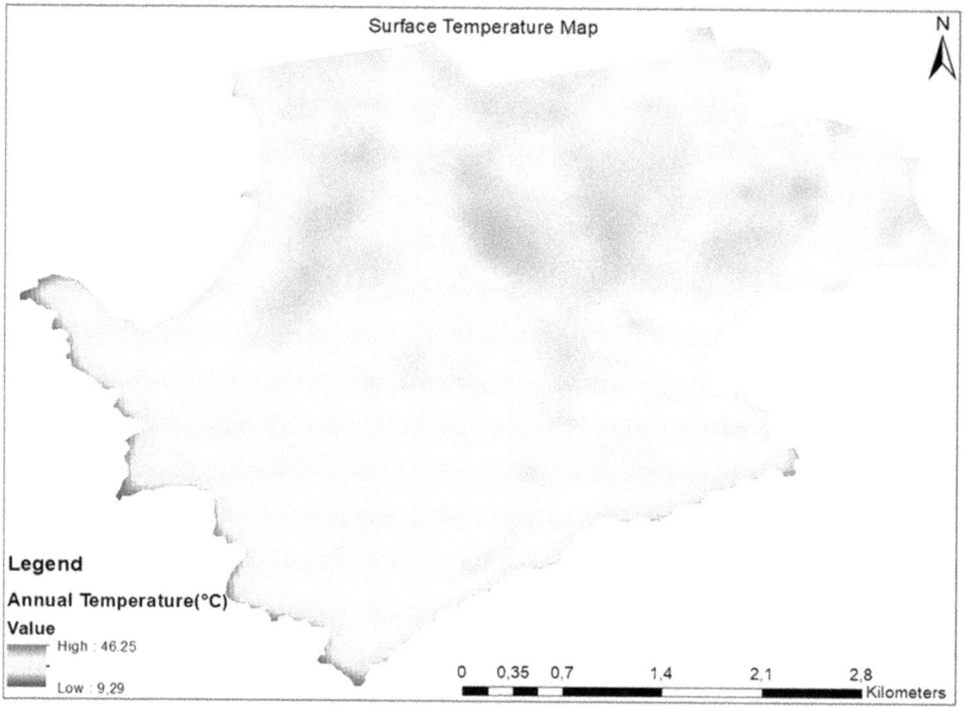

Figure 12.7: Annual temperature map.
(*Source:* Authors' own work using ArcGIS 10.8.1)

Figure 12.8 presents a solar irradiation map of the study area, providing crucial insights into the spatial distribution of solar energy potential. Solar irradiation data were sourced from SolarGIS (2023), an open-source platform that provides high-resolution solar resource data. The dataset was subsequently clipped to the boundaries of the study area to focus on the relevant spatial characteristics.

The map employs a gradient colour scheme to represent variations in solar irradiation intensity.

- Light yellow shades indicate areas with moderate solar irradiation, suggesting reasonable solar energy potential but possibly requiring additional considerations for optimal energy generation.

- Darker yellow shades, which are predominantly concentrated in the top-centre portion of the map, represent areas with higher solar irradiation levels, making them more favourable for the installation of a PV solar farm.

This solar irradiation analysis is critical for identifying optimal locations for PV solar infrastructure within the SMA. Areas with higher solar irradiation are more likely to maximise energy output, thereby improving the efficiency and cost-effectiveness of a potential solar energy system.

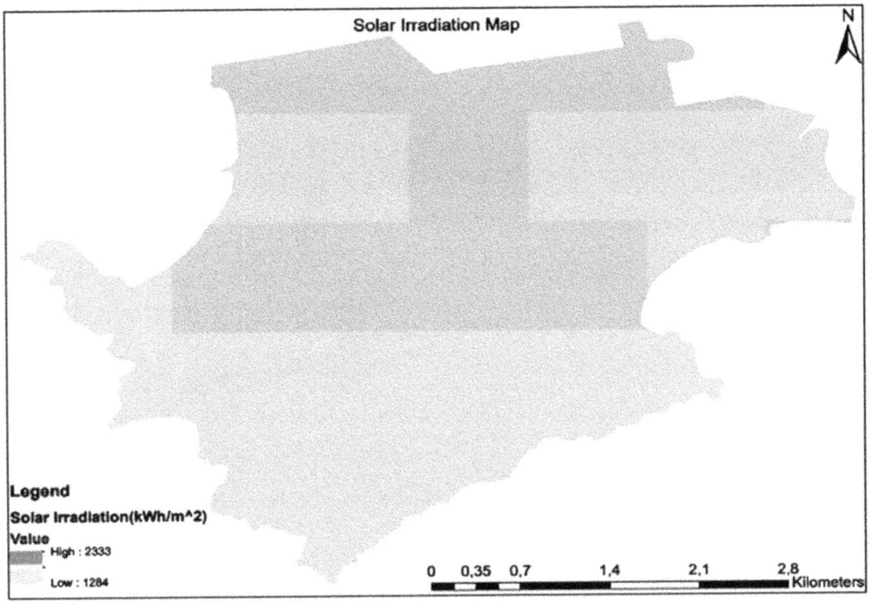

Figure 12.8: Annual solar irradiation map.
(*Source:* Authors' own work using ArcGIS 10.8.1)

The results of this analysis highlight the different types of land use/land cover and the associated restrictions within the selected study area. Figure 12.9 presents a land use/land cover map generated using data from the Esri Sentinel-2 Land Cover Observer 2022. A supervised classification approach was applied to accurately map land use/land cover and identify restricted areas within the study region (Esri, 2022). Certain land-use categories are unsuitable for the construction of community-scale PV solar farms because of environmental, legal, and infrastructural constraints. The restricted areas include:

- **Protected areas**. These regions have been designated for conservation and biodiversity protection. Any form of infrastructure development, including solar farms, can disrupt ecosystems, wildlife habitats, and legally protected zones.
- **Water bodies**. Rivers, lakes, and reservoirs are unsuitable due to flood risks, unstable terrain, and the need to preserve water resources. Additionally, solar farm installations in or near water bodies can impact aquatic ecosystems and water quality.
- **Residential areas**. Urban and suburban regions are not viable for large-scale PV solar installations because of space limitations, zoning regulations, and potential conflicts with existing infrastructure and communities.

By identifying restricted areas, this analysis ensures that the site selection process aligns with environmental sustainability, legal compliance, and optimal land-use planning. The insights provided by the land use/land cover map (Figure 12.9) contribute to a more informed and strategic approach to determining suitable locations for PV solar farm development within the SMA.

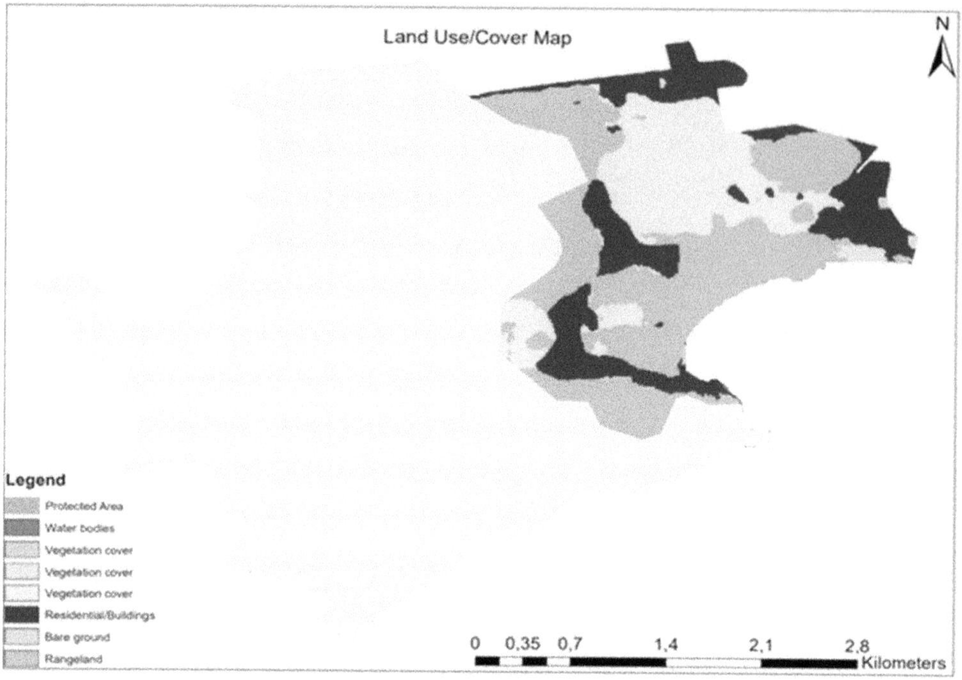

Figure 12.9: Land use/cover map.
(*Source:* Authors' own work using ArcGIS 10.8.1)

The final product of this analysis was the Land Suitability Index (LSI) map (Figure 12.10), which serves as a decision-making tool for identifying optimal locations for the construction of a community-scale PV solar farm within the SMA. This was accomplished through a multi-criteria evaluation process using ArcGIS, integrating seven key thematic maps (Figures 12.3–12.9) that influence land suitability.

To systematically assess the suitability of different areas, a weighted overlay analysis tool in ArcGIS was used. This method allowed for the combination of multiple spatial datasets, each assigned a relative weight based on its importance in determining the feasibility of a solar farm. The result was a composite suitability map (Figure 12.10) that classified the land into three distinct suitability indices:

- **Index 1 (Red): Least suitable areas**. These regions are characterised by significant constraints, such as steep slopes, unsuitable land cover, low solar irradiation, and proximity to restricted zones. Owing to these limitations, constructing a PV solar farm in these areas would be impractical and inefficient.

- **Index 2 (Yellow): Moderately suitable areas**. These zones exhibit conditions that are partially favourable for solar farm development. While they meet some essential criteria (such as adequate solar irradiation and proximity to infrastructure), they may also present challenges (such as suboptimal slopes or land-use restrictions), requiring further assessment and potential mitigation strategies before development.

- **Index 3 (Green): Highly suitable areas.** These locations possess the most favourable conditions for PV solar farm installation, including high solar irradiation, relatively flat terrain, proximity to roads and substations, and appropriate land cover. These areas offer the best opportunities for efficient and cost-effective solar energy generation in the future.

The LSI map provides crucial insights into land suitability, enabling informed decision-making regarding the placement and feasibility of PV solar farms in the SMA. The results indicate that while much of the area is unsuitable for solar development, specific portions meet the necessary criteria and could be effectively utilised for renewable energy infrastructure.

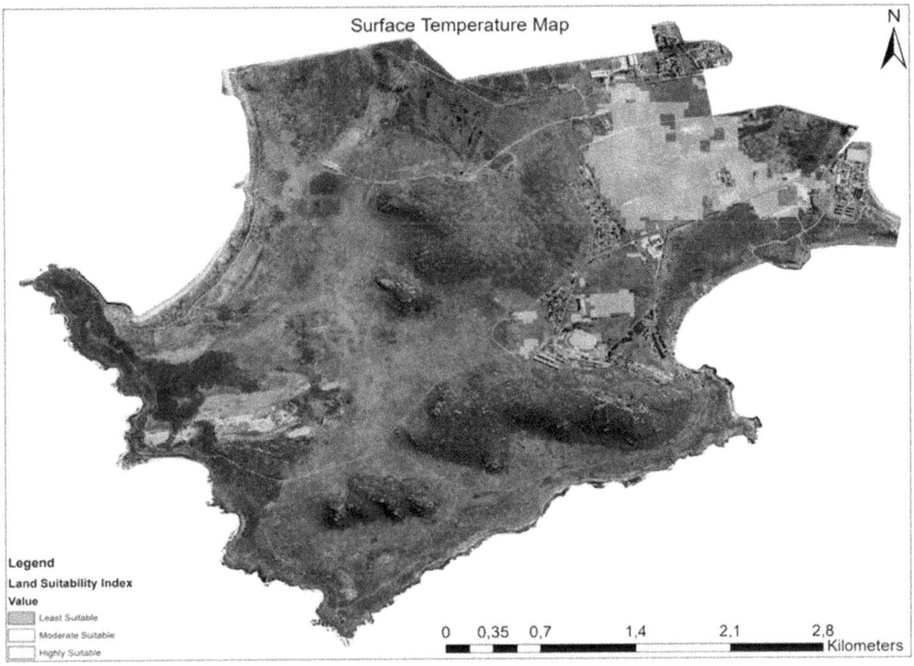

Figure 12.10: Land Suitability Index map.
(*Source:* Authors' own work using ArcGIS 10.8.1)

Conclusion

The application of the AHP-GIS-based method provides a comprehensive assessment of land suitability for the construction of a community-scale PV solar farm (Shrestha et al., 2022). The results indicate that only a small portion of the SMA is suitable for development, whereas most of the land is unsuitable for development.

It is crucial to recognise the significance of solar energy in reducing carbon emissions and enhancing sustainable electricity generation. The AHP-GIS-based method has proven to be an effective and cost-efficient tool for land suitability analyses. The factors considered in this study included annual solar irradiation, annual surface temperature, slope, aspect, land use/land cover, proximity to roads, and distance from miniature substations. The proposed methodology was successfully implemented to identify suitable locations for constructing a PV solar farm within the SMA.

This research holds value for the South African Military Academy as it offers a sustainable alternative to costly and environmentally harmful diesel-powered generators, particularly during load shedding periods. Moreover, given that the Saldanha Bay area hosts an Industrial Development Zone and Special Economic Zone, where energy demand is high, adopting solar power could significantly mitigate the region's carbon footprint. Consequently, the Saldanha Bay Municipality can leverage this study to promote eco-friendly energy solutions through the development of PV solar farms.

Despite its contributions, this study encountered two key limitations which provide opportunities for further research. First, the solar irradiation data were sourced from SolarGIS without cross-validation against data from the South African Weather Service, which could have potentially improved the study's accuracy. Second, the study was conducted on a small scale (limited to the SMA), posing challenges in data collection. A broader-scale study covering a larger geographic area could yield more robust and generalisable findings.

References

Ahmad, S. & Tahar, R.M. 2014. Selection of renewable energy sources for sustainable development of electricity generation system using analytic hierarchy process: A case of Malaysia. *Renewable Energy*, 63: 458–466. https://doi.org/10.1016/j.renene.2013.10.001

Akinci, H. & Özalp, A.Y. 2022. Optimal site selection for solar photovoltaic power plants using geographical information systems and fuzzy logic approach: A case study in Artvin, Turkey. *Arabian Journal of Geosciences*, 15(9). https://doi.org/10.1007/s12517-022-10129-y

Akkas, O.P., Erten, M.Y., Cam, E. & Inanc, N. 2017. Optimal site selection for a solar power plant in the Central Anatolian region of Turkey. *International Journal of Photoenergy*, 2017: 1–13. https://doi.org/10.1155/2017/7452715

Aly, A., Jensen, S.S. & Pedersen, A.B. 2017. Solar power potential of Tanzania: Identifying CSP and PV hot spots through a GIS multicriteria decision making analysis. *Renewable Energy*, 113: 159–175. https://doi.org/10.1016/j.renene.2017.05.077

Aruldoss, M., Lakshmi, T.M. & Venkatesan, V.P. 2013. A survey on multi-criteria decision-making methods and their applications. *American Journal of Information Systems*, 1(1): 31–43.

Asakereh, M., Omid, R., Alimardani, R. & Sarmadian, F. 2014. Developing a GIS-based fuzzy AHP model for selecting solar energy sites in Shodirwan region in Iran. *International Journal of Advanced Science and Technology*, 68: 37–48. https://doi.org/10.14257/ijast.2014.68.04

Axelson, E. 1977. A summary of the history of human settlement at Saldanha Bay. *Transactions of the Royal Society of South Africa*, 42(3–4): 215–221. https://doi.org/10.1080/00359197709519914

Bezuidenhout, J. 2012. Mapping of historical human activities in the Saldanha Bay Military Area. *Scientia Militaria: South African Journal of Military Studies*, 40(2): 89–101. https://doi.org/10.5787/40-2-998

Böhm, J. & Peterson, S. 2021. Fossil fuel subsidy inventories vs. net carbon prices: A consistent approach for measuring fossil fuel price incentives (No. 2186). Kiel Working Paper. https://www.kielinstitut.de/publications/fossil-fuel-subsidy-inventories-vs-net-carbon-prices-a-consistent-approach-for-measuring-fossil-fuel-price-incentives-6819 [28 March 2025].

Chang, K.T. 2008. *Introduction to Geographic Information Systems*. Boston, MA: McGraw-Hill.

Colak, H.E., Memisoglu, T. & Gercek, Y. 2020. Optimal site selection for solar photovoltaic (PV) power plants using GIS and AHP: A case study of Malatya Province, Turkey. *Renewable Energy*, 149: 565–576. https://doi.org/10.1016/j.renene.2019.12.078

Doorga, J.R., Rughooputh, S.D. & Boojhawon, R. 2019. Multi-criteria GIS-based modelling technique for identifying potential PV solar farm sites: A case study in Mauritius. *Renewable Energy*, 133: 1201–1219. https://doi.org/10.1016/j.renene.2018.08.105

Dubey, S., Sarvaiya, J.N. & Seshadri, B. 2013. Temperature dependent photovoltaic (PV) efficiency and its effect on PV production in the world–a review. *Energy Procedia*, 33: 311–321. https://doi.org/10.1016/j.egypro.2013.05.072

Eskom. 2013. *Eskom Interim Integrated Report*. Johannesburg: Eskom Holdings Limited. https://www.eskom.co.za/wp-content/uploads/2022/01/IIR2013.pdf [30 March 2025].

Esri [Environmental Systems Research Institute]. 2022. Sentinel-2 Land Cover Explorer. Available at: https://livingatlas.arcgis.com/landcoverexplorer/#mapCenter=17.93306%2C-33.02176%2C13.474082634089331&mode=step&timeExtent=2017%2C2023&year=2022 [12 October 2023].

Goldberg, A. 2015. The economic impact of load shedding: The case of South African retailers. Doctoral dissertation. Pretoria: University of Pretoria.

Henrico, I. 2021. Optimal interpolation method to predict the bathymetry of Saldanha Bay. *Transactions in GIS*, 25(4): 1991–2009. https://doi.org/10.1111/tgis.12783

Hizam, W.R., Sukojo, B.M. & Jaelani, L.M. 2017. Development of land surface temperature retrieval algorithm from Landsat 8 TIRS and long-term in-situ data. *Modern Environmental Science and Engineering*, 3(3): 200–213. https://10.15341/mese(2333-2581)/03.03.2017/008

IEA [International Energy Agency]. 2022. *World Energy Outlook 2022*. https://www.iea.org/reports/world-energy-outlook-2022 [30 March 2025].

Kengpol, P., Rontlaong, P. & Tuominen, M. 2012. Design of a decision support system for site selection using fuzzy AHP: A case study of a solar power plant in northeastern parts of Thailand. *Proceedings of PICMET 2012: Technology Management for Emerging Technologies*. 734–743.

Lee, A.H., Kang, H.Y. & Liou, Y.J. 2017. A hybrid multiple-criteria decision-making approach for photovoltaic solar plant location selection. *Sustainability*, 9(2). https://doi.org/10.3390/su9020184

Malczewski, J. & Rinner, C. 2015. *Multicriteria Decision Analysis in Geographic Information Science*. Volume 1. New York, NY: Springer. https://doi.org/10.1007/978-3-540-74757-4

Manieniyan, V., Thambidurai, M. & Selvakumar, R. 2009. Study on energy crisis and the future of fossil fuels. *Proceedings of SHEE*, 10: 2234–3689.

Pretorius, I., Piketh, S.J. & Burger, R.P. 2015. The impact of the South African energy crisis on emissions. *WIT Transactions on Ecology and the Environment*, 198: 255–264. https://doi.org/10.2495/AIR150211

Rodríguez Mañay, L.O., Guaita-Pradas, I. & Marques-Perez, I. 2022. Measuring the supply chain performance of the floricultural sector using the SCOR model and a multicriteria decision-making method. *Horticulturae*, 8(2). https://doi.org/10.3390/horticulturae8020168

Rossouw, D. 2008. A technical risk evaluation of the Kantienpan volcanic-hosted massive sulphide (VHMS) deposit and its financial viability. Doctoral dissertation. Pretoria: University of Pretoria.

Ruiz-Fernández, M.D., Ramos-Pichardo, J.D., Ibáñez-Masero, O., Cabrera-Troya, J., Carmona-Rega, M.I. & Ortega-Galán, Á.M. 2020. Compassion fatigue, burnout, compassion satisfaction, and perceived stress in healthcare professionals during the COVID-19 health crisis in Spain. *Journal of Clinical Nursing*, 29(21–22): 4321–4330. https://doi.org/10.1111/jocn.15469

Sánchez-Lozano, J.M., Teruel-Solano, J., Soto-Elvira, P.L. & García-Cascales, M.S. 2013. Geographical Information Systems (GIS) and Multi-Criteria Decision Making (MCDM) methods for the evaluation of solar PV farm locations: Case study in southeastern Spain. *Renewable and Sustainable Energy Reviews*, 24: 544–556. https://doi.org/10.1016/j.rser.2013.03.019

Shrestha, R.R., Acharya, A., Karmacharya, N., Sapkota, M., Paudyal, B., Basnet, P., Timilsina, S. & Shakya, H.D. 2022. Application of AHP and GIS for optimal solar site identification in Madhesh Province, Nepal. *International Journal of the Analytic Hierarchy Process*, 14(3). https://doi.org/10.13033/ijahp.v14i3.986

SolarGIS. 2023. Most accurate data and software for PV plant investments. https://files.solargis.com/public/doc/Collaterals/Brochures-leaflets/Company_Brochure-EN.pdf [12 October 2023].

Sozen, A., Mirzapour, M. & Çakir, M.T. 2015. Selection of the best location for solar plants in Turkey. *Journal of Energy in Southern Africa*, 26(4): 72–80. https://doi.org/10.17159/2413-3051/2016/v26i4a2093

Tahri, M., Hakdaoui, M. & Maanan, M. 2015. The evaluation of PV solar farm locations applying Geographic Information System and Multi-Criteria Decision-Making methods: Case study in southern Morocco. *Renewable and Sustainable Energy Reviews*, 51: 1354–1362. https://doi.org/10.1016/j.rser.2015.07.054

Tariq, M.I., Mian, N.A., Sohail, A., Alyas, T. & Ahmad, R. 2020. Evaluation of the challenges in the Internet of Medical Things with multicriteria decision making (AHP and TOPSIS) to overcome its obstruction under fuzzy environment. *Mobile Information Systems*, 2020: 1–19. https://doi.org/10.1155/2020/8815651

Tunç, A., Tuncay, G., Alacakanat, Z. & Sevimli, F.S. 2019. GIS-based solar power plant site selection using Analytic Hierarchy Process (AHP) in Istanbul, Turkey. *The International Archives of the Photogrammetry, Remote Sensing and Spatial Information Sciences*, 42: 1353–1360. https://doi.org/10.5194/isprs-archives-XLII-2-W13-1353-2019

Uyan, M. 2013. GIS-based solar PV farms site selection using Analytic Hierarchy Process (AHP) in Karapinar region, Konya/Turkey. *Renewable and Sustainable Energy Reviews*, 28: 11–17. https://doi.org/10.1016/j.rser.2013.07.042

Wiguna, K.A., Sarno, R. & Ariyani, N.F. 2016. Optimisation of PV solar farm site selection using multi-criteria decision-making fuzzy AHP and PROMETHEE: Case study in Bali. *2016 International Conference on Information & Communication Technology and Systems (ICTS)*. 237–243. https://doi.org/10.1109/ICTS.2016.7910305

Williams, D.K. 2022. Differences between north- and south-facing slopes. Sciencing. https://www.sciencing.com/differences-between-north-southfacing-slopes-8568075/ [30 January 2025].

Xin, Z., Chen, L., Yang, Y., Miao, K. & Li, Z. 2019. Application of AHP-DEA model in comprehensive evaluation of distribution network planning reliability. *iSPEC 2019 IEEE Sustainable Power and Energy Conference: Grid Modernization for Energy Revolution, Proceedings*. 290–294. https://doi.org/10.1109/iSPEC48194.2019.8975139

Yushchenko, A., de Bono, A., Chatenoux, B., Patel, M.K. & Ray, N. 2018. GIS-based assessment of photovoltaic (PV) and concentrated solar power (CSP) generation potential in West Africa. *Renewable and Sustainable Energy Reviews*, 81: 2088–2103. https://doi.org/10.1016/j.rser.2017.06.021

Zoghi, M., Ehsani, A.H., Sadat, M., Amiri, M.J. & Karimi, S. 2017. Optimization of solar site selection by fuzzy logic model and weighted linear combination method in arid and semi-arid regions: A case study in Isfahan, Iran. *Renewable and Sustainable Energy Reviews*, 68: 986–996. https://doi.org/10.1016/j.rser.2015.07.014

CHAPTER 13

Modelling the maximum potential wave energy zones in North Bay, Saldanha, for small-scale green energy

L.L. Matiwane and R.L. Uys

Abstract

The pursuit of renewable energy sources has become increasingly significant in addressing environmental challenges and reducing dependence on fossil fuels. A promising approach in this context is the exploitation of wave energy, which offers substantial, clean, and sustainable power potential. This study examined the North Bay region in Saldanha Bay to model zones of maximum potential for wave energy suitable for small-scale green energy generation. Utilising oceanographic and geographic data, including wave height, wave period, coastal bathymetry, wind speed, and wind direction, this study identified areas within North Bay that exhibit optimal conditions for wave energy exploitation. These potential zones are characterised by high wave energy densities and appropriate water depths for small-scale wave energy converters. By integrating these datasets and employing advanced modelling tools such as SWAN, MATLAB, and ArcGIS, the research maps locations in North Bay with the highest wave energy potential. These zones are identified as areas where the wave conditions consistently demonstrate favourable characteristics for energy extraction. The results offer valuable insights into the spatial distribution of wave energy resources and provide a foundation for further analysis and decision-making regarding small-scale green energy projects. These findings contribute to the expanding body of knowledge on wave energy resource assessment, specifically for the Saldanha region. The identified maximum potential wave energy zones offer critical insights for energy developers, facilitating informed decision-making and the sustainable utilisation of wave energy resources. The integration of small-scale green energy generation in North Bay can significantly contribute to a more sustainable and environmentally friendly energy landscape.

Keywords: small-scale green energy, wave energy, wave energy zones, wave modelling

Introduction

Governments bear the fundamental responsibility of ensuring a secure environment for their citizens while advancing economic and social programmes. Electricity is a vital component of the developmental trajectory of any nation. In South Africa, coal-fuelled technology is the primary foundation and power source for electricity generation. However, the South African power sector is experiencing supply stress due to its intermittent inability to meet the increasing demand for electricity, exacerbated by climate change induced by gas emissions from fossil fuel combustion. This challenge is further compounded by ageing power plants, necessitating a significant transition for utility companies from a coal-dependent generation model to one that incorporates alternative energy sources (Simon et al., 2021). Among the emerging renewable energy sources, wave-generated electricity is anticipated to grow substantially, with global potential comparable to that of wind and hydropower. Several wave power technologies are currently being commercialised, raising concerns regarding their environmental impact (Langhamer et al., 2010).

Saldanha Bay, located on the West Coast of South Africa in the Western Cape province, northwest of Cape Town, is characterised by its protected seas, sand dunes, mudflats, salt marshes, seagrass beds, and intertidal zones. The region experiences an annual reversal in wind direction and speed, with south-westerly onshore winds prevailing during summer and north-easterly offshore winds during winter (Flemming, 2015; Henrico & Bezuidenhout, 2020). The study area, as illustrated in Figure 13.1, contains several notable features. First, Malgas Island, which is 7 m high and situated 1.2 km east of the North Head Lighthouse, marks the northern entrance to Saldanha Bay. This island is officially designated as a marine-protected guano island. On its northern side, there are several white-painted huts and a flagstaff, surrounded by shoals and underwater rocks extending 370 m and beyond from the shore. Second, North Bay features alternating sandy beaches and rocky outcrops. The eastern half of the bay, exposed to the dominant swell, experiences heavy breakers far from the shore, whereas the western portion is generally well-protected from the swell and north-westerly winds. Third, Marcus Island, although no longer an island, remains classified as a guano island. It is connected to the mainland by a breakwater, which includes a spending beach on its seaward side, designed to absorb wave energy (SAN HO-22, 2014).

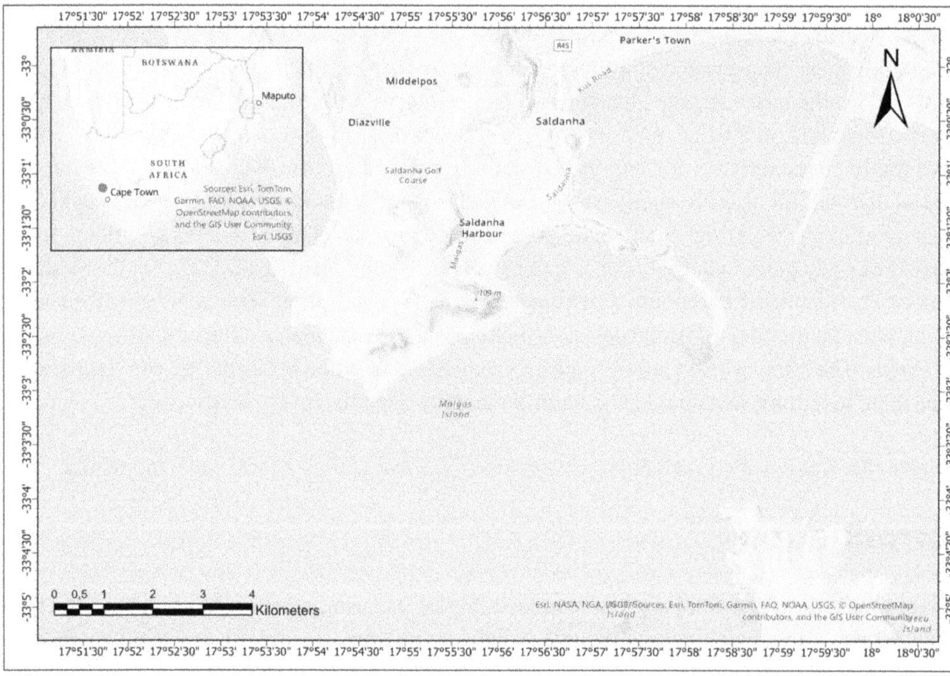

Figure 13.1: The study area is indicated by a red dot representing Saldanha Bay on the map of South Africa.
(*Source:* Authors' own work using ArcGIS 10.8.1)

The escalating global demand for energy, coupled with the environmental repercussions of fossil fuel dependency, has intensified the pursuit of renewable energy alternatives such as biofuels. Wave energy, characterised as a clean and sustainable resource, offers a promising solution to these challenges (Ellabban et al., 2014). This study examined the wave energy potential of North Bay, Saldanha Bay, with a focus on identifying optimal sites for small-scale green energy generation. By integrating oceanographic and geographic data, such as wave height, wave period, bathymetry, wind speed, and direction, with advanced modelling tools, including SWAN, MATLAB, and ArcGIS, this research aims to map and characterise the zones of maximum potential wave energy within North Bay. The findings of this study will provide valuable insights for energy developers and contribute to the sustainable utilisation of wave energy resources in the Saldanha region, ultimately facilitating a transition towards a more sustainable energy future.

Research methodology

This study aimed to model the maximum potential wave energy zones in North Bay for small-scale green energy production. The objective of this study was to identify the optimal locations for establishing a wave energy farm. This research used three software programmes – ArcGIS, MATLAB, and SWAN – integrated at various stages from data analysis to result presentation.

ArcGIS, developed by Esri, is a comprehensive geographic information system (GIS) software package that facilitates the creation, management, analysis, and visualisation of geographic data. This enables users to interact with spatial data, including geographic and descriptive information. The software offers numerous functionalities, such as data editing, spatial analysis, geo-processing, mapping, and data visualisation. Users can access various data sources, including pre-existing databases, satellite imagery, GPS devices, and aerial photography (Chang, 2019). MATLAB provides a high-level programming language and environment for numerical computing, data analysis, and visualisation (Chapra, 2008).

The third-generation numerical wave model, Simulating Waves Nearshore (SWAN), is frequently employed for wave analysis (Folley, 2016). SWAN is computationally efficient in shallow water and is considered to be cost-effective. The model results can be generalised to closely match field observations (Booij et al., 1999). SWAN served as the primary modelling tool in this study to assess wave conditions using bathymetric, wind, and wave boundary data.

Data collection

To effectively conduct modelling using SWAN, the model requires only the essential inputs typical of numerical wave propagation models: boundary conditions, such as offshore deep-water wave parameters (Hs, Tp, and MWD), in addition to the site's bathymetry (Roberts et al., 2014). Data pertaining to wave height, wave period, wind speed, and wind direction for 2020 were obtained from the CSIR. Bathymetric data were provided by the South African Navy Hydrographic Office.

Data processing

Processing bathymetric data and establishing study area coordinates are the initial steps in wave modelling. Bathymetric data are typically presented with irregular spacing in digital x, y, and z formats linked to a geographic coordinate system. It is feasible to interpolate unevenly spaced bathymetric data into grid data (Rusu et al., 2008). The raw bathymetric data collected were in MS Excel format and organised in x-y-z coordinates. The original Excel (.xlsx) spreadsheet was converted and saved as a comma-separated values (CSV) file (.csv) to ensure compatibility with ArcGIS. Coastline shapefiles were then created in ArcGIS, and the bathymetric data were uploaded.

The inverse distance weighting interpolation technique was employed because of its superior performance in mapping the bathymetry of Saldanha Bay (Henrico, 2021). As illustrated in Figure 13.2, the interpolation of bathymetric points was conducted using the inverse distance weighting technique, resulting in the creation of a raster file with contour lines. This file was integrated with a fishnet grid to obtain the data points. A conversion tool from raster to ASCII files was applied. The resulting ASCII file was then exported to Excel, where data cleaning was performed, and subsequently saved as a *.xlsx file *.bot (bottom) file for use with SWAN. Determining the appropriate cell size (resolution) is crucial, as a coarse grid can lead to information loss at higher frequencies and provide less information on a rough computational grid. In the simulation with a coarse grid cell size of 100 × 100 m, there was no provision for energy at frequencies greater than 0.02 Hz. With such an imperfect grid, the model can only replicate the slowest wave groups because it has limited points to represent the faster wave groups. Following several simulation steps (Steps 1 to 3, see below), an accurate grid with an average grid cell size of 30 × 30 m was selected (Soler, 2006). This provided the most accurate results when the energy spectra were plotted.

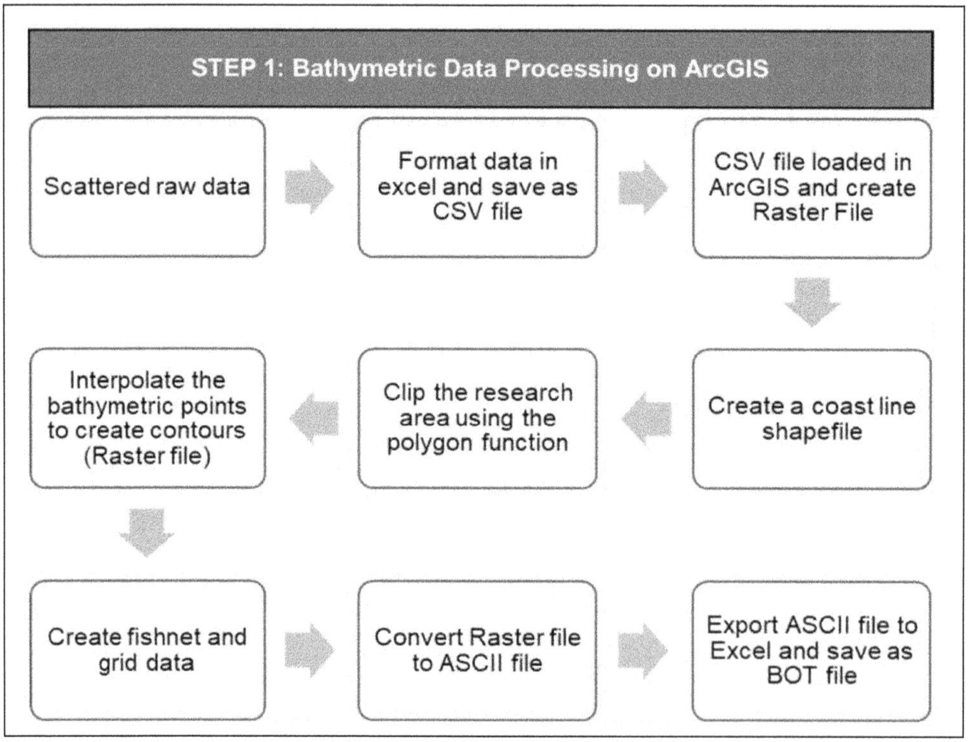

Figure 13.2: Bathymetric data processing in ArcGIS.

The SWAN input file was constructed according to the specifications outlined in the bathymetric driver, with the bottom grid serving as the foundation for establishing the computational grid used in the modelling process (Rusu et al., 2008). In Step 2 (Figure 13.3), the computational grid was developed to align with the bottom grid using bathymetric data generated in the *.bot file. In SWAN, output files (*.mat and text spectrum files) are produced as inputs for MATLAB, facilitating the plotting process.

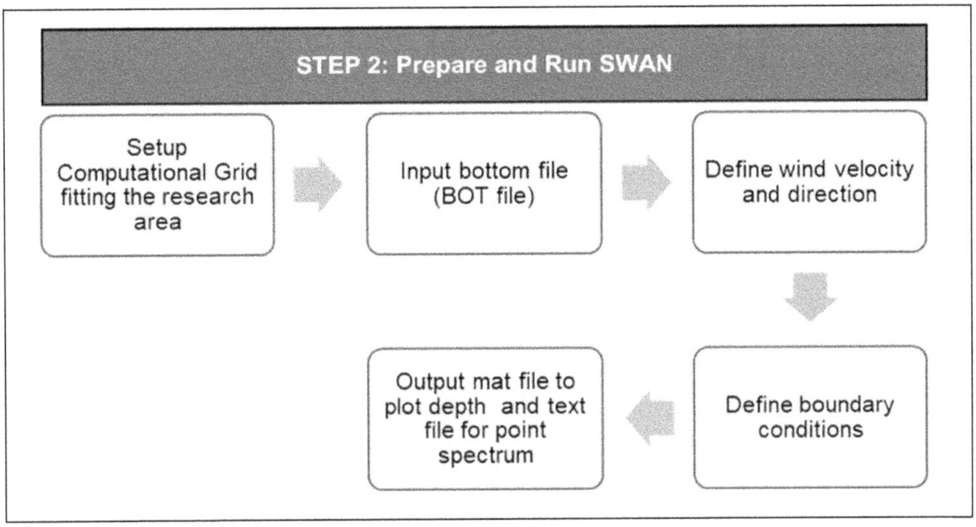

Figure 13.3: Preparation and run of SWAN.

Once the *.mat files were generated by SWAN, and MATLAB was employed to plot the energy spectrum, directional spreading, and energy dissipation, following the procedures outlined in Step 3 (Figure 13.4).

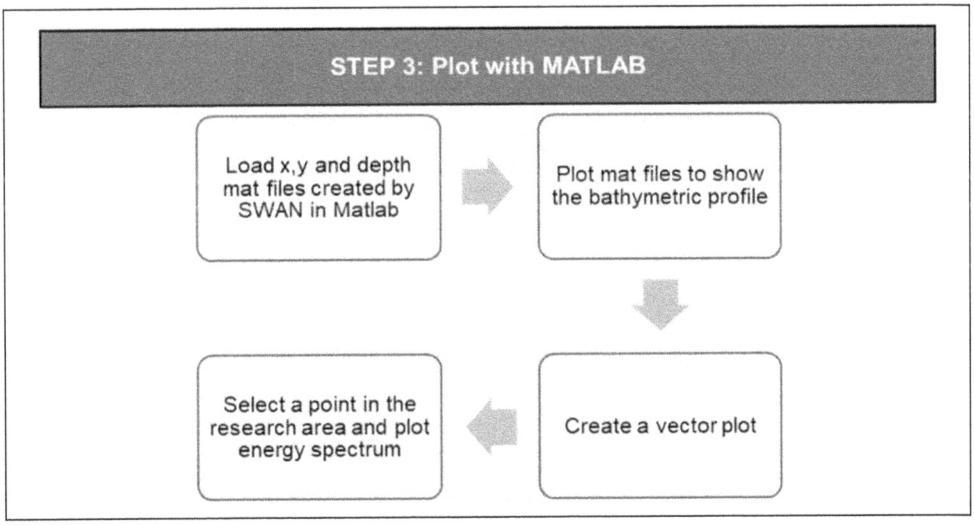

Figure 13.4: Plotting using MATLAB.

Results and findings

Ocean waves undergo various modifications as they propagate from deep to intermediate and shallow waters. These waves interact with the seabed topography, resulting in changes in their shape and velocity, which subsequently lead to variations in wavelength, direction, and wave height until the waves ultimately break. The term 'wave transformation' refers to these modifications that occur as waves travel from deep-water zones to shallow waters (Bosboom & Stive, 2021). Consequently, the wave spectra exhibit significant differences between the deep and shallow water regions, owing to the aforementioned interactions with the seabed. Therefore, it is essential to understand the alterations in wave characteristics nearshore. The SWAN model can simulate the effects of wave conditions in shallow waters, thereby facilitating the study of wave transformation (Holthuijsen, 2007).

Wave characteristics

The ratio of the wave height to its wavelength is referred to as wave steepness. A wave that exhibits a gradual slope from trough to crest is characterised by modest wave steepness. Conversely, a wave with moderate steepness displays a more pronounced crest than trough. A high wave steepness indicates a steep wave with a sharp peak, which may become unstable and break (Bosboom & Stive, 2021). The modelled wave steepness in North Bay is illustrated in Figure 13.5, where the steepness ranges from 0.02–0.14. According to Holthuijsen (2007), a wave steepness of 0.14 is associated with white-capping, which is visually observed in this region.

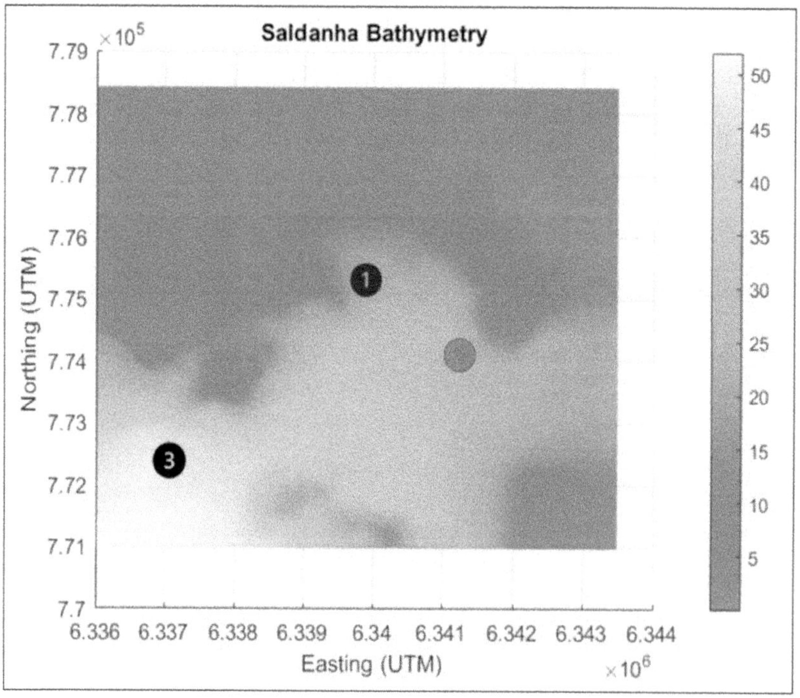

Figure 13.5: Bathymetry and three prospective sites for wave energy harvesting, identified as locations 1, 2, and 3, respectively.
(*Source:* Authors' own work using MATLAB R2024b)

Shoaling is a process that influences wave energy, resulting in changes in wave height, steepness, and breaking as waves approach shallower waters near the shore. The increase in wave height is attributed to interactions with the seabed, which cause the waves to bunch up as their speed decreases (Bosboom & Stive, 2021). The energy of a wave is directly proportional to its height, specifically to the square of its height. Consequently, if the wave height doubles, the energy quadruples (MacHutchon & Bosman, 2006). In shallower waters, waves may steepen as crests travel more rapidly than troughs. The interaction with the seabed, which dissipates some of the wave energy, leads to wave breaking as they near the coast (Bosboom & Stive, 2021).

Wave characteristics are altered over and around obstructions, thereby affecting the wave direction and energy distribution. Refraction occurs when waves change direction owing to variations in water depth, thereby affecting the distribution of wave energy. Refraction can concentrate wave energy near points and headlands, creating regions with focused wave energy, known as wave concentrations. The dispersion of wave energy over a large area reduces the wave height and energy. This dispersion can occur due to refraction and diffraction. Diffraction occurs when waves spread around obstructions or coastal features, distributing energy along the shoreline. When diffracted waves interact, they can interfere constructively or destructively, influencing wave energy levels at specific locations (Bosboom & Stive, 2021; Gweba, 2018; Holthuijsen, 2007; Soler, 2006).

Significant wave height and wave direction

The significant wave height is a critical parameter for assessing the energy potential of ocean waves. This metric is typically represented as the average height of the highest one-third of waves within a specified time frame and serves as a statistical measure of wave height. This approximately aligns with the visual estimation of the average wave height by mariners (Holthuijsen, 2007). Waves with greater significant wave heights are associated with an increased energy potential. Analysing the distribution and magnitude of significant wave heights in North Bay can facilitate the identification of areas with the highest wave energy concentration, which may be suitable for wave energy farming (Bosboom & Stive, 2021; Soler, 2006).

The efficiency of energy capture is significantly influenced by the alignment between the wave direction and the orientation of the energy converters. Several wave energy devices are designed to extract the maximum energy when waves approach a specific direction. Optimising the placement of these devices necessitates the determination of the prevailing wave directions. By analysing these predominant directions, developers can allocate resources more effectively and focus on regions with the highest energy fluxes. Strategic resource placement maximises the energy production potential. Additionally, wave direction affects coastal erosion, sediment transport, and marine ecosystem health. Therefore, considering wave direction is crucial for planning sustainable projects and assessing potential environmental impacts (Bosboom & Stive, 2021; Soler, 2006).

Joint North Sea Wave Project (JONSWAP)

The Joint North Sea Wave Project (JONSWAP) is a prominent wave spectrum model used in oceanography to characterise the spectral properties of ocean surface waves. It is also a requisite component of the SWAN model, indicating the spectral model to be used. The JONSWAP spectrum was developed by analysing wave data collected in the North Sea by several European countries in the late 1970s. Within a sea wave field, the JONSWAP spectrum represents the distribution of wave energy across various frequencies, providing a mathematical description of how the wave energy content varies with wave frequency (Holthuijsen, 2007).

The Pierson-Moskowitz (PM) and JONSWAP spectral forms are among the most prevalent empirical spectral shapes. The PM spectrum is applicable to deep water, where the depth significantly exceeds the wavelength, and is based on the assumption of a steady state and a fully developed sea state. Conversely, the JONSWAP spectrum is frequently employed in nearshore and coastal applications, proving particularly useful for defining sea conditions with moderate to fully developed waves (Gweba, 2018; Holthuijsen, 2007; Joubert & Van Niekerk, 2013).

According to Holthuijsen (2007), the peak enhancement factor (γ) in the JONSWAP spectrum is critical. It is used to modify the PM spectrum to represent more accurately the properties of the observed waves under realistic conditions. As noted by Joubert and Van Niekerk (2013), the ratio of the peak energy densities of the PM and JONSWAP spectra is used to determine the value of γ; thus, a JONSWAP spectrum with a γ-value of 1 corresponds to a PM spectrum. Gweba (2018) reports that the mean γ-value in the Saldanha Bay area is 2.2, which was the γ-value employed in their study.

Directional spreading

Directional spreading refers to the distribution of wave energy across various wave directions. This distribution can fluctuate owing to irregularities in the sea state, which is essential for understanding the efficiency of wave energy capture and enhancing the design of energy conversion devices. The energy of waves originating from different directions varies. Factors such as prevailing winds, fetch (the duration and distance over which the wind exerts force), and coastal features can result in certain orientations exhibiting waves with greater intensity. Accurate modelling of directional spreading is instrumental in identifying the most favourable wave directions for energy extraction (Holthuijsen, 2007; Joubert & Bosman, 2008; Joubert & Van Niekerk, 2013).

Energy spectra and directional spreading

Three potential sites were identified, as depicted in Figure 13.5: locations 1, 2, and 3. The selection of these sites was informed by factors such as bathymetry, wave direction, the proposed aquaculture site situated centrally within the bay, and the necessity to avoid obstructing vessel traffic in the bay. The energy spectra and directional spreading at the three selected sites are illustrated in Figure 13.6.

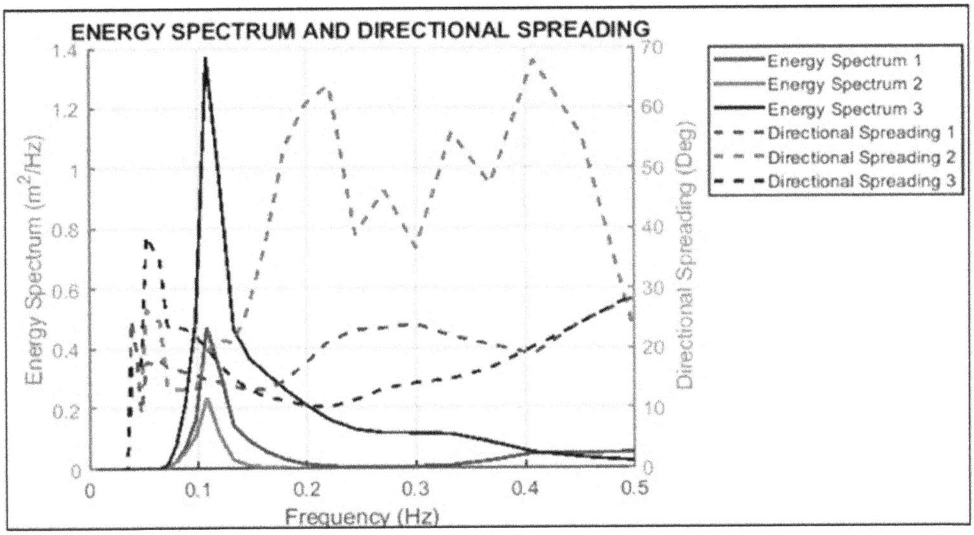

Figure 13.6: The energy spectrum and directional spreading for the three selected locations, with locations 1, 2, and 3 in blue, red, and black, respectively.

In shallow waters, waves can break when they become sufficiently steep. The collapse of the wave crest and the ensuing turbulence in the water result in rapid energy dissipation from the breaking waves (Bosboom & Stive, 2021). As illustrated in Figure 13.5, the wave steepness was most pronounced at location 1 among the three locations, leading to reduced energy due to dissipation. Conversely, location 3 exhibited an average wave steepness, which facilitated better energy conservation. Location 2 demonstrated minimal wave steepness, with energy reduction primarily attributed to spreading loss.

As previously mentioned, diffraction can redistribute wave energy along coastal features, such as islands. Waves often extend into the shadow zones behind the obstructions. Depending on the shoreline geometry, this can result in areas with either increased or decreased wave energy. Refraction influences the spatial distribution of wave energy by concentrating the waves at specific locations. Wave energy concentration can shift in areas with varying water depths, as waves tend to refract towards shallower waters (Bosboom &Stive, 2021; Gweba, 2018; Holthuijsen, 2007; Soler, 2006). Locations 1 and 2 experienced diffraction and refraction owing to their position in the shallower regions of the bay, leading to significant energy dissipation. In contrast, location 3, which is situated in a deeper region with fewer obstructions, exhibits higher energy levels.

In Figure 13.6, three spectra are depicted: location 1 (blue), location 2 (red), and location 3 (black). The black spectrum displays the highest energy peak, approximately 1.3 m^2/Hz, at approximately 0.1 Hz. A larger area under the curve, indicating variance, suggests dominant energy in low-frequency waves. The blue spectrum shows a relatively broad energy distribution compared to location 2, indicating mixed wave conditions owing to shallow conditions and diffraction. The red spectrum had the lowest energy peak, suggesting a low-energy system caused by refraction and shallow water. According to Holthuijsen (2007), the peak at approximately 0.1 Hz indicates the presence of swell, which was evident at location 3.

The useful energy range, as depicted in Figure 13.6, lies between approximately 0.08 and 0.2 Hz. Only location 2 exhibited an increase in directional spreading over these frequencies, indicating a spreading loss in this area. Location 1 and 3 showed a slight initial decrease in directional spreading. The minimum spreading loss is typically expected at maximum energy levels but is observed here to have shifted to the right (lower energy levels). The reason for this shift is unclear. The shape of the directional spreading graph is not well understood but is speculated to have a maximum value in the dominant wind direction (Holthuijsen, 2007).

Conclusion

In North Bay, three distinct locations characterised by varying wave properties and bathymetric features were selected for the analysis. The energy spectra and directional spreading at these sites were plotted and compared to each other. Location 1 exhibited the highest wave steepness; however, the shallowness of the area resulted in significant energy dissipation, refraction, and diffraction. Location 2 demonstrated the lowest wave steepness, likely because of the presence of a breakwater wall extending to Marcus Island, which was engineered to absorb wave energy. Consequently, Location 2 recorded the lowest energy levels among the sites, which was attributed to energy dissipation, refraction, and diffraction. Location 3, situated southwest of Malgas Island, experienced moderate wave steepness and a higher significant wave height than the other locations. This site encountered reduced refraction due to its depth and diminished diffraction owing to minimal coastal obstruction, leading to higher energy levels than those at the other locations.

This study aimed to facilitate the implementation of small-scale green energy through wave farming by examining the intricate geography of wave energy potential in North Bay. Using innovative research methodologies, including various software, programming, and analytical techniques, this study produced illustrative results that identified the most suitable location for wave energy farming. The modelling conducted in North Bay enabled the identification of the optimal area for small-scale wave energy farming in the region. Based solely on physical oceanographic factors, location 3 was determined to be the most suitable.

References

Booij, N., Ris, R.C. & Holthuijsen, L.H. 1999. A third-generation wave model for coastal regions: 1. Model description and validation. *Journal of Geophysical Research: Oceans*, 104(C4): 7649–7666. https://doi.org/10.1029/98JC02622

Bosboom, J. & Stive, M.J.F. 2021. *Coastal Dynamics*. TU Delft OPEN Books. https://doi.org/10.5074/T.2021.001

Chang, K. 2019. *Introduction to Geographic Information Systems*. Ninth edition. New York, NY: McGraw-Hill.

Chapra, S.C. 2008. *Applied Numerical Methods with MATLAB for Engineers and Scientists*. Second edition. New York, NY: McGraw-Hill.

Ellabban, O., Abu-Rub, H. & Blaabjerg, F. 2014. Renewable energy resources: Current status, future prospects, and their enabling technology. *Renewable and Sustainable Energy Reviews*, 39: 748–764. https://doi.org/10.1016/j.rser.2014.07.113

Flemming, B.W. 2015. Depositional processes in Saldanha Bay and Langebaan Lagoon (Western Cape, South Africa). CSIR Research Report 362. National Research Institute for Oceanology, Stellenbosch, South Africa.

Folley, M. 2016. *Numerical Modelling of Wave Energy Converters: State-of-the-Art Techniques for Single Devices and Arrays*. Oxford: Academic Press.

Gweba, B. 2018. Developing a methodology for the assessment of wave energy along the South African Coast. Master's dissertation. Stellenbosch: Stellenbosch University.

Henrico, I. 2021. Optimal interpolation method to predict the bathymetry of Saldanha Bay. *Transactions in GIS*, 25(4): 1991–2009. https://doi.org/10.1111/tgis.12783

Henrico, I. & Bezuidenhout, J. 2020. Determining the change in the bathymetry of Saldanha Bay due to the harbour construction in the seventies. *South African Journal of Geomatics*, 9(2): 236–249. https://doi.org/10.4314/sajg.v9i2.16

Holthuijsen, L.H. 2007. *Waves in Oceanic and Coastal Waters*. Cambridge: Cambridge University Press. https://doi.org/10.1017/CBO9780511618536

Joubert, J.R. & Bosman, M.D.E. 2008. An investigation of the wave energy resource on the South African Coast, focusing on the spatial distribution of the South West Coast. Master's dissertation. Stellenbosch: Stellenbosch University.

Joubert, J.R. & Van Niekerk, J.L. 2013. *South African Wave Energy Resource Data: A Case Study*.

Langhamer, O., Haikonen, K. & Sundberg, J. 2010. Wave power – Sustainable energy or environmentally costly? A review with special emphasis on linear wave energy converters. *Renewable and Sustainable Energy Reviews*, 14(4): 1329–1335. https://doi.org/10.1016/j.rser.2009.11.016

MacHutchon, K.R. & Bosman, D.E. 2006. The characterisation of South African sea storms. Master's dissertation. Stellenbosch: Stellenbosch University.

Roberts, J.D., Chang, G. & Jones, C. 2014. *Investigation of Wave Energy Converter Effects on Near-shore Wave Fields: Model Generation Validation and Evaluation – Kaneohe Bay HI*. Sandia report (SAND2014-17400). Albuquerque, NM: Sandia National Laboratories. https://doi.org/10.2172/1156602

Rusu, E., Conley, D. & Ferreira-Coelho, E. 2008. A hybrid framework for predicting waves and longshore currents. *Journal of Marine Systems*, 69(1–2): 59–73. https://doi.org/10.1016/j.jmarsys.2007.02.009

SAN HO-22. 2014. *South African Sailing Directions SAN HO-22. Volume 2 - The Coast of Namibia and the Republic of South Africa (RSA) from Kunene River to Cape Agulhas*. Fifth edition. The Hydrographer, SA Navy.

Simon, G., Dikeledi, G. & Warlimont, J. 2021. The Challenges of Power Generation, Transmission, and Distribution in the South African Power Sector. *"Ovidius" University Annals, Economic Sciences Series*, XXI (1).

Soler, C. 2006. Swell and wave-groups at Saldanha Bay. Master's dissertation. Delft: Delft University of Technology.

CHAPTER 14

Geospatial tools and techniques in military exercise planning: Insights from the South African Military Academy's Exercise Trans Enduro

M. Mathoho and H.A.P. Smit

Abstract

The South African Military Academy (SAMA) is a military institution of higher education and professional military development with a legacy of training military officers for more than 70 years. It also houses the Faculty of Military Science at Stellenbosch University. This chapter presents a case study focusing on the application of geospatial techniques and data to accomplish reconnaissance tasks during the planning phase of the annual Exercise Trans Enduro conducted by the SAMA. The aim was to explore the suitability and effectiveness of geospatial techniques in replacing or assisting with traditional physical reconnaissance, which incurs significant annual costs of approximately R170,000 for the SAMA. The exercise, facilitated by military instructors and supported by Stellenbosch University, aims to develop leadership qualities essential for modern military officers in achieving specific objectives through teamwork and community outreach through the Masiza program. Traditionally, the exercise planning phase involved a team of approximately 23 individuals travelling from Saldanha to Gqeberha to conduct reconnaissance in various towns along National Route number 2 (N2).

However, the results of this study reveal that geospatial techniques can efficiently accomplish most of the reconnaissance tasks required for exercise planning. Coupling this with physical reconnaissance by a small group to verify the results can lead to substantial cost savings. The study acquired data, such as shapefiles and maps, from reputable sources such as OpenStreetMap and USGS, and accessed essential statistics from government databases to identify disadvantaged areas suitable for the community outreach program (Masiza). The acquired data were effectively visualised and mapped using ArcGIS, enabling the study to achieve the desired results and explore the efficacy of geospatial techniques. The results indicate the feasibility of accomplishing reconnaissance tasks, thus enabling the creation of geospatial PDFs, resulting in significant cost savings, and shedding light on the benefits and implications of adopting geospatial techniques for reconnaissance in military exercise planning. Moreover, this research serves as a groundbreaking development for instructors and students, offering valuable time and cost-saving opportunities. By utilising geospatial techniques, this study showcases a potential shift towards more efficient and economical methods in reconnaissance for planning purposes, making it an invaluable contribution to the field of military operations and geospatial applications.

Keywords: cost-effective, data visualisation, exercise planning, geospatial PDF, geospatial techniques, reconnaissance, Trans Enduro

Introduction

Democracy in South Africa came with the integration of the armed forces; as a result, the South African National Defence Force (SANDF) was formed (Nathan, 1998). Furthermore, under civilian oversight, a defence review and the White Paper on Defence were drawn to redefine and broaden the functions and goals of the SANDF. Among the broadened functions, in support of the policy and the constitution, the SANDF now takes on non-traditional military roles, such as humanitarian intervention exercises and peacekeeping operations. The SANDF conducts humanitarian operations, which also require thorough route planning in cases of mobilisation and/or disaster relief operations and community outreach programmes, such as the South African Military Academy's (SAMA) annual Exercise Trans Enduro (Le Roux, 2005). Figure 14.1 illustrates a few activities that took place during the exercise in 2021, that is, the launching parade, the mayor's address during the parade, running, cycling, and sea teams busy with their daily activities, and members of the transport team preparing vehicles before moving out.

Figure 14.1: Exercise Trans Enduro 2021.
(*Source:* Photographs provided courtesy of the South African Military Academy)

The origins of Exercise Trans Enduro can be traced back to 1982, when a group of young officers at SAMA discovered that they had no leave days after their examinations and that they would have to stay at SAMA during the winter break (Helfrich, 2021). According to Helfrich (2021), these officers persuaded the then commandant of the SAMA to allow them to compete in a ten-day endurance race using two inflatable rubber boats from Gqeberha (formerly known as Port Elizabeth) to Saldanha. This initial effort evolved, and the exercise currently consists of nine teams. The endurance teams on land include the running and cycling teams (covering between 10 and 50 km per day) and the sea team (covering between 20 and 50 sea miles per day depending on sea conditions). Exercise Trans Enduro also has support teams, of which one (the land team) solely focuses on fundraising, and the others focus on media, signals, logistics, and '*kaaswors*' (for beverages and entertainment). Apart from raising funds and developing endurance as primary objectives on the Eastern and Western Cape routes, Exercise Trans Enduro participants (primarily SAMA students) have decided to include community-related relief work among disadvantaged communities in their activities. This includes cleaning streets, painting schools, and visiting retirement homes and villages, among other activities (Helfrich, 2018).

Planning Exercise Trans Enduro

Every year, a new committee is formed to plan and lead the exercise, and this committee is elected immediately after the previous year's exercise is completed. After the election, the committee begins preparing and planning the exercise. One of the prerequisites for the newly elected committee is gathering data concerning the towns that will be visited during the exercise, which is accomplished through physical reconnaissance. According to information acquired from the 2021 Exercise Trans Enduro commander, the committee, which consists of nine team leaders and their seconds-in-command, together with military instructors, travel in two 22-seater buses and two Land Cruiser 4x4 vehicles to Gqeberha to inspect all the towns that will be visited during the exercise (Captain D.D. Oss 2023, personal communication, 15 June). During this reconnaissance, the members must be accommodated and fed, the vehicles need fuel, and the group must pay for tollgates and secure parking. Approximately R170,000 or more must be spent on the exercise's reconnaissance each year. Thus, this is a strain on the SAMA's financial resources, and any way in which it can be reduced should be investigated.

Military operations rely heavily on route planning to conduct exercises effectively; thus, it is an essential component of military tactical skills that contribute to productive intelligence (Tarapata, 2003). Reconnaissance is a prerequisite for all military operations, as the military relies on it in most cases. However, physical reconnaissance is expensive, time-consuming, delays planning, and can endanger the lives of soldiers, which may compromise the mission (Kopuletý & Palasiewicz, 2018). Physical reconnaissance requires an enormous amount of financial input and other resources, as well as physical labour, for information gathering. Modern militaries use new technology, such as planes, robots, drones, unmanned aerial vehicles (UAVs), or Unmanned Ground Vehicles (UGVs), to conduct reconnaissance, but not all militaries can acquire, employ, and maintain the technology that is necessary for this type of reconnaissance. This is especially true in developing and/or third-world countries. Furthermore, training and the process of technology procurement add to the time and financial constraints of the mission (Henrico et al., 2020). Equally important, there is a risk of compromising missions because this technological equipment can still be detected (McKearney, 2000). Oron-Gilad & Parmet (2017) reiterated that using equipment such as UGVs affects soldiers' ability to make quick decisions. Hence, depending on the nature of the mission, militaries still rely on traditional ways of doing reconnaissance.

Geospatial tools and techniques provide a cheaper and quicker way to acquire information as they are readily available to plan military exercises (Henrico et al., 2020). Among many other applications of geospatial tools within the defence industry, these tools can be used for intelligence data integration, flight or path planning, reconnaissance, and base construction planning. The Military Academy's Department of Military Geography has resources (computers and software) that can be used to gather the information needed to plan an exercise such as Exercise Trans Enduro. The research reported here used geospatial techniques to conduct most of the reconnaissance required to plan Exercise Trans Enduro so that SAMA could save some of the costs of conducting physical reconnaissance. The information gathered in this manner can be augmented by a small team verifying the information during physical reconnaissance.

Geospatial tools and techniques in exercise planning

During peacetime, the military conducts exercises other than war, such as humanitarian interventions, disaster relief, outreach programmes, and 'show of force', or, in this case, the SAMA's Exercise Trans Enduro. Military exercises are essential for preparing military forces for combat scenarios and maintaining their readiness (Frazier & Hutto, 2017). Equally important, reconnaissance is an aspect of military operations that involves collecting and analysing information about the enemy and the battlefield (Young & Adam, 2014). Despite the rapid advancement in reconnaissance technology, including geospatial tools and techniques, physical reconnaissance is still performed for some missions.

Geospatial tools have become increasingly important in the planning and execution of military exercises worldwide because of their ability to provide valuable information about terrain, infrastructure, and potential threats (Ngeno, 2021). Based on Singh et al. (2021), geospatial techniques can be defined as the use of specialised software and hardware to collect, analyse, and visualise geographic data. In addition, one of the key applications of geospatial tools and techniques in military reconnaissance is the creation of maps because of their ability to provide real-time information integrated from different sources. Furthermore, GIS enables planners of military operations to make well-informed decisions on the movement of soldiers (route planning), logistical support, and the effects of environmental factors and their influence on the exercise.

The use of geospatial tools in exercise planning has become increasingly popular because of their ability to improve the efficiency and efficacy of the planning process. According to Yang et al. (2014), geospatial tools such as GIS and remote sensing have been utilised in various aspects of exercise planning, including site selection, hazard analysis, resource allocation, and risk assessment. For example, GIS allows for the integration of various data sources, such as topography, population density, and infrastructure, to identify suitable locations for exercise activities (Abdalla, 2016). Furthermore, remote sensing can provide valuable information on environmental conditions, such as vegetation coverage and elevation, that can affect the planning process. Through the integration of these tools and techniques, exercise planners can create detailed and comprehensive plans that consider various factors that may affect the success of the exercise. The use of geospatial techniques can also aid in the analysis of past exercises to identify areas for improvement and develop more effective strategies for future exercises.

Other applications of geospatial tools and techniques within the military realm have been in counter-insurgency operations, peacekeeping operations, and the fight against rhino poaching in South Africa. These techniques provide military planners with a powerful tool to collect, analyse, and visualise geographic data, enabling them to make more informed decisions, allocate resources more effectively, and respond more quickly to changing conditions. Based on the literature by Namufohamba (2019), it is important to note that there can be challenges with the use of geospatial tools and other technologies used for information gathering in the military, based on the need to produce accurate and up-to-date data. Karakaya (2014) sheds light on how information gathering using new technologies such as UAVs or other GIS tools can be costly due to reasons such as the need for training, as they require certain expertise to operate and maintain such equipment. Furthermore, some of these technologies can be detected or intercepted, thereby compromising the mission.

A study conducted by Henrico et al. (2020) emphasised the importance of studying terrain for military-related operations. Their study was a comparative analysis of the use of open-source desktop QGIS, ArcGIS, and GIS products developed by Esri. It evaluated the effectiveness of utilising open-source GIS rather than physical reconnaissance for military purposes, especially with a limited budget or at short notice. Lakomy (2022) supports the notion that the usage of open-source programmes, especially for intelligence gathering, has become more common in the military environment. However, some studies have reported that the use of open-source products can be problematic. Best (2011) emphasised the confidentiality issue and the possibility of unauthorised access to records of information when using open sources, mentioning that there can be a lack of trust and credibility for information being used through the use of open-source software.

Derek and Sikora (2019) conducted a study on bicycle route planning based on a multi-criteria GIS analysis in urban and suburban areas. Their study aimed to provide a new approach to multi-criteria road analysis. ArcGIS software was used to analyse and visualise the most suitable bicycle routes in Imotski, Croatia. A similar study was conducted by Hsu and Lin (2011) to evaluate routes in terms of bicycle use. A multi-criteria model was developed to assess the suitability of designing bicycle routes using GIS. Another similar study that appraised the effectiveness of GIS as a tool for bicycle route planning was conducted by Huang and Ye (1995), who developed a procedure using GIS to select routes in the city of Berkeley. The procedure includes developing a database, finding the most desirable route between each origin and destination pair, and aggregating the results to identify the most desirable routes in the city.

As outlined above, geospatial tools provide a robust basis for determining, planning, and evaluating the most suitable routes and locations for an exercise.

Study area

Exercise Trans Enduro is a tradition of the SAMA and is planned and executed by the members of the institution. SAMA is a military unit of the SANDF that houses the Faculty of Military Science of Stellenbosch University. Overlooking Saldanha Bay from Malgaskop, the SAMA is located in the Saldanha Military Area on the West Coast of the Western Cape, South Africa. The exercise, during the execution phase, commences at Gqeberha, in the Eastern Cape province, and the activities take place along the N2 running through some of South Africa's major coastal towns/cities, namely Jeffrey's Bay, Plettenberg Bay, George, Stilbaai, Struisbaai, Hermanus, and Bloubergstrand, ending at Saldanha Bay in the Western Cape (Figure 14.2). At a length of approximately 2,250 km, the N2 is the main highway along South Africa's coast, bordered by the Indian Ocean, with Exercise Trans Enduro only covering approximately 1,000 km of the distance.

The area where Exercise Trans Enduro takes place is characterised by variations in elevation and topography, ranging from fairly flat to mountainous areas. The average elevation of the Exercise Trans Enduro route ranges between sea level and approximately 40 m above sea level (Pauw, 2009). The N2 is a scenic route with natural and cultural beauty, used mainly as a holiday route but is also ideal for tourist drives (Ramukumba, 2019).

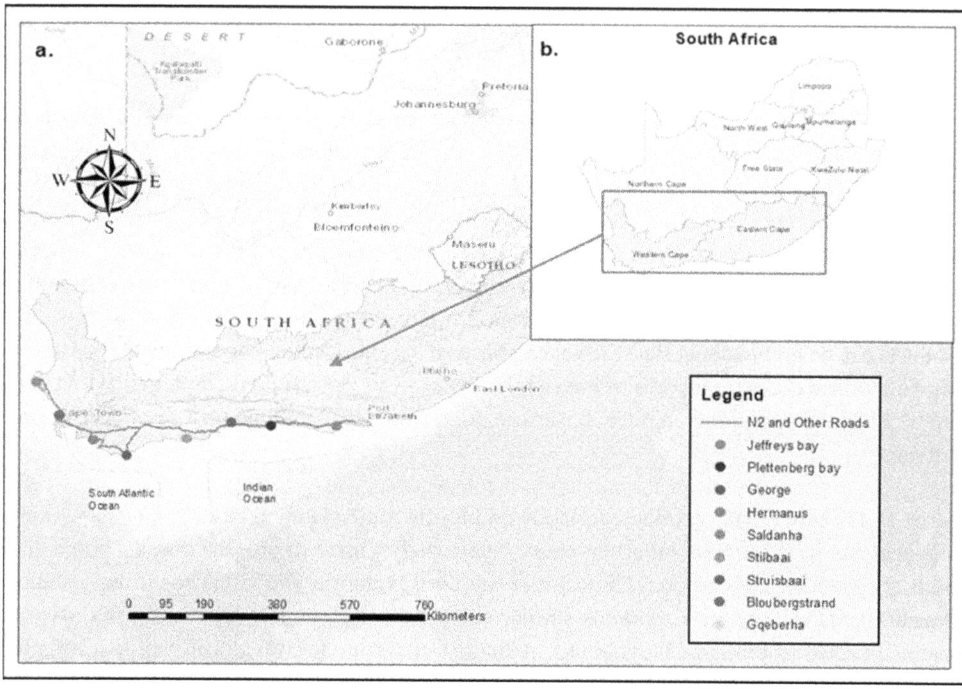

Figure 14.2: A map illustrating the towns/cities and route of Exercise Trans Enduro (a) and the location of the study area within South Africa (b).
(*Source:* Authors' own work using ArcGIS 10.8.1)

Methodology

Planning a military operation should not add too much to the costs of executing the actual exercise (Winkler & Steinberg, 1999). Geospatial techniques and open-source data have proven effective in planning military operations (Henrico et al., 2020). This research adopted a case study design to provide an in-depth understanding of how geospatial techniques can partly replace traditional physical reconnaissance in military exercise planning, focusing on SAMA's Exercise Trans Enduro (Figure 14.3). According to Toyon (2023), a case study is a detailed study of a specific subject, such as a person, group, place, event, organisation, or phenomenon. A case study research design usually involves qualitative methods and is best for describing, comparing, evaluating, and understanding the different aspects of a research problem.

This study also followed a qualitative approach, which is more descriptive and flexible in terms of interpreting data. Furthermore, descriptive data were presented as numbers and the location of the exercise (Febriantoko et al., 2023). A logical sequence is used to deduce the premises that lead to firm conclusions. The study also followed a pragmatic research paradigm because it allowed a mix of quantitative and qualitative methods (Feilzer, 2010).

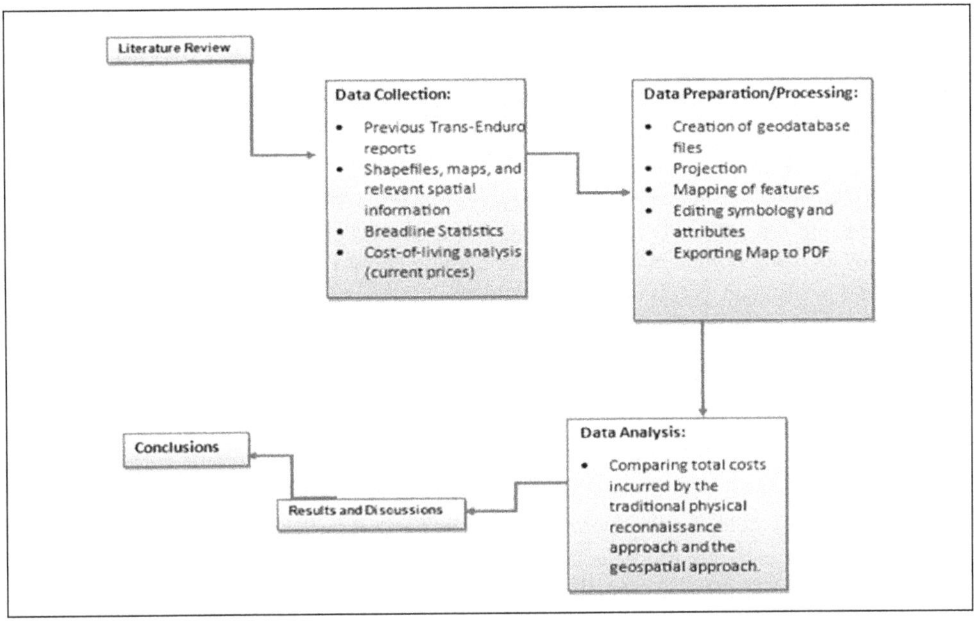

Figure 14.3: Methodology workflow.

This research used existing qualitative data, such as previous Exercise Trans Enduro exercise reports and planning data available at the unit's media department, to understand the participants' experiences and perspectives to determine challenges related to traditional reconnaissance methods and the potential benefits of geospatial techniques. Geospatial data were acquired from reliable sources such as OpenStreetMap, USGS, and Google Earth Pro. These data include shapefiles, maps, and other relevant spatial information. Furthermore, government databases and breadline statistics were viewed as information about disadvantaged areas suitable for the community outreach programme (Masiza).

The data were prepared and processed to ensure accuracy and consistency. This involved data transformation, projection adjustment of data for each town because they fall under different Universal Transverse Mercator (UTM) zones, and addressing gaps or errors. ArcGIS software (ArcMap and ArcGIS Pro) was used for digitisation and editing, which included mapping and visualisation of the acquired geospatial data to identify possible launching sites, accommodation options, proximity to military bases for logistical support and recovery, hospitals, shopping centres, running and cycling routes, sea team routes, and replenishing points for the sea team. When mapping was performed, ArcGIS was used to create an interactive geospatial PDF. In addition, the costs associated with the traditional reconnaissance method were calculated based on the current cost of living analysis (current prices) in South Africa obtained from reliable sites such as GlobalPetrolPrices and then compared with those of the geospatial approach (Expatistan, 2023).

This methodology thus utilised a systematic approach to assess the suitability and effectiveness of geospatial techniques in replacing traditional reconnaissance methods in military exercise planning at the SAMA.

Results

Geospatial PDF

Interactive geospatial PDFs were developed using ArcGIS. According to Neustaedter et al. (2024), a geospatial PDF is a portable document format (PDF) file that contains extra information that ties the image to coordinates on a map. It is interactive in nature and can be used as a standard PDF file but has the added capacity of limited mapping features. It is important to note that before exporting a map into a PDF on ArcMap or ArcGIS Pro, one must digitise and create different layers using points, polylines, and polygons. These features represent real objects on the ground. For instance, points in this study were used to represent point features such as malls or shopping centres, caravan parks/camping sites, hospitals, military bases, and replenishment points for the sea team. Furthermore, each type of object has an icon that closely relates to and represents the type of feature it is, such as a launching site with a 'chequered flag' or a shopping centre with a 'shopping cart' icon.

Figure 14.4 illustrates the results with all the features after digitisation, which were identified based on the requirements for the exercise on Google Earth Pro. Thereafter, the files were saved as a .kmz file and converted to layers for further digitisation, data processing, and editing to produce a map that has all the information needed to plan Exercise Trans Enduro before exporting to an interactive geospatial PDF.

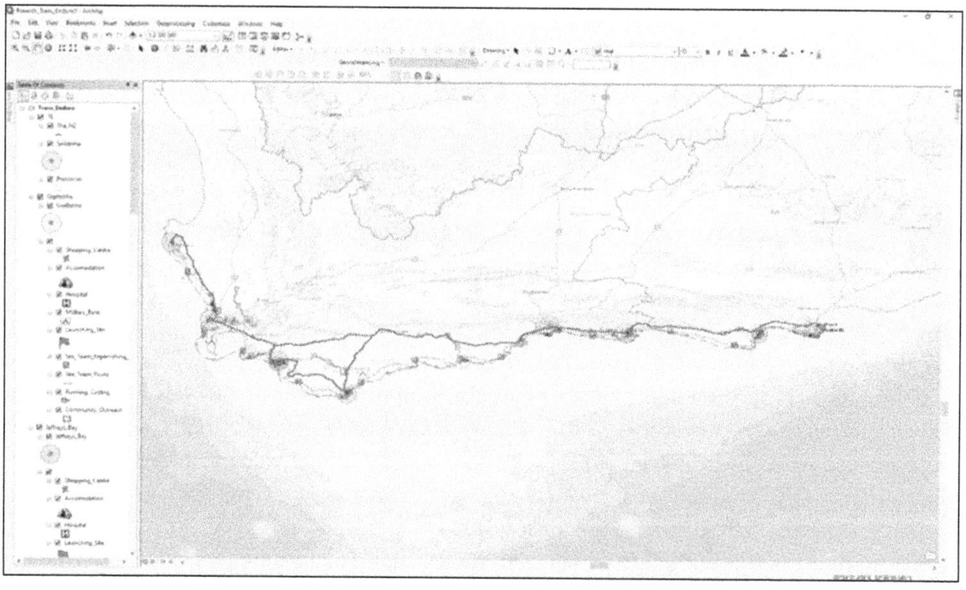

Figure 14.4: ArcMap product before exporting to PDF.
(*Source:* Authors' own work using ArcGIS 10.8.1)

Figure 14.5 illustrates the main results of this study, which is the geospatial PDF exported using ArcGIS.

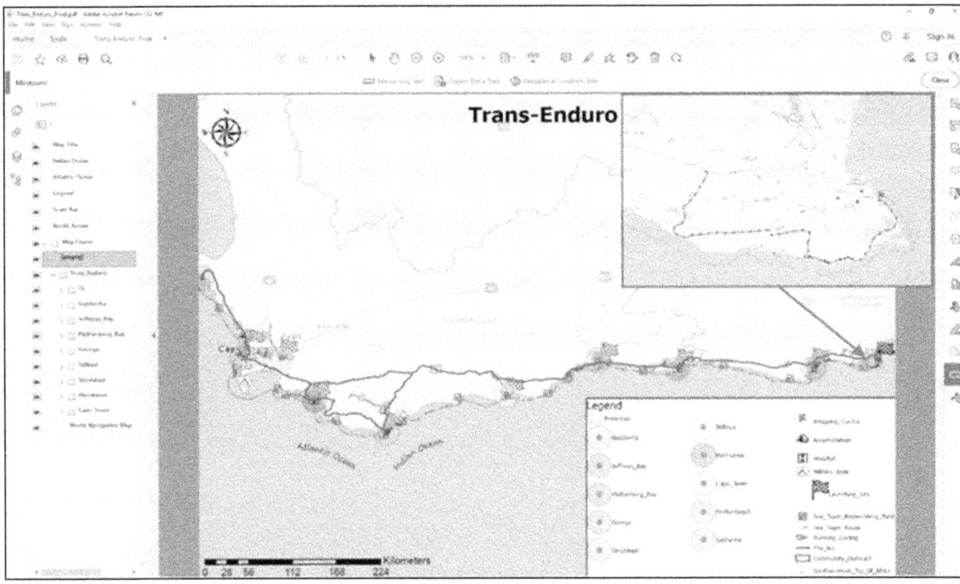

Figure 14.5: Geospatial PDF of the Exercise Trans Enduro map with a subset of Gqeberha. (*Source:* Authors' own work using ArcGIS 10.8.1 and the GeoPDF functionality in Adobe Acrobat Reader)

This geospatial PDF includes multiple map layers that can be toggled on or off, as shown on the left side of Figure 14.5, under 'Layers' in the table of contents. From the table of contents, there is a main layer which displays everything on the map titled 'Trans-Enduro', wherein after extracting it, you get sub-layers which are the different towns visited during Exercise Trans Enduro. Upon extracting the town layers, all the features that appear on the map and are found within that specific town can be accessed.

The PDF includes a 'legend' that explains what each symbol and colour on the map represents, which is shown in the table of contents as a layer that can be switched on and off. The features in the legend represent various aspects of the reconnaissance data that were digitised based on the information required and gathered, such as launching sites represented by a chequered flag icon, places of accommodation represented by a tent icon since it is either a camping site or caravan park, military bases represented by a military base icon, hospitals represented by an H or hospital icon, shopping centres represented by a shopping cart icon, running and cycling routes represented by a cycling route icon, sea team routes represented by a ferry route icon, and a mechanic icon for sea team replenishment points.

In addition, it is possible to zoom in and out of the map and pan to explore different areas, as illustrated by the red-bordered subset 'Gqeberha' in Figure 14.5. This functionality allows the user to focus on specific regions or zoom out for a broader perspective. This also gives users the ability to customise views of the exercise area and add their annotations or highlights to the PDF. It also helps when marking specific points of interest or making notes about them during exercise planning.

Geospatial PDFs allow for interactive elements, such as clickable icons, markers, and links. When you click on these elements, they reveal additional information or details about specific locations or features on the map. One of the most important tools for exploring Exercise Trans Enduro planning is the measuring tool. Figure 14.6 illustrates the functions that can be performed using the measuring tool. The measuring tool function allows the user to measure distances, the perimeter, or the area of features on the map, as indicated by the red-bordered elements, where a straight-line distance was measured using the red arrow, which is the distance between Stilbaai and Struisbaai in kilometres. Equally important, the unit of measurement can be changed at any time by the user.

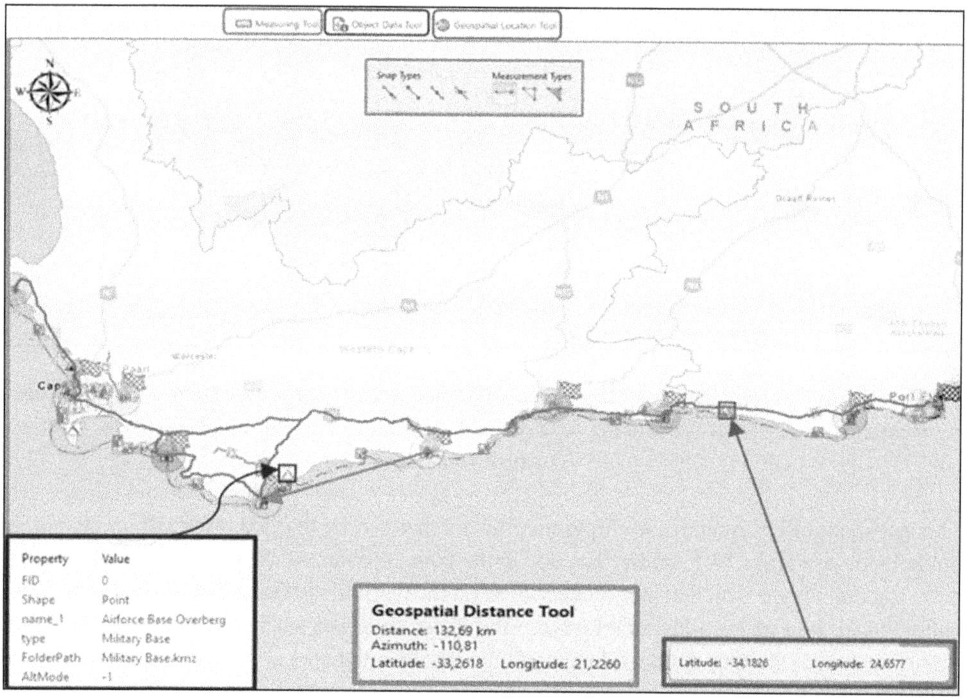

Figure 14.6: Measuring tools in geospatial PDF.
(*Source:* Authors' own work using ArcGIS 10.8.1 and the GeoPDF functionality in Adobe Acrobat Reader)

The object data tool, illustrated by the dark blue-bordered elements, allows the user to view attributes of any object/feature selected on the map. For instance, the dark, blue-bordered feature selected in Figure 14.6 indicates that the selected feature is a military base and that the name of that base is the Air Force Base Overberg. The geospatial location tool, on the other hand, simply provides the geographical coordinates of any object/feature selected on the map, as illustrated by the purple-bordered elements in Figure 14.6.

Cost savings

Table 14.1 illustrates the total costs incurred by the traditional method of conducting reconnaissance at the SAMA. The values were calculated based on the current rates of cost of living in South Africa obtained from various websites (Expatistan, 2023; GlobalPetrolPrices.

com, 2023; SAFacts, 2022). The cost of living in South Africa is relatively affordable compared to other countries; however, it has been a different story for South Africans (Expatistan, 2023). The relatively high cost of living calls for solutions that result in saving funds for institutions, civilian and government, to still be able to render services without compromising standards.

Table 14.1: Total costs incurred by Exercise Trans Enduro reconnaissance based on different methods

	Accommodation	Fuel	Meals per day	Tollgate fees
Full committee (23 members + 4 vehicles)	R98 325	R31 733.05	R35 075	R894 (2 x minibuses + 2 x Land Cruisers)
No 2ICs (13 members + 2 vehicles)	R41 162.50	R15 866.53	R17 537.50	R447 (1 x minibus + 1 x Land Cruisers)
The committee (Geo tools + telephone)	R153,60			
Total (full committee)	R166 027.05			
Total (No 2ICs)	R75 013.53			
Total (GeoPDF)	R153.60			

The calculations for the amounts shown in Table 14.1 were conducted as follows.

- The average overnight accommodation price in South Africa, according to Cowling (2023), is between R855 and R1,301 per night; therefore, R855 was multiplied by the number of members in the Exercise Trans Enduro committee (including the accompanying military instructors), and the product was multiplied by the number of days that the team will need overnight accommodation during the five-day reconnaissance.

- The fuel was calculated because SAMA uses diesel-powered minibuses for trips and Land Cruiser vehicles, and according to manufacturers' specifications, the average fuel consumption of the vehicles is approximately 15 litres per 100 km (Cars.co.za, 2023). The total distance travelled during reconnaissance from Saldanha to Gqeberha, if all the towns in the study area are visited, will have an addition of at least 50 km per town, which will give a total distance of at least 2,100 km. According to GlobalPetrolPrices.com (2023), the average price of diesel per litre is R25. The total money for the fuel required for reconnaissance in rands was calculated by taking the total distance that ought to be travelled, dividing it by the average fuel consumption, multiplying it by the average price for diesel, and multiplying it by the number of vehicles.

- According to SAFacts (2022), the average cost of meals (breakfast, lunch, and supper) per person per day is R305. The total cost of meals is the average cost of meals per day multiplied by the number of members doing reconnaissance and the number of days for reconnaissance, which is five.

- There is only one tollgate on the Garden Route between Saldanha and Gqeberha (Tsitsikama Toll Plaza), and the fees for light vehicles are R63.50 and R160 for medium or heavy vehicles as of March 2023 (Toll Road Tariffs, 2023). Each vehicle must pay fees for a return trip. Therefore, the total was calculated by multiplying the number of light vehicles by the toll fees for light vehicles multiplied by two, plus the number of medium or heavy vehicles multiplied by their toll fees multiplied by two.

- According to SAFacts (2022), the average cost of a phone call in South Africa is R0.32 per minute for landline calls. Assuming that the total time spent making arrangements with the different stakeholders (municipalities, accommodation sites, shopping centre managers, etc.) for each town telephonically is approximately 60 minutes. The calculation is the average cost of a phone call per minute multiplied by 60 minutes, multiplied by the number of towns, which gives the estimated cost incurred by conducting reconnaissance using geospatial tools and techniques.

- In addition, this study ignores the salaries of the members doing the reconnaissance as a cost factor for this exercise since their salaries are not in any way affected or changed when they are involved in this exercise.

Evaluating the effectiveness of geospatial techniques in SAMA'S Exercise Trans Enduro reconnaissance and planning

The study demonstrates that utilising geospatial tools and techniques for reconnaissance significantly saves costs compared with the traditional method of sending a team to various towns along the N2. These results suggest that SAMA can allocate its resources more efficiently, saving between R70,000 and R170,000 annually. These cost savings can be redirected to other critical areas within SAMA or used to expand Exercise Trans Enduro.

The results of this study indicate that geospatial tools, such as Google Earth Pro and ArcGIS with its extensions, allow for efficient data collection and visualisation. This efficiency streamlines the reconnaissance process and saves time and effort. This implies that SAMA can plan exercises faster, cheaper, and with less manpower than humans. The geospatial approach provides a comprehensive reconnaissance or view of the towns to be visited during the Exercise Trans Enduro by providing details such as suitable areas for launching sites, accommodation options, proximity to military bases, hospitals, shopping centres, running and cycling routes, sea team routes, and replenishment points for the sea team. Furthermore, these results suggest that geospatial techniques enable a thorough understanding of the target area, enhancing the planning process and potentially improving the exercise outcomes.

The study's assertion that the research serves as a game changer for instructors and students implies that the adoption of geospatial tools can fundamentally transform the way reconnaissance and exercise planning are conducted at SAMA. This suggests that this shift offers significant advantages in terms of time and cost savings, which can be crucial for military training institutions, especially if expanded to the planning of other types of activities. Incorporating geospatial techniques into military training offers educational benefits for students, as they can gain practical experience with modern technology and data analysis methods. This can better prepare them for real-world military operations, where geospatial intelligence is becoming increasingly important.

It is common for a research study to have limitations that can affect the scope and generalisability of its findings. Some limitations include the fact that the study is focused on the specific context of SAMA and the Exercise Trans Enduro, and the findings may not be directly applicable to other military institutions or exercises with different objectives and requirements. Thus, the use of this geospatial PDF is limited to SAMA military instructors, students, and Exercise Trans Enduro planners. This study relied on data sources such as Google Earth Pro and other open sources. These sources may have limitations in terms of accuracy, currency, or coverage, which can affect the quality of reconnaissance data and subsequent analyses. Hence, telephone engagement is of utmost importance when selecting areas of interest based on exercise requirements.

The success of the geospatial approach depends on the GIS software, and it is important to note that technical issues, such as the inability to load large amounts of data at a time, could result in the software crashing or failing to save work progress in time. The success of the exercise can also be jeopardised by the ongoing national power crisis with unpredictable load shedding. Equally important, the renewal of the yearly product licence or software limitations could impact the feasibility and effectiveness of this approach. It is also important to note that the findings may be time-sensitive and may not account for evolving technologies, changes in exercise requirements, or shifts in military training practices. Therefore, the PDF needs to be updated periodically due to changes in the local governance of the different towns, changes in the names of establishments such as shopping centres and accommodation options, or even street names. External factors, such as changes in government policies, funding, or military priorities, could impact the feasibility and sustainability of implementing geospatial techniques in military training. While the study highlights educational benefits, it may not comprehensively assess the long-term impact of geospatial training on students' performance and readiness as military officers. It is important to acknowledge these limitations to recognise areas where further research or adjustments may be needed for a more comprehensive understanding of the integration of geospatial techniques into military training and planning.

Overall, using geospatial PDFs involves actively engaging with their interactive features to gain insights into reconnaissance data, exercise planning, and the exercise area itself. It allows users to explore, analyse, and extract valuable information for decision-making and planning. In addition, the study's results indicate that the integration of geospatial techniques and data into the reconnaissance and planning phases of military exercises, such as Exercise Trans Enduro, can result in cost savings, increased efficiency, enhanced understanding of exercise areas, and improved educational experiences for military officers in training. This shift can have far-reaching implications for military institutions such as the SAMA, offering valuable opportunities to improve and optimise their training processes.

Conclusion

This study demonstrated the transformative potential of incorporating geospatial techniques and data into the reconnaissance and planning phases of the SAMA Exercise Trans Enduro. Reconnaissance, which had previously relied exclusively on a fairly large team of personnel travelling to the study area to perform in-situ reconnaissance, incurred high annual costs and resource usage. This research has the potential to usher in a new era of efficiency, cost-effectiveness, and increased planning skills by utilising geospatial tools and data sources, specifically Google Earth Pro and ArcGIS.

One of the key results of this study is the significant cost reduction achieved through the geospatial technique. SAMA can transfer resources to other essential parts of military education and development by eliminating the need for physical reconnaissance. An approximate R170,000 annual cost reduction is a concrete benefit that allows for more efficient resource allocation. The use of geospatial techniques has the potential to provide a comprehensive understanding of the towns to be visited during the Exercise Trans Enduro, as evidenced by the detailed map(s) created using the geospatial data. These maps include critical information, such as launching sites, accommodation options, proximity to military bases, hospitals, shopping centres, running and cycling routes, sea team routes, and sea team replenishment points. This increased situational awareness considerably enhances the planning and execution of exercises. ArcGIS enables the creation of interactive geographic PDFs, and these dynamic documents enable users to engage with exercise planning information in an easy-to-use and interesting manner.

The demonstrated feasibility of conducting reconnaissance tasks using geospatial tools represents newfound efficiency and cost-effectiveness, as well as a paradigm shift in SAMA exercise planning. It is a transformation that saves time rather than being an incremental improvement only. Finally, the use of geospatial tools and data in the Exercise Trans Enduro planning constitutes a significant improvement. It shows how innovation may improve not only the cost-effectiveness of military operations but also the educational experience of future military leaders.

The findings and recommendations of this study provide a solid foundation for the continuous integration of geospatial tools and techniques into SAMA's curriculum and operational practices, ensuring that the institution remains at the forefront of military education in South Africa. Although geospatial techniques cannot entirely replace the need for physical reconnaissance, they significantly reduce its scope. For instance, a small team (e.g. four personnel in one vehicle) could validate selected sites, achieving substantial cost savings while maintaining operational validity. This pragmatic approach highlights the practical value of geospatial tools in enhancing efficiency and modernising reconnaissance for planned military operations.

References

Abdalla, R. 2016. Geospatial data integration. In: R. Abdalla (ed.). *Introduction to Geospatial Information and Communication Technology (GeoICT)*. Cham: Springer. 105–124. https://doi.org/10.1007/978-3-319-33603-9_6

Best, C. 2011. Challenges in open source intelligence. *2011 European Intelligence and Security Informatics Conference*. *IEEE*,. 58–62. https://doi.org/10.1109/EISIC.2011.41

Bubear, R. 2023. Pricing! Toyota Land Cruiser 70 Series 2.8 GD-6 to gain manual 'box. *Cars.co.za*, 9 December. https://www.cars.co.za/motoring-news/pricing-toyota-land-cruiser-70-series-2-8-gd-6-to-gain-manual-box/260390/ [10 June 2023].

Cowling, N. 2023. Average daily rate (ADR) of hotels in South Africa from 2012 to 2023 (in South African Rand). *Statista*. https://www.statista.com/statistics/800460/hotel-adr-south-africa/ [27 July 2023].

Derek, J. & Sikora, M. 2019. Bicycle route planning using multiple criteria GIS analysis. *2019 International Conference on Software, Telecommunications and Computer Networks (SoftCOM)*. *IEEE*. 1–5. https://doi.org/10.23919/SOFTCOM.2019.8903800

Expatistan. 2023. Cost of living in South Africa. https://www.expatistan.com/cost-of-living/country/south-africa [10 September 2023].

Febriantoko, J., Indriasari1, D., Putra, M.S.G.P. & Marpen, R. 2023. Evaluation of tourism business activities and multi-sectoral income optimization: Case in Pagaralam City, South Sumatra. *Proceedings of 6th FIRST T3 2022 International Conference (FIRST-SS 2022)*. https://doi.org/10.2991/978-2-38476-026-8_15

Feilzer, M.Y. 2010. Doing mixed methods research pragmatically: Implications for the rediscovery of pragmatism as a research paradigm. *Journal of Mixed Methods Research*, 4(1): 6–16. https://doi.org/10.1177/1558689809349691

Frazier, D.V. & Hutto, J.W. 2017. The socialization of military power: Security cooperation and doctrine development through multinational military exercises. *Defence Studies*, 17(4): 379–397. https://doi.org/10.1080/14702436.2017.1377050

GlobalPetrolPrices.com. 2023. South Africa diesel prices. https://www.globalpetrolprices.com/South-Africa/diesel_prices/ [25 September 2023].

Helfrich, K. 2018. Trans Enduro 2018. *DefenceWeb*. https://www.defenceweb.co.za/sa-defence/sa-defence-sa-defence/trans-enduro-2018/ [13 March 2023].

Helfrich, K. 2021. Mil Academy Trans Enduro 2021 kicking off. *DefenceWeb*. https://www.defenceweb.co.za/sa-defence/sa-defence-sa-defence/mil-academy-trans-enduro-2021-kicking-off/ [1 March 2023].

Henrico, S., Coetzee, S. & Cooper, A. 2020. Is open source GIS feasible in military operations? Evaluation by applying a use case. *Scientia Militaria: South African Journal of Military Studies*, 48(1): 41–60. https://doi.org/10.5787/48-1-1259

Hsu, T.P. & Lin, Y.T. 2011. A model for planning a bicycle network with multi-criteria suitability evaluation using GIS. *WIT Transactions on Ecology and the Environment*, 148: 243–252. https://doi.org/10.2495/RAV110231

Huang, Y. & Ye, G. 1995. Selecting bicycle commuting routes using GIS. *Berkeley Planning Journal*, 10(1): 55–75. https://doi.org/10.5070/BP310113062

Karakaya, M. 2014. UAV route planning for maximum target coverage. *Computer Science & Engineering: An International Journal (CSEIJ)*, 4(1). https://doi.org/10.5121/cseij.2014.4103

Kopuletý, M. & Palasiewicz, T. 2018. Advanced military robots supporting engineer reconnaissance in military operations. *Modelling and Simulation for Autonomous Systems: 4th International Conference, MESAS 2017*. 285–302. https://doi.org/10.1007/978-3-319-76072-8_20

Lakomy, M.M. 2022. Assessing the potential of OSINT on the internet in supporting military operations. *Bezpieczeństwo. Teoria i Praktyka*, 3(2022): 297–309. https://doi.org/10.48269/2451-0718-btip-2022-3-022

Le Roux, L. 2005. The post-apartheid South African military: Transforming with the nation. In: M. Rupiya (ed.). *Evolutions and Revolutions: A Contemporary History of Militaries in Southern Africa*. Institute for Security Studies. 235–268.

McKearney, T.J. 2000. Collaborative planning for military operations: Emerging technologies and changing command organizations. *Proceedings of the Command and Control Research and Technology Symposium, Naval Postgraduate School, Monterey, CA*.

Namufohamba, M. 2019. Challenges of geospatial technology on national security in the 21st century: The case of the Namibia Defence Force. Doctoral dissertation. Windhoek East: University of Namibia.

Nathan, L. 1998. The 1996 defence white paper. In: J. Cock & P. Mckenzie (eds.). *From Defence to Development: Redirecting Military Resources in South Africa*. Cape Town: David Philip. 41–54.

Neustaedter, E.R., Kemna, R.F., Padilla, A.J. & Otarod, D. 2024. Geospatial PDF map of the compilation of GIS data for the mineral industries and related infrastructure of Africa. (No. 2024-1041). US Geological Survey. https://doi.org/10.3133/ofr20241041

Ngeno, K.S. 2021. Documenting the use of geospatial technologies in defence and related intelligence. B.Sc. dissertation. Nairobi: University of Nairobi.

Oron-Gilad, T. & Parmet, Y. 2017. Close target reconnaissance: A field evaluation of dismounted soldiers utilizing video feed from an unmanned ground vehicle in patrol missions. *Journal of Cognitive Engineering and Decision Making*, 11(1): 63–80. https://doi.org/10.1177/1555343416675077

Pauw, J. 2009. Challenges to sustainability in the Garden Route: Water, land, and economy. Nelson Mandela Metropolitan University, Port Elizabeth.

Ramukumba, T. 2019. Analysing tourism stakeholders' perceptions of the attractiveness and competitiveness of the Garden Route as a tourist destination. *African Journal of Hospitality, Tourism and Leisure*, 8(1): 1–9.

SAFacts. 2022. Average cost of food per day in South Africa. https://safacts.co.za/average-cost-of-food-per-day-in-south-africa/ [15 August 2023].

Singh, R.K., Kumar, P., Mukherjee, S., Suman, S., Pandey, V. & Srivastava, P.K. 2021. Application of geospatial technology in agricultural water management. In: P.K. Srivastava, M. Gupta, G. Tsakiris & N.W. Quinn (eds.). *Agricultural Water Management: Theories and Practices*. London: Academic Press. 31–45. https://doi.org/10.1016/B978-0-12-812362-1.00003-5

Tarapata, Z. 2003. Military route planning in battlefield simulation: Effectiveness problems and potential solutions. *Journal of Telecommunications and Information Technology*, (4): 47–56. https://doi.org/10.26636/jtit.2003.4.201

Toll Road Tariffs (effective from 1 March 2023). 2023. www.roadtransportexecutive.co.za [27 July 2023].

Toyon, M.A.S. 2023. Introduction to research: Mastering the basics. *Scholars Journal of Research in Social Science (SJRSS)*, 3(1): 1–24. https://doi.org/10.5281/zenodo.7724719

Winkler, J.D. & Steinberg, P.S. (1999) Making military education more effective and affordable. https://doi.org/10.7249/RB3007

Yang, L., Qi, J., Xiao, J. & Yong, X. 2014. A literature review of UAV 3D path planning. *Proceedings of the 11th World Congress on Intelligent Control and Automation, IEEE*. 2376–2381. https://doi.org/10.1109/WCICA.2014.7053093

Young, A.B. & Adam, B. 2014. Employing intelligence, surveillance, and reconnaissance: Organizing, training, and equipping to get it right. *Air & Space Power Journal*, 28(1): 26–44.

ABOUT THE EDITORS

Ivan Henrico is the Vice-Dean for Research and Internationalisation in the Faculty of Military Science at Stellenbosch University, where he also serves as an associate professor in the Department of Military Geography. A C2-rated researcher (NRF, 2024), he holds a PhD in Geoinformatics from the University of Pretoria and specialises in military geography, geopolitics, remote sensing, and geographic information systems (GIS). He has published widely in accredited journals and scholarly book collections, presented his work at leading international conferences, and leads several collaborative research initiatives with global partners, including the United States Military Academy at West Point and Charles University in Prague. An award-winning educator, he has supervised numerous postgraduate students and developed specialist GIS and remote-sensing training programmes. His research focuses on integrating geospatial technologies with strategic, defence, and security applications.

[https://orcid.org/0000-0002-1080-5107]
Email: ivanh@sun.ac.za

Susan Henrico is a senior lecturer in the Department of Defence Intelligence Studies at the Faculty of Military Science, Stellenbosch University. She holds a PhD in geography from the University of Pretoria, specialising in the acceptance and use of open-source geospatial software. Her research interests span military geography, geosciences, GIS, FOSSGIS, geospatial intelligence, and cyber studies. A former section head of geospatial training in the South African National Defence Force, she has extensive experience in defence intelligence, curriculum development, and capacity building across Africa. Dr Henrico has published widely in accredited journals, presented at leading international conferences, and collaborates globally on projects addressing geospatial applications in civil and military domains.

[https://orcid.org/0000-0001-5312-0813]
Email: susanh@sun.ac.za

Rikus le Roux is an associate professor in Technology and the Head of the Department of Military Technology at the Faculty of Military Science, Stellenbosch University. He holds a PhD in computer and electronic engineering from North-West University (2015) and specialises in radiation instrumentation, detection and measurement, geophysics, and environmental radioactivity. He lectures in radar systems, electronic warfare, information warfare, and numerical techniques, and has an extensive publication record in national and international journals and conference proceedings. Prof le Roux maintains active research collaborations with the International Atomic Energy Agency (IAEA), Ghana Atomic Energy Commission (GAEC), University of Nairobi, University of Ghana, the Joint Institute for Nuclear Research (JINR), Medusa Explorations in Groningen, Netherlands, and leading South African research institutions. An experienced postgraduate supervisor and peer reviewer for scientific journals, he continues to advance research at the nexus of detection systems and applied science.

[https://orcid.org/0000-0001-9039-7010]
Email: rikusr@sun.ac.za

www.ingramcontent.com/pod-product-compliance
Ingram Content Group UK Ltd.
Pitfield, Milton Keynes, MK11 3LW, UK
UKHW051118290926
15471UKWH00010B/1735

9 780639 889429